# Advanced Textile Based Polymer Composites: Synthesis, Characterization and Applications

# Advanced Textile Based Polymer Composites: Synthesis, Characterization and Applications

Editors

**Muhammad Tayyab Noman**
**Michal Petrů**

MDPI • Basel • Beijing • Wuhan • Barcelona • Belgrade • Manchester • Tokyo • Cluj • Tianjin

*Editors*

| | |
|---|---|
| Muhammad Tayyab Noman | Michal Petrů |
| Department of Machinery Construction | Department of Machinery Construction |
| Technical University of Liberec | Technical University of Liberec |
| Liberec | Liberec |
| Czech Republic | Czech Republic |


*Editorial Office*
MDPI
St. Alban-Anlage 66
4052 Basel, Switzerland

This is a reprint of articles from the Special Issue published online in the open access journal *Polymers* (ISSN 2073-4360) (available at: www.mdpi.com/journal/polymers/special_issues/adv_text_polym_compos).

For citation purposes, cite each article independently as indicated on the article page online and as indicated below:

LastName, A.A.; LastName, B.B.; LastName, C.C. Article Title. *Journal Name* **Year**, *Volume Number*, Page Range.

**ISBN 978-3-0365-7325-0 (Hbk)**
**ISBN 978-3-0365-7324-3 (PDF)**

# Contents

**Preface to "Advanced Textile Based Polymer Composites: Synthesis, Characterization and Applications"** . . . . . . . . . . . . . . . . . . . . . . . . . . . . . . . . . . . . . . . . . . . . . **vii**

**Muhammad Tayyab Noman, Nesrine Amor, Michal Petru, Aamir Mahmood and Pavel Kejzlar**
Photocatalytic Behaviour of Zinc Oxide Nanostructures on Surface Activation of Polymeric Fibres
Reprinted from: *Polymers* **2021**, *13*, 1227, doi:10.3390/polym13081227 . . . . . . . . . . . . . . . . **1**

**Muhammad Zeeshan, Hong Hu and Ehsan Etemadi**
Geometric Analysis of Three-Dimensional Woven Fabric with in-Plane Auxetic Behavior
Reprinted from: *Polymers* **2023**, *15*, 1326, doi:10.3390/polym15051326 . . . . . . . . . . . . . . . . **19**

**Joana C. Antunes, Inês P. Moreira, Fernanda Gomes, Fernando Cunha, Mariana Henriques and Raúl Fangueiro**
Recent Trends in Protective Textiles against Biological Threats: A Focus on Biological Warfare Agents
Reprinted from: *Polymers* **2022**, *14*, 1599, doi:10.3390/polym14081599 . . . . . . . . . . . . . . . . **37**

**Nga-Wun Li, Kit-Lun Yick, Annie Yu and Sen Ning**
Mechanical and Thermal Behaviours of Weft-Knitted Spacer Fabric Structure with Inlays for Insole Applications
Reprinted from: *Polymers* **2022**, *14*, 619, doi:10.3390/polym14030619 . . . . . . . . . . . . . . . . . **69**

**Shi Hu, Dan Wang, Aravin Prince Periyasamy, Dana Kremenakova, Jiri Militky and Maros Tunak**
Ultrathin Multilayer Textile Structure with Enhanced EMI Shielding and Air-Permeable Properties
Reprinted from: *Polymers* **2021**, *13*, 4176, doi:10.3390/polym13234176 . . . . . . . . . . . . . . . . **85**

**Ion Sandu, Claudiu Teodor Fleaca, Florian Dumitrache, Bogdan Alexandru Sava, Iuliana Urzica and Iulia Antohe et al.**
Shaping in the Third Direction; Synthesis of Patterned Colloidal Crystals by Polyester Fabric-Guided Self-Assembly
Reprinted from: *Polymers* **2021**, *13*, 4081, doi:10.3390/polym13234081 . . . . . . . . . . . . . . . . **101**

**Annie Yu, Sachiko Sukigara and Miwa Shirakihara**
Effect of Silicone Inlaid Materials on Reinforcing Compressive Strength of Weft-Knitted Spacer Fabric for Cushioning Applications
Reprinted from: *Polymers* **2021**, *13*, 3645, doi:10.3390/polym13213645 . . . . . . . . . . . . . . . . **117**

**Nesrine Amor, Muhammad Tayyab Noman and Michal Petru**
Prediction of Methylene Blue Removal by Nano $TiO_2$ Using Deep Neural Network
Reprinted from: *Polymers* **2021**, *13*, 3104, doi:10.3390/polym13183104 . . . . . . . . . . . . . . . . **129**

**Chang-Pin Chang, Cheng-Hung Shih, Jhu-Lin You, Meng-Jey Youh, Yih-Ming Liu and Ming-Der Ger**
Preparation and Ballistic Performance of a Multi-Layer Armor System Composed of Kevlar/Polyurea Composites and Shear Thickening Fluid (STF)-Filled Paper Honeycomb Panels
Reprinted from: *Polymers* **2021**, *13*, 3080, doi:10.3390/polym13183080 . . . . . . . . . . . . . . . . **141**

**Nesrine Amor, Muhammad Tayyab Noman and Michal Petru**
Classification of Textile Polymer Composites: Recent Trends and Challenges
Reprinted from: *Polymers* **2021**, *13*, 2592, doi:10.3390/polym13162592 . . . . . . . . . . . . . . . . **153**

**Aamir Mahmood, Muhammad Tayyab Noman, Miroslava Pechočiaková, Nesrine Amor, Michal Petrů and Mohamed Abdelkader et al.**
Geopolymers and Fiber-Reinforced Concrete Composites in Civil Engineering
Reprinted from: *Polymers* **2021**, *13*, 2099, doi:10.3390/polym13132099 . . . . . . . . . . . . . . . . **181**

# Preface to "Advanced Textile Based Polymer Composites: Synthesis, Characterization and Applications"

In contemporary society, the use of textiles is far more advanced than ever before. Technical textiles, textile-reinforced composites, and textile use in the fashion industry are common examples. The demand for textiles as a reinforcement for composites has grown exceptionally as a result of the promotion of green chemistry and cost effectiveness. The fabrication of textiles and polymer-based composites as advanced and multifunctional materials in applied industries is a primary aim of this book. Textile materials, their structures, surface-treated textiles, nanocoated textiles, mathematical modeling, and the use of artificial intelligence (machine learning) for the prediction of most important parameters; textile material reinforcement in composites and polymer composites; and the synthesis, characterization, and applications of textile-based polymer composites for economic and environmental sustainability are some of the topics covered in this reprint.

This reprint covers the following research topics:

- Synthesis and characterization of functional textiles—natural, synthetic, blended, etc.;
- Synthesis and characterization of polymer composites—natural, synthetic, hybrid, inorganic, etc.;
- Analysis of developed composites—interfacial, mechanical, thermal, physical, etc.;
- Surface treatment of textile-based polymer composites—coating, sorption processes. etc.;
- Modelling and simulation of textile-based polymer composites—artificial intelligence, machine learning, process optimization, statistical analysis, etc.

Researchers, technologists, and students who are interested in textile-based polymer composites would find this reprint useful. I would like to take this opportunity to thank all the authors for their work and support. I would like to thank the production team of the publishing company for their continued support in getting this reprint to its current form.

**Muhammad Tayyab Noman and Michal Petrů**
*Editors*

polymers

Review

# Photocatalytic Behaviour of Zinc Oxide Nanostructures on Surface Activation of Polymeric Fibres

**Muhammad Tayyab Noman [1,*], Nesrine Amor [1], Michal Petru [1], Aamir Mahmood [2] and Pavel Kejzlar [3]**

[1] Department of Machinery Construction, Institute for Nanomaterials, Advanced Technologies and Innovation (CXI), Studentská 1402/2, 461 17 Liberec 1, Technical University of Liberec, 46117 Liberec, Czech Republic; nesrine.amor@tul.cz (N.A.); michal.petru@tul.cz (M.P.)
[2] Department of Material Engineering, Faculty of Textile Engineering, Studentská 1402/2, 461 17 Liberec 1, Technical University of Liberec, 46117 Liberec, Czech Republic; aamir.mahmood@tul.cz
[3] Department of Material Science, Faculty of Mechanical Engineering, Studentská 1402/2, 461 17 Liberec 1, Technical University of Liberec, 46117 Liberec, Czech Republic; pavel.kejzlar@tul.cz
* Correspondence: muhammad.tayyab.noman@tul.cz; Tel.: +420-776396302

**Abstract:** Zinc oxide (ZnO) in various nano forms (nanoparticles, nanorods, nanosheets, nanowires and nanoflowers) has received remarkable attention worldwide for its functional diversity in different fields i.e., paints, cosmetics, coatings, rubber and composites. The purpose of this article is to investigate the role of photocatalytic activity (role of photogenerated radical scavengers) of nano ZnO (nZnO) for the surface activation of polymeric natural fibres especially cotton and their combined effect in photocatalytic applications. Photocatalytic behaviour is a crucial property that enables nZnO as a potential and competitive candidate for commercial applications. The confirmed features of nZnO were characterised by different analytical tools, i.e., scanning electron microscopy (SEM), field emission SEM (FESEM) and elemental detection spectroscopy (EDX). These techniques confirm the size, morphology, structure, crystallinity, shape and dimensions of nZnO. The morphology and size play a crucial role in surface activation of polymeric fibres. In addition, synthesis methods, variables and some of the critical aspects of nZnO that significantly affect the photocatalytic activity are also discussed in detail. This paper delineates a vivid picture to new comers about the significance of nZnO in photocatalytic applications.

**Keywords:** nZnO; photocatalytic activity; polymeric fibres; cotton; stabilization

**Citation:** Noman, M.T.; Amor, N.; Petru, M.; Mahmood, A.; Kejzlar, P. Photocatalytic Behaviour of Zinc Oxide Nanostructures on Surface Activation of Polymeric Fibres. *Polymers* **2021**, *13*, 1227. https://doi.org/10.3390/polym13081227

Academic Editor: Nathanael Guigo

Received: 22 March 2021
Accepted: 7 April 2021
Published: 10 April 2021

**Publisher's Note:** MDPI stays neutral with regard to jurisdictional claims in published maps and institutional affiliations.



## 1. Introduction

The last two decades are eye witness to a prestigious revolution made by nano science, as researchers enlarge their circle for nanomaterials especially metal oxide-based nanomaterials, i.e., zinc oxide (ZnO). Nano ZnO (nZnO) composed of different forms of nanostructures i.e., nanoparticles, nanorods, nanowires, nanobelts, nanosheets and nanoflowers, has gained significant attention from researchers for the fabrication of sensors [1–3], medical devices [4–7], composites [8–11] and photocatalysts [12–14] for various applications [15–18]. nZnO is a fascinating material that possesses and reveals exceptional physicochemical properties when used in photocatalytic applications. As a semiconductor, ZnO has high thermal conductivity, high exciton binding energy (60 m eV), high electron mobility and wide band gap, i.e., 3.2–3.4 eV [19]. The potential of photocatalytic activity of nZnO expands its scope in biomedical [20], industrial [21,22], catalysis [23,24], coatings [25,26], sensors [27,28], textiles [29,30] and energy conversion devices, i.e., fuel and solar cells [31]. nZnO has also been used in personal (sunscreen) and beauty care (cosmetics) products due to excellent ultraviolet (UV) absorption properties. UV rays are considered as the primary cause of skin diseases, i.e., wrinkles, skin cancer, aging and sunburns. On the nanoscale, ZnO shows significantly high optical, electronic and antimicrobial properties and due to these properties—nZnO provides great protection against the breaking down of skin collagen and the skin regeneration mechanism. Therefore, the addition of nZnO in the formulation of cosmetics not only protects our skin from harmful rays (UV-B and UV-A) but also enhances the attractiveness of beauty care. The photocatalytic activity of nZnO and its effects on the morphology of fibrous surface is a topic of great interest that should be addressed in detail. The literature discussed above only explains the application of photocatalytic activity of nZnO in different fields. However, the role of radical scavengers produced during the process of photocatalysis and their typical effect on the morphology and surface topography of polymeric fibrous material is still an area of investigation for the explanation of enhanced photocatalytic activity. As well as the authors searched, there was no such literature available on this hot issue. Therefore, our investigation provides an up-to-date knowledge on the role of photogenerated charge carriers (radical scavengers) and their induced potential for enhanced photocatalytic activity of nZnO and their combined effect on surface topography of polymeric fibrous materials. This investigation delineates a gateway for newcomers and experienced researchers in their respective areas as a valuable reference. In order to achieve our objectives, this review paper discusses the latest literature under two categories, i.e., the development of nZnO coated textiles and the role of photocatalytic activity on surface properties.

Polymeric fibres (cotton, polyester, jute, polyamide, wool and polypropylene) have been widely used in many fields of life due to versatility in their different properties. Surface roughness, porosity and capillary imbibition action in porous media are some of the very important properties of polymeric fibrous materials from textiles and composites point of view. Xiao et al. reported the application of a fractal model for capillary flow through a single tortuous capillary with surface roughness in porous media. The derived model is tested against imbibition mass and imbibition height. The results revealed that both of these characteristics decreased with an increase in relative surface roughness. It was also observed that the equilibrium time for rough surfaces decreased with an increment in relative roughness [32].

In a different study, Xiao et al. studied the fluid transport through fibrous porous media under a fractal model with dimensionless permeability and Kozeny–Charman constant. The obtained results were in good agreement with the previously performed studies and with the experimental data. The results explain that the physical phenomenon of fluids transport through porous media can be elucidated in a better way with this fractal model as the values of the Kozeny–Charman constant increases with an increase in porosity, tortuosity fractal dimension, relative roughness and pore dimensions [33]. Polysaccharides are a class of renewable and sustainable natural polymeric carbohydrates including chitosan, cellulose, gum, starch, alginate, pectin and chitin. The applications (energy storage devices, medical devices, composites, sorbents, nano catalysts and light weight porous materials) of these polymeric carbohydrates have increased tremendously during the last few years due to their biodegradability, non-toxicity, sustainability and environmentally friendly benefits [34–37]. In a recent study, Ahmed et al. discussed the benefits of chitosan and chitin in the fabrication of carbon-based composites for waste water treatment. They reported that due to mechanical properties and easy handling, these biopolymers not only show excellent compatibility for all kind of carbonaceous materials like carbon nanotubes, graphene, biochar, activated carbon and graphene oxide but also provide excellent results for the adsorption of water pollutants [38]. In a review article, Nasrollahzadeh et al. elucidated the role of bio polysaccharides, i.e., chitosan, pectin, cellulose and alginate in waste water treatment. Natural biopolymers are excellent candidates for the elimination of aqueous contaminants and aquatic pollution when utilised as nano sorbents or nano catalysts in the composition of nanobiocomposites. A list of the most used biopolymers or polymeric carbohydrates is presented in Figure 1 [39].

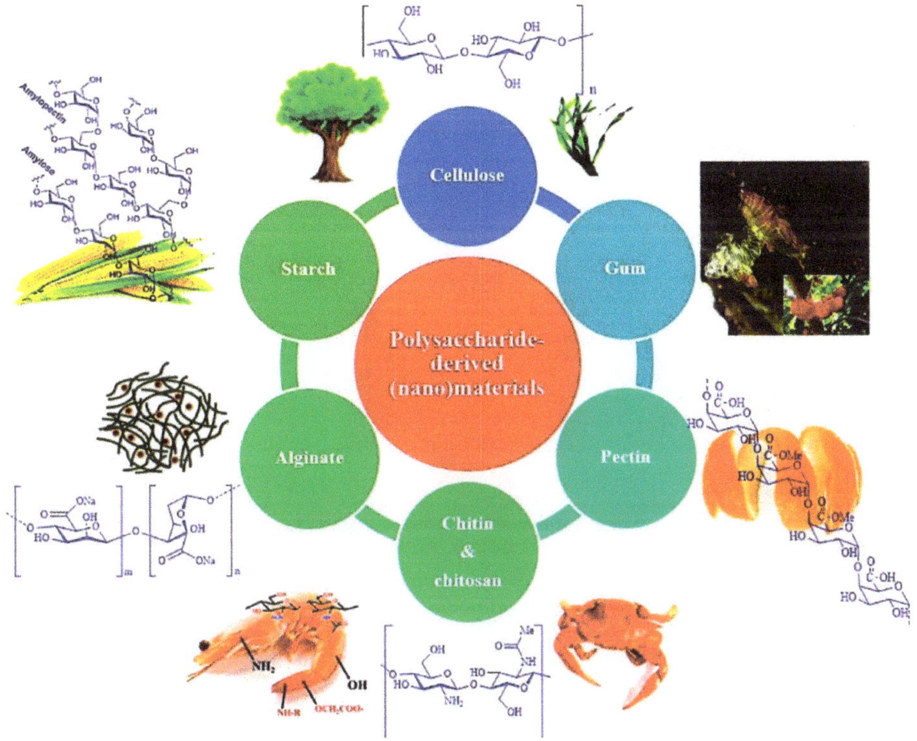


**Figure 1.** A group of sustainable and environmentally friendly polymeric carbohydrates (polysaccharides). Reprinted with permission from Reference [39]. Copyright 2021, with permission from Elsevier.

In our study, the range of polymeric fibres covers from cotton, jute, flax and their combinations with other fibres for a comparative analysis of photocatalytic activity of nZnO coated and noncoated samples. In a recent article, Theerthagiri et al. reviewed biological and energy applications (electrochemical supercapacitors, lithium-ion batteries, dye sensitised solar cells, photocatalyst, bioimaging, sensors, drug delivery, toxicity and gene delivery) of nZnO and explained that physicochemical properties are responsible for a dramatic change in nZnO behaviour and differentiates nZnO from its bulk counterpart. ZnO associated photocatalysis is a type of heterogeneous photocatalysis where the reactants, photocatalysts and the products are in different phases during the reaction mechanism. They discussed the effects of crystallinity, particle size and surface morphology on the performance efficiency of nZnO [40]. Nada et al. reported a successful allocation of Zn element in the polymer matrix of chitosan gelatine-based hybrid nanobiocomposites. Acid catalysed amino addition reaction was observed during the incorporation of zinc, inside the chitosan matrix and later on, the nanofibers of zinc incorporated chitosan were developed by an electrospinning process. The results elucidate that chitosan nanofibers with the addition of zinc element showed excellent antimicrobial properties [41]. However, it is still believed that our approach towards the critical aspects of photocatalytic performance of nZnO is exceptional and provides a vivid analysis for future studies. In addition, this review paper uncovers new horizons for researchers working with textiles and provides new ideas about the utilization of nZnO in photocatalytic applications.

In recent years, plenty of articles have been published explaining the significance of nZnO as a photocatalyst [42]. However, there is still a vacant place to dig up for dimensions, synthesis routes and variation in size, that can explain the photocatalytic behaviour of nZnO in a better way. The recent literature explains the selection of a suitable synthesis route and that is very important to control the size and to harness maximum photocatalytic performance. Therefore, we started with synthesis routes and important variables involved during the synthesis, then discussed their critical exigency for better photocatalytic activity, elaborated on individual articles that provided new insights and ended up with photocatalytic applications of nZnO coated textiles. Synthesis routes and parameters elucidated a crucial impact on the dimensional stability of nZnO. In an experimental study, Abramova et al. developed medically proven antimicrobial textiles by coating sol-gel synthesised nZnO on the surface of fabric through ultrasonic irradiations. They explained that the fabricated samples coated with zinc oxide in combination with titanium dioxide exhibited 99.99% suppression level of Escherichia coli (*E. coli*). These results reveal that sol-gel is an excellent method to deposit nanostructures on the surface of textiles [43]. Barreto et al. used a microwave-assisted method for the fabrication of nZnO and found that selected variables and reagents, i.e., temperature, additives, time and microwave power impart significant impact on growth, morphology, shape and size of nZnO. They explained that the addition of surfactants enhances the photocatalytic performance of nZnO to a significant level [44]. In a recent study, Noman et al. synthesised ZnO nanoparticles (ZnO NPs) via sonication and coated them on cotton and polyester fabrics with varying thickness. The results reveal that a smooth nZnO coating on both fabrics significantly depends on ultrasonic irradiation's time and intensity. Moreover, longer irradiation time enhances the porosity that allows nZnO to go deeper and creates a smooth layer on both fabrics. These results depict that a higher amount and a smooth coating of nZnO significantly enhance the photocatalytic performance of investigated samples [45]. It is observed from the previous literature that the structure of ZnO plays a crucial role in the augmentation of photocatalytic activity. ZnO exists in two main crystal structures, i.e., cubic blend and hexagonal wurtzite. However, hexagonal wurtzite is the most common and most stable form of ZnO at ambient conditions. Krol et al. explained the lattice parameters, i.e., a = 0.325 nm and c = 0.521 nm, with three growth directions of hexagonal wurtzite crystal where every tetrahedral Zn atom is surrounded by four O atoms. These parameters of the wurtzite structure are responsible for higher photocatalytic efficiency than the cubic zinc blend. A typical wurtzite crystal structure with primary growth directions is illustrated in Figure 2 [46].

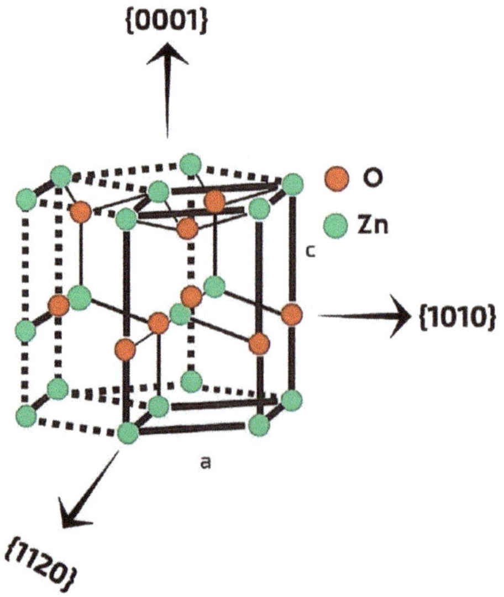


**Figure 2.** A typical wurtzite crystal structure of ZnO with indicated directions. Reprinted with permission from Reference [46]. Copyright 2017, with permission from Elsevier.

However, in a previous study, Pala and Metiu evaluated the density functions of ZnO-based thin films to calculate the amount of energy required during the formation of oxygen vacancy and structures of plain, hybrid and thin films used in pure and hybrid forms for photocatalytic applications. The results elucidate that after the removal of oxygen vacancy, the wurtzite crystal structure solely depends on the thickness of thin layers of ZnO as presented in Figure 3 [47].

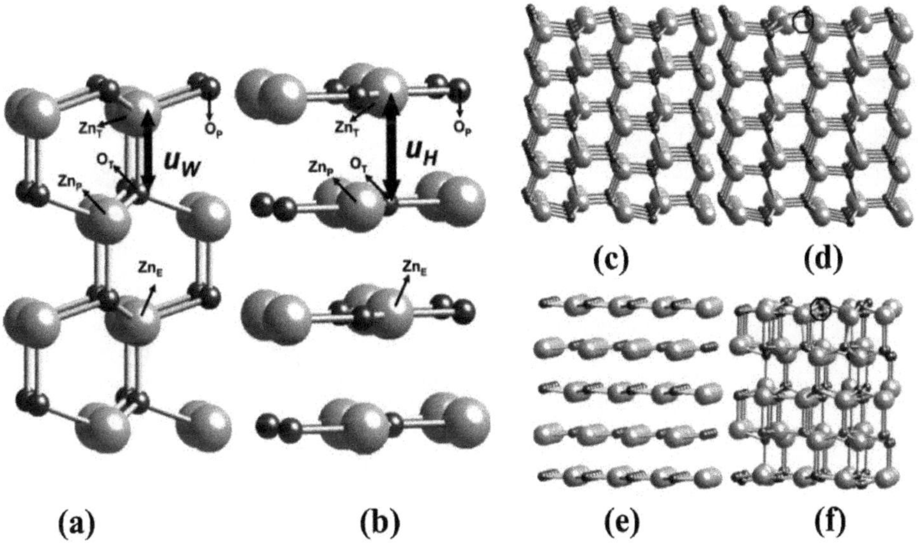


**Figure 3.** (**a**) Ball and stick model of ZnO with wurtzite structure. The light spheres denote Zn atoms while dark spheres denote O atoms. Zn and O atoms are in different planes (**b**) ZnO with hexagonal structure. Both O and Zn atoms are in the same plane (**c**) ZnO stoichiometric structure with five layers, (**d**) ZnO stoichiometric structure after removing O atom, (**e**) Single layer ZnO stoichiometric structure, and (**f**) Single layer ZnO stoichiometric structure after removing O atom. Reprinted with permission from Reference [47]. Copyright 2007, with permission from the American Chemical Society.

Ong et al. thoroughly reviewed and explained the synthesis mechanism of nZnO and described the properties responsible for higher photocatalytic activity. They categorised ZnO as a green eco-friendly material that can be efficiently utilised for organic pollutants removal during waste-water treatment and other purification processes. They claimed ZnO is a cost effective, non-toxic and more efficient photocatalyst than $TiO_2$ as it absorbs a greater fraction of the solar spectrum. Moreover, they classified nZnO into several categories [48]. Generally, nanomaterials are classified (on the basis of their structure and shape) from 0 to 3 in dimensions. Spherical nanoparticles are zero dimensional (0D); nanorods, nanowires, nanotubes and nanobelts are one dimensional (1D); nanosheets, nanolayers, nanofilms, graphene and nanocoatings are two dimensional (2D); porous nanostructures (nanoflowers) are three dimensional (3D) and further subdivision gives us quantum dot arrays, respectively [49]. Figure 4 illustrates the field emission scanning electron microscopy (FESEM) images with different morphologies and different dimensions of nZnO collected by various researchers [50–53].

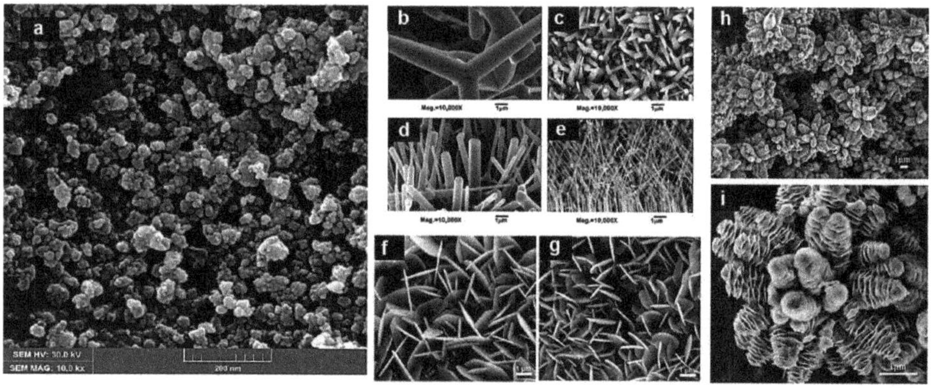

**Figure 4.** Scanning electron microscopy (SEM) and field emission SEM (FESEM) images for: (**a**) 0D (nanoparticles); (**b–e**) 1D (nanorods); (**f,g**) 2D (nanofilms); and (**h,i**) 3D (nanoflowers), respectively. Reprinted with permission from References [50–53]. Copyright 2009, 2013, and 2014, respectively, with permission from Elsevier.

From a medical point of view, zinc is an important element that exists in tissues all through the human body. Zinc plays a vibrant role during the synthesis of proteins, nucleic acid, neurogenesis and hematopoiesis by taking part in metabolism as zinc is the major component of different enzymatic systems [54]. nZnO has been used in food additives as it is easily absorbed or digested by the human body. The US Food and Drug Administration (FDA) graded ZnO as a safe material [55]. Due to availability, non-toxicity and being less expensive as compared to other metal oxides, nZnO has been given preferences in anti-cancer, bioimaging and drug-delivery applications [18,56,57]. Jiang et al. reviewed latest developments in the fabrication of nZnO for biomedical applications. They defined some valuable characteristics, i.e., biocompatibility, less toxicity, easy access and economic efficiency that enable nZnO to be a suitable and competent material for biomedical applications. They explained that crystal growth is an important factor for biomedical application and in the case of the sol-gel method, the crystal growth of nZnO is much more significant as compared to other methods, i.e., chemical precipitation, pyrolytic process, biological process and solution free method. Their results show that nZnO (due to larger surface area) enhances the intercellular generation of reactive oxygen species (ROS) and damages the cancer cells. Moreover, they reported a potential antimicrobial effect of nZnO against a number of Gram-positive and Gram-negative pathogens. The schematic representation of antimicrobial activity of nZnO is illustrated in Figure 5 [58]. The main factor for excellent performance of nZnO is the excessive generation and induction of ROS. Moreover, the performance is based on the accumulation of nZnO in the cytoplasm of bacterial cell walls or in the outer membranes and the release of $Zn^{+2}$ ions. This mechanism disintegrates the cell membrane and damages the membrane protein that results in killing of bacteria.

It has been revealed from the discussion that physical, mesoporous and chemical properties of nZnO are very critical and fascinating as these properties provide a theoretical description of key parameters, i.e., size, elasticity, energy distribution and thermodynamics of nZnO. Therefore, this paper summarises the important findings of research carried out in recent years on nZnO coated textiles and their enhanced photocatalytic activity.

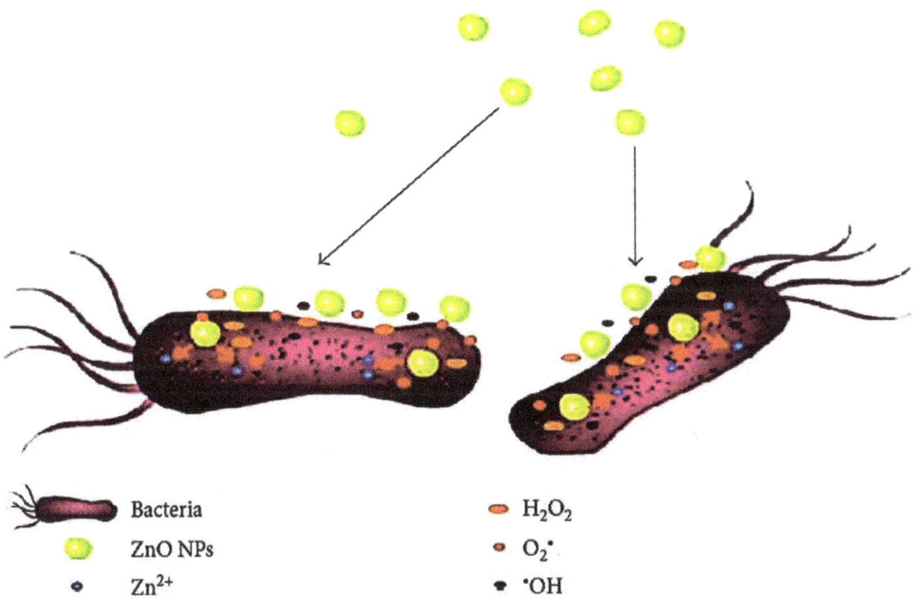


**Figure 5.** Graphical representation of antimicrobial activity of nZnO. Reprinted with permission from Reference [58]. Copyright 2018, with permission from Hindawi.

## 2. Synthesis of nZnO and Coating Process on Textiles

In principle, there are two approaches, i.e., top–down and bottom–up, used for the synthesis of nanostructures. The first one uses grinding, cutting, slicing and milling processes to get nanostructures from bulk material. Conversely, the bottom–up approach arranges atoms and molecules either by physical methods (sonication, physical vapour deposition and thermal evaporation); chemical methods (sol-gel, hydrothermal, chemical vapour deposition, precipitation and solvothermal) or biological methods (controlled deposition and growth) to create nanostructures. The top–down approach is not suitable for photocatalytic performance as this technique causes serious damage to the crystal structure of nanomaterials, that significantly affects overall properties. Therefore, bottom–up techniques are the best to fabricate photocatalytically active nanomaterials or functional nanomaterials.

Water-based chemical methods (often called wet chemical methods) offer several benefits, i.e., low energy input, uncomplicated equipment, cheap, easy to handle reagents and being environmentally friendly [59,60]. Furthermore, they make tailoring of synthesis variables easy throughout the process that helps in gaining control over the size, shape, structure and composition of the resulting nanomaterials. Therefore, in this section, the most used chemical methods such as sol-gel, hydrothermal and coprecipitation with their influence on reaction variables that affect the crystallization kinetics, morphology, particle size distribution and facet formation of nZnO will be discussed in detail.

The sol-gel method is considered to be one of the most versatile methods to fabricate advanced and novel materials especially nanomaterials [61–63]. The sol-gel method initializes with hydrolyzation, polymerization and ends with condensation reactions. This method comprises the preparation of a colloidal solution (generally known as sol) that is further converted into gel. The precursors generally used for the sol-gel method are metal alkoxides and chloride salts. For nZnO, the most common precursors are zinc chloride ($ZnCl_2$), zinc nitrate hexahydrate ($Zn(NO_3)_2{\cdot}6H_2O$) and zinc acetate $Zn(CH_3CO_2)_2$. The most important variables that affect the size, shape and dimensions of nZnO in sol-gel synthesis are the concentration of solvents, nature of solvents, temperature and molar ratio.

Sui and Charpentier provided a detail explanation about sol-gel nano synthesis mechanism of metal oxides for supercritical fluid applications. They explained the benefits of supercritical fluids and supercritical drying in the synthesis of solid products. It is already known that at ambient conditions, some inevitable shrinkage of the solid leads to the collapse of the microstructure that results in low specific surface area. The formation of sols in the liquid phase gets an infinite level of viscosity when converted into a gel results in the formation of xerogel and aerogel with ambient drying and supercritical drying respectively. The results explain the overall chemistry of metal oxide-based nanomaterials synthesised by the sol-gel process and suggest the use of supercritical fluids i.e., $H_2O$, $CO_2$ and organic solvents in nanofabrication. The graphical representation of the mechanism of the sol-gel synthesis is illustrated in Figure 6 [64].

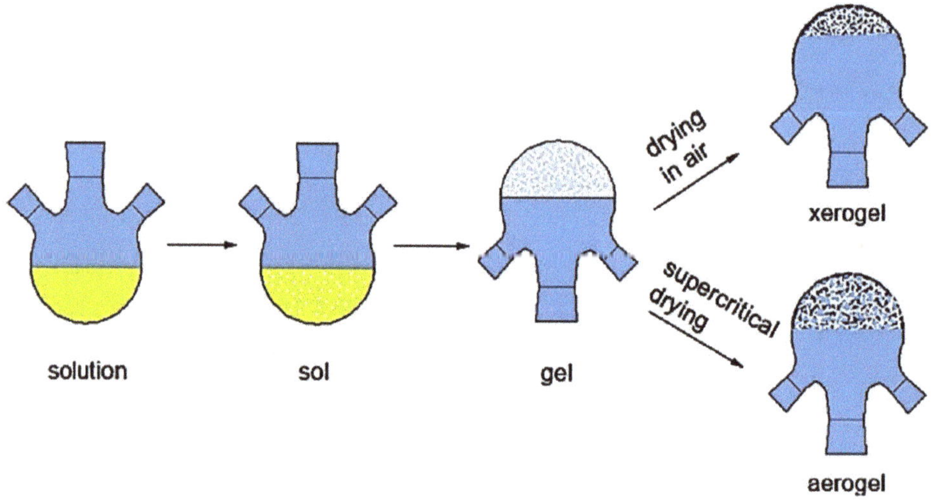


**Figure 6.** Graphical illustration of sol formation in liquid phase, infinite viscosity of gel, shrinkage of xerogel and an aerogel without shrinkage. Reprinted with permission from Reference [64]. Copyright 2012, with permission from American Chemical Society.

Alias et al. reported sol-gel synthesis of ZnO NPs under different pH conditions. The resulting nano powders showed agglomeration at pH 6 to 7 (acidic and neutral conditions). However, alkaline pH conditions (pH = 9) were more favourable in obtaining nZnO powders. The sizes of the resulting nanoparticles were lower in acidic medium. However, the optical properties were better in alkaline conditions. The sizes were significantly uniform at pH 9 to 11 and the size range was 37 to 50 nm at pH 9 and 11, respectively. Chemical composition of as synthesised nZnO explained the existence of methanol solvent near the x-axis origin. The results of energy dispersive X-ray (EDX) analysis confirm the Zn and O elemental peak for all pH levels as illustrated in Figure 7. The results also explain the growth mechanism of nZnO in a precise manner. For sol-gel synthesis, hydrolysis and polycondensation (nucleation) are the primary steps to prepare particles, while for growth of nZnO with high crystallinity, a sufficient amount of $OH^-$ ions are necessary. The results of FESEM for surface topography, structural and morphological analysis at different pH conditions are presented in Figure 8 [65]. Gupta et al. achieved a uniform nZnO coating on cotton by the sol-gel method and fabricated smart wearable electromagnetic interference (EMI) shielding electrically conductive textiles that protect the human body from electromagnetic pollution [66].

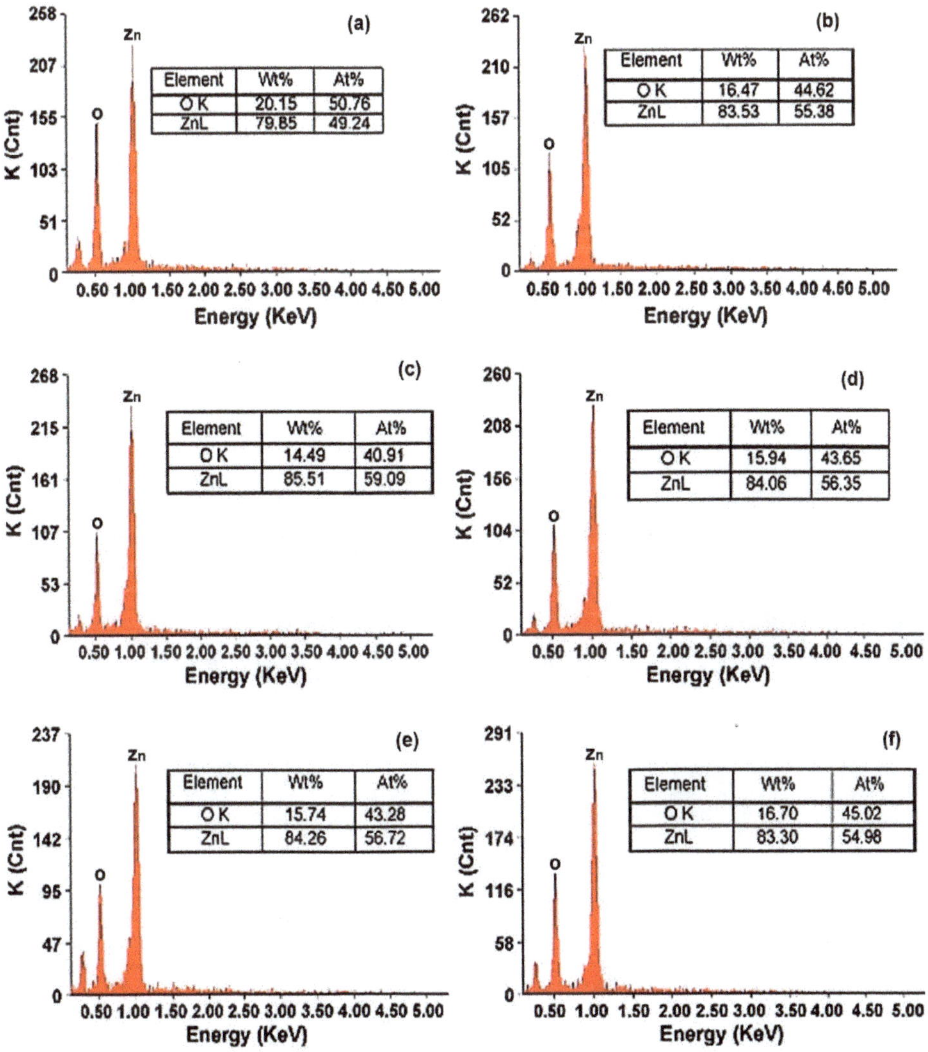


**Figure 7.** (**a**–**f**) Elemental detection spectroscopy (EDX) analysis of sol-gel synthesised nZnO at varies pH from 6 to 11, respectively. Reprinted with permission from Reference [65]. Copyright 2010, with permission from Elsevier.

Hydrothermal is another widely used wet chemical method that uses autoclaves under controlled conditions. In autoclaves, the temperature rises above 100 °C and reaches to saturated vapor pressure. This method is generally followed in order to achieve small size particles. Commonly used precursors for hydrothermal synthesis are zinc nitrate hexahydrate and zinc sulphate heptahydrate ($ZnSO_4{\cdot}7H_2O$). The variables that affect the size, shape and dimensions of nZnO in hydrothermal synthesis are pH, calcination temperature and heating time. In an experimental study, Gong et al. incorporated hydrothermally synthesised nZnO over the optical fibre surface and evaluated the growth mechanism. The results reveal that an augmentation in preferred growth and less oxygen vacancy are mainly due to higher UV irradiation power [67]. In a recent study, Koutavarapu et al. reported hydrothermally synthesised ZnO nanosheets based hybrid nanoribbons. Photocatalytic performance was evaluated against tetracycline, a toxic organic pollutant. The results show 98% photodegradation of tetracycline within 90 min [68].

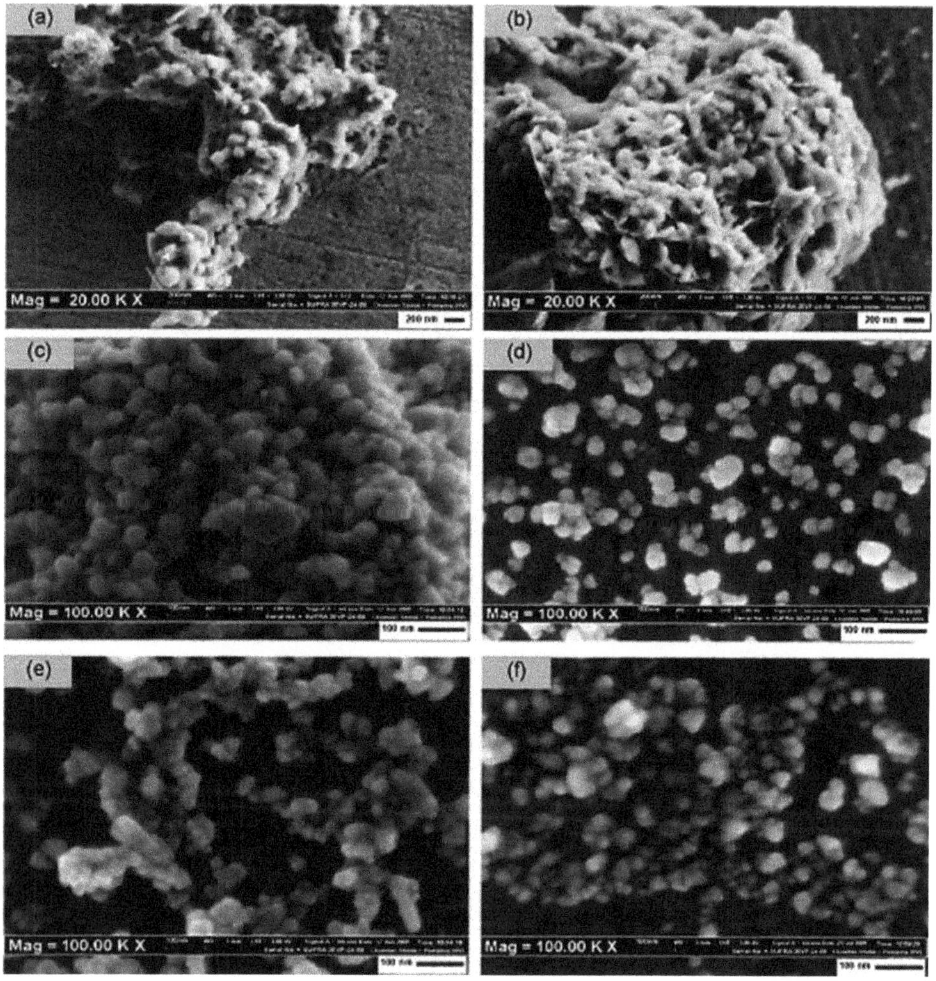


**Figure 8.** (**a–f**) FESEM analysis of sol-gel synthesised nZnO at varies pH from 6 to 11, respectively. Reprinted with permission from Reference [65]. Copyright 2010, with permission from Elsevier.

Precipitation and coprecipitation methods are also very renowned to fabricate nZnO and have gained attention in recent years. These methods involve a reaction of zinc salts (zinc precursors normally $ZnSO_4 \cdot 7H_2O$, $Zn(CH_3CO_2)_2$, $Zn(NO_3)_2 \cdot 6H_2O$ and $ZnCl_2$) with alkaline solutions i.e., NaOH, KOH, LiOH and $NH_4OH$. The reaction starts between $Zn^{2+}$ and $OH^-$ that follows aggregation and a stable colloidal suspension of nZnO is formed in the presence of alcohol. nZnO with different morphologies are obtained by controlling the typical variables, i.e., pH of the solution, concentration of the solution, type of precursor, calcination temperature and type of alkali, of these processes. By using the precipitation method, Shetti et al. reported the synthesis of nZnO taking 0.2 M $Zn(NO_3)_2 \cdot 6H_2O$ and 0.4 M KOH and further applied it on carbon—nZnO coated carbon electrode designed to detect Molinate, a thiocarbamate herbicide, by cyclic and voltametric methods [69].

The wide band gap of ZnO allows this material to function only in the UV range. In addition, surface oxygen vacancy and a higher recombination rate of photo generated charge carriers decreases the photocatalytic efficiency of ZnO. The defects in ZnO structure defined by nonbonding electrons in polarization, band gap and elastic modulus. The crystal size is associated with annealing temperature. The surface defects, interatomic and atomic interactions and variations in electronic distribution are the dominant parameters in determining the photocatalytic performance. Doping is an excellent method to overcome these issues by tailoring the optical properties of nZnO. Doping comprises the insertion of a specific element (ions) into the crystal lattice of ZnO that modifies and controls the band gap of ZnO with a direct consequence on the photocatalytic activity of ZnO. The physicochemical properties of doped ZnO significantly depend on dopant type. For nZnO, commonly used doping agents are cerium, titanium dioxide, calcium, boron, cobalt and magnesium. Doped ZnO is generally considered as a hybrid multifunctional metal oxide and is significantly used for photocatalytic, sensing, energy and biomedical applications. In a recent study, Kim and Yong fabricated boron doped ZnO and reported a significantly higher photocatalytic hydrogen production. They observed that boron doped ZnO illustrates type 2 alignment of band structures whereas undoped ZnO exhibits z-scheme band. Due to this dramatic change in band structures, boron-doped ZnO shows a 2.9-times higher $H_2$ production rate than undoped ZnO [31]. The reason for excellent photocatalytic activity of nZnO is the formation of ROS. Photocatalytic reaction initializes when a light beam strikes the surface of nZnO. The formation of excess numbers of ROS (due to larger surface area of nanostructures) with their strong power of decomposition degrade organic pollutant that cause staining and health issues [70]. The photocatalytic activity of nZnO in bare and doped forms for photodegradation of organic pollutants and solar driven water splitting is graphically illustrated in Figures 9 and 10, respectively.

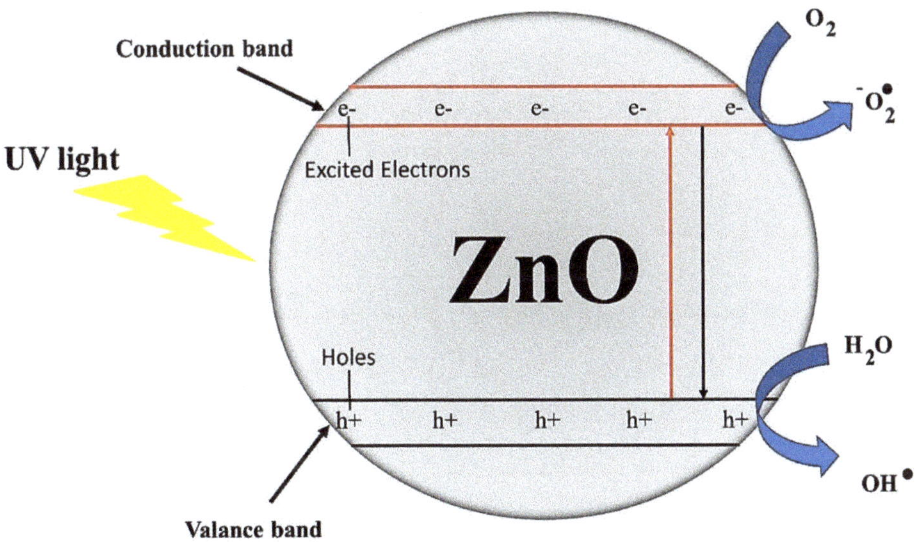


**Figure 9.** A typical mechanism of photocatalytic activity of pure nZnO. Reprinted with permission from Reference [70]. Copyright 2021, with permission from Elsevier.

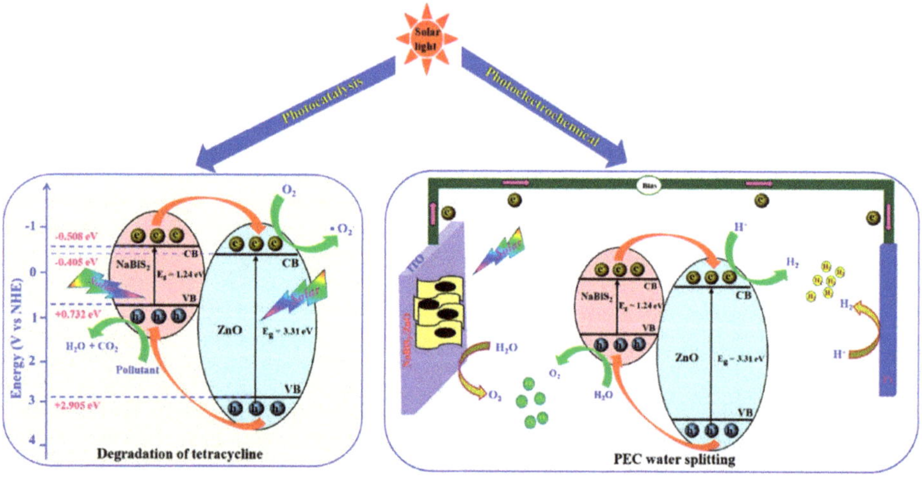


**Figure 10.** A comparative analysis of photocatalytic activity of doped nZnO for the photodegradation of organic pollutant (tetracycline) and for photocatalytic water splitting. Reprinted with permission from Reference [68]. Copyright 2021, with permission from Elsevier.

Table 1 summarises a brief overview of other synthesis methods used in the fabrication of nZnO.

**Table 1.** Overview of some synthesis methods for nZnO.

| Method | Used Precursors | Conditions | Structure; Size | References |
|---|---|---|---|---|
| Ultrasonics | $Zn(NO_3)_2{\cdot}6H_2O$, $C_6H_8O_7{\cdot}H_2O$ | Sonication for 0.5–2 h, room temperature | Hexagonal, spherical particles | [71] |
| Chemical bath deposition (CBD) | $Zn(CH_3CO_2)_2{\cdot}2H_2O$, $C_2H_4(NH_2)_2$ | 80 °C temperature, magnetic stirring up to 3 h | Core shell rod like structure, size 500 nm | [72] |
| Sol-gel | $Zn(CH_3CO_2)_2$, $C_2H_5OH$, $(C_6H_9NO)n$ | 100 °C temperature, stirring time 2 h, calcination at 300 °C | Size 140 nm | [73] |
| Sol-gel | $Zn(CH_3CO_2)_2{\cdot}2H_2O$, $C_2H_5OH$, NaOH | 60 °C temperature, time 17 h, annealing at 200 °C | Hexagonal wurtzite, spherical, size 22 nm | [74] |
| Co-precipitation | $Zn(CH_3CO_2)_2{\cdot}2H_2O$, NaOH, $C_2H_5OH$, $NH_4OH$ | Agitation time 1 h, annealing at 60 °C, for 6 h | Nanorods, wurtzite crystal structure | [75] |
| Successive ionic layer adsorption and reaction (SILAR) | $ZnSO_4{\cdot}7H_2O$, $NH_4OH$, $CuSO_4{\cdot}5H_2O$ | Dipping cycles 30, annealing 500 °C for 2 h | Hexagonal wurtzite, spherical, size 34 nm | [76] |
| Hydrothermal | $Zn(NO_3)_2{\cdot}6H_2O$, NaOH, $Zn(CH_3CO_2)_2{\cdot}2H_2O$, $C_2H_5OH$, | Heating 50 °C, stirring 30 min, oven drying 150 °C for 6 h | Nanorods, wurtzite crystal structure | [77] |
| Two step sonication modified sol-gel | $Zn(CH_3CO_2)_2{\cdot}2H_2O$, $NH_4OH$, KOH, Tetraethyl orthosilicate (TEOS) | 60 °C temperature, time 1.5 h, sonication time 1 h | Hexagonal wurtzite, spherical nanoparticles | [78] |
| Biological | Tabernaemontana divaricata leaves extract, $Zn(NO_3)_2{\cdot}6H_2O$ | 80 °C temperature, continuous stirring, drying at 450 °C for 2 h | Hexagonal wurtzite, spherical, size 20–50 nm | [79] |
| | Abutilon indicum leaves extract, $Zn(NO_3)_2{\cdot}6H_2O$ | Temperature of furnace at 200 °C for 3 min., calcination for 2 h | Hexagonal wurtzite, spheroid rod like shape, size 16–20 nm, band gap 3.36 eV. | [80] |

## 3. Photocatalytic Applications

For photocatalytic applications, higher photocatalytic activity and optical properties are key features that enable nZnO a good choice to be utilised as a potential (photocatalyst) material. nZnO is an extensively used material as a photocatalyst in the photodegradation of various organic pollutants. Many researchers have utilised nZnO in the textile sector. Ghayempour and Montazer reported a coating mechanism of nZnO on a cotton surface with the help of tragacanth gum—a natural biopolymer—and ultrasonic energy. The results reveal a wurtzite structure with star like shape of as synthesised nZnO that shows good to excellent photocatalytic efficiency against the photodegradation of methylene blue (MB) [81]. In another study, Chakrabarti and Banerjee coated sonochemically synthesised nZnO on cotton fabric by padding to achieve multifunctional characteristics. The results explain that developed samples show significantly enhanced photocatalytic properties and degrade 69% trypan blue, a direct azo dye, under solar self-cleaning [82]. In the photodegradation of organic pollutants, $OH^{\bullet}$ radicals take part in a redox reaction and convert the long chain molecules into $H_2O$ and $CO_2$. In a recent study, Noman and Petru reported a sonochemical coating of nZnO on cotton fabric under controlled conditions and compared the comfort performance against $TiO_2$. In an in-situ coating, $ZnCl_2$ was used as a precursor and an average particle size of 30 nm was successfully achieved. The results depict that nZnO coated fabric exhibits higher properties than $TiO_2$ coated samples [83]. Both doped and undoped ZnO has been used for photocatalytic applications and preference is given to the type of application, i.e., for photodegradation of organic pollutants, mostly pure nZnO is used while for photocatalytic water splitting, energy and sensing applications, doped nZnO is preferably used. The previous literature elucidates that pure nZnO works better in the degradation applications and doped nZnO in other applications. Nair et al. worked with undoped and cobalt-doped nZnO and performed photocatalytic experiments against MB solution and reported that photocatalytic efficiency of pure ZnO is significantly higher than cobalt doped ZnO. The reasons for higher photocatalytic activity are zinc vacancies and interstitial oxygen atoms from acceptor states and oxygen vacancies and interstitial zinc atom from donor states [84]. In another study, Kaur and Singhal synthesised various metal doped nZnO and evaluated photocatalytic performance against methyl orange (MO). The results show that MO degradation follows first order reaction kinetics and undoped nZnO performs much better than metal-doped ZnO [85]. Photocatalysis is a dynamic process that triggers a series of redox (reduction and oxidation) reactions and convert long chain molecules into less toxic materials, i.e., $CO_2$ and $H_2O$. The primary oxidizing species that work during photocatalysis are superoxide anions and hydroxyl radicals. It is observed from all above discussed literature that different morphologies of nZnO perform differently during photocatalytic applications and significantly affect the photocatalytic performance. One dimensional nanorods, nanoneedles and nanowires provide better results during photocatalytic applications as compared to nanoparticles because they provide more surface area than nanoparticles. FESEM images of nZnO with various morphologies like nanoparticles, nanorods, nanoflowers and nanoneedles are provided in Figure 11.

For hydrogen generation, electrochemical and biosensors, doped nZnO has opened a new research area because doping of suitable element enhances the efficiency of nZnO to a significant level for these applications. In a recent study, Bukkitgar et al. reported excellent electrocatalytic sensing performance of Mg-doped ZnO nanoflakes for an anti-inflammatory drug, mefenamic acid. The results reveal that pH and selection of dopant are important variables to control the sensing performance of doped nZnO as well as to develop durable and highly efficient sensors. Dopant selection significantly affects the electrochemical detection limit and offers better results under low detection limit [86]. In their previous study, barium was loaded as a dopant and Ba doped ZnO was used to detect mefenamic acid under cyclic voltammogram. The results show good recovery values and suggest the developed sensors for practical use [87]. However, the role of bare/undoped metal oxide semiconductors especially ZnO for solar photocatalytic applications is still undeniable. In a recent review article, Karthikeyan et al. elucidated the pivotal aspects of utilising different types of metal oxide semiconductors as a photocatalyst in hydrogen production, dye sensitised solar cells, energy storage batteries, water splitting, electrodes and sensors. They covered zinc oxide, titanium dioxide, copper oxide, tungsten oxide, tin oxide and their nanostructures and specified the vibrant role of nano dimensions for solar photocatalytic applications [88].

**Figure 11.** Illustration of different morphologies of nZnO taken by FESEM analysis: (**a**) ZnO nanoparticles; (**b–e**) ZnO nanorods; (**f,g**) ZnO nanoflowers; and (**h,i**) ZnO nanoneedles. Reprinted with permission from References [89,90]. Copyright 2010, 2013 respectively, with permission from Elsevier.

## 4. Future Direction

The contributions in this review paper can lead new comers to new lines of inquiry about the role of photogenerated charge carriers (ROS, radical scavengers) on the surface morphology of polymeric fibrous material and on photocatalytic applications. The discussed literature clarifies that synthesis methods and the selection of precursors are very important in order to get better results for photocatalytic applications. The latest progress and discussion particularly in these areas have been elucidated in this thematic issue in order to understand the variables and their combined effect on better photocatalytic activity. Zinc oxide nanostructures as a photocatalyst attract enormous attention due to their potentials of harnessing solar energy directly for the production of hydrogen and solar fuels, and degrade toxic pollutants as well. However, the efficiency of nZnO is reduced due to faster electron-hole recombination rate and low light consumption. The fabrication of high-quality photocatalysts on largescale for real time applications is still a challenge. Advancement is also necessary for the fabrication of an efficient, sustainable, cost-effective and facile photocatalyst. Therefore, there is still a paucity to go deeper and investigate other dimensions that have emerged in light of the findings presented here.

## 5. Summary

Zinc oxide nanostructures have high specific surface area, optical properties and flexibility that elucidate their potential for excellent photocatalytic applications especially photodegradation of organic pollutants, self-cleaning and antimicrobial efficiency. Higher photocatalytic activity significantly depends on synthesis routes, precursors type, selected variables and dopant type. Doped nZnO shows excellent results for water splitting, $H_2$ production, sensing and electrochemical detection of various materials. It was observed that doping shifts the bandgap and provides more radical scavengers, i.e., $OH^{\bullet}$ radicals. The discussed literature elucidates that the synthesis methods control the crystallinity, dimensional stability, particle size and overall photocatalytic activity of nZnO. Moreover, an extensive investigation of photocatalytic activity and the role of photogenerated charge carriers (ROS) provide a comprehensive knowledge of reaction kinetics and crystal structure for next generation materials and devices.

**Author Contributions:** M.T.N. conceived, designed, performed experiments, analysed the results and wrote the manuscript. N.A. and A.M. performed experiments. P.K. analysed the results. M.P. analysed the results, supervised and acquired funding. All of the authors participated in critical analysis and preparation of the manuscript. All authors have read and agreed to the published version of the manuscript.

**Funding:** This work was supported by the Ministry of Education, Youth and Sports of the Czech Republic and the European Union (European Structural and Investment Funds-Operational Programme Research, Development and Education) in the frames of the project "Modular platform for autonomous chassis of specialised electric vehicles for freight and equipment transportation", Reg. No. CZ.02.1.01/0.0/0.0/16_025/0007293.

**Institutional Review Board Statement:** Not applicable.

**Informed Consent Statement:** Not applicable.

**Conflicts of Interest:** The authors declare no conflict of interest.

## References

1. Ilager, D.; Shetti, N.P.; Malladi, R.S.; Shetty, N.S.; Reddy, K.R.; Aminabhavi, T.M. Synthesis of Ca-doped ZnO nanoparticles and its application as highly efficient electrochemical sensor for the determination of anti-viral drug, acyclovir. *J. Mol. Liq.* **2020**, *322*, 114552. [CrossRef]
2. Kulkarni, D.R.; Malode, S.J.; Prabhu, K.K.; Ayachit, N.H.; Kulkarni, R.M.; Shetti, N.P. Development of a novel nanosensor using Ca-doped ZnO for antihistamine drug. *Mater. Chem. Phys.* **2020**, *246*, 122791. [CrossRef]
3. Shanbhag, M.M.; Shetti, N.P.; Kulkarni, R.M.; Chandra, P. Nanostructured Ba/ZnO modified electrode as a sensor material for detection of organosulfur thiosalicylic acid. *Microchem. J.* **2020**, *159*, 105409. [CrossRef]
4. Wasim, M.; Ashraf, M.; Tayyaba, S.; Nazir, A. Simulation and synthesis of ZnO nanorods on AAO nano porous template for use in a mems devices. *Dig. J. Nanomater. Biostruct.* **2019**, *14*, 559–567.
5. Liu, H.; Liu, J.; Xie, X.; Li, X. Development of photo-magnetic drug delivery system by facile-designed dual stimuli-responsive modified biopolymeric chitosan capped nano-vesicle to improve efficiency in the anesthetic effect and its biological investigations. *J. Photochem. Photobiol. B Biol.* **2020**, *202*, 111716. [CrossRef] [PubMed]
6. Andryukov, B.G.; Besednova, N.N.; Romashko, R.V.; Zaporozhets, T.S.; Efimov, T.A. Label-free biosensors for laboratory-based diagnostics of infections: Current achievements and new trends. *Biosensors* **2020**, *10*, 11. [CrossRef]
7. Souza, J.M.T.; de Araujo, A.R.; de Carvalho, A.M.A.; Amorim, A.d.G.N.; Daboit, T.C.; de Almeida, J.R.d.S.; da Silva, D.A.; Eaton, P. Sustainably produced cashew gum-capped zinc oxide nanoparticles show antifungal activity against Candida parapsilosis. *J. Clean. Prod.* **2020**, *247*, 119085. [CrossRef]
8. Manjula, N.; Chen, S.-M. One-pot synthesis of rod-shaped gadolinia doped zinc oxide decorated on graphene oxide composite as an efficient electrode material for isoprenaline sensor. *Compos. Part B Eng.* **2021**, *211*, 108631. [CrossRef]
9. Chen, Z.; Zhang, D.; Zhang, Y.; Zhang, H.; Zhang, S. Influence of multi-dimensional nanomaterials composite form on thermal and ultraviolet oxidation aging resistances of SBS modified asphalt. *Constr. Build. Mater.* **2021**, *273*, 122054. [CrossRef]
10. Islam, S.E.; Hang, D.-R.; Chen, C.-H.; Chou, M.M.; Liang, C.-T.; Sharma, K.H. Rational design of hetero-dimensional C-ZnO/MoS2 nanocomposite anchored on 3D mesoporous carbon framework towards synergistically enhanced stability and efficient visible-light-driven photocatalytic activity. *Chemosphere* **2021**, *266*, 129148. [CrossRef] [PubMed]
11. Aaryashree, A.; Mandal, B.; Biswas, A.; Bhardwaj, R.; Agarwal, A.; Das, A.K.; Mukherjee, S. Mesoporous Tyrosine Functionalized BTC-ZnO Composite for Highly Selective Capacitive CO Sensor. *IEEE Sens. J.* **2020**, *21*, 2610–2617. [CrossRef]
12. Gadisa, B.T.; Appiah-Ntiamoah, R.; Kim, H. In-situ derived hierarchical ZnO/Zn-C nanofiber with high photocatalytic activity and recyclability under solar light. *Appl. Surf. Sci.* **2019**, *491*, 350–359. [CrossRef]
13. He, J.; Zhang, Y.; Guo, Y.; Rhodes, G.; Yeom, J.; Li, H.; Zhang, W. Photocatalytic degradation of cephalexin by ZnO nanowires under simulated sunlight: Kinetics, influencing factors, and mechanisms. *Environ. Int.* **2019**, *132*, 105105. [CrossRef] [PubMed]
14. Messih, M.A.; Shalan, A.E.; Sanad, M.F.; Ahmed, M. Facile approach to prepare ZnO@ $SiO_2$ nanomaterials for photocatalytic degradation of some organic pollutant models. *J. Mater. Sci. Mater. Electron.* **2019**, *30*, 14291–14299. [CrossRef]
15. Ahmad, R.; Majhi, S.M.; Zhang, X.; Swager, T.M.; Salama, K.N. Recent progress and perspectives of gas sensors based on vertically oriented ZnO nanomaterials. *Adv. Colloid Interface Sci.* **2019**, *270*, 1–27. [CrossRef] [PubMed]
16. Azeem, M.; Noman, M.T.; Wiener, J.; Petru, M.; Louda, P. Structural design of efficient fog collectors: A review. *Environ. Technol. Innov.* **2020**, *20*, 101169. [CrossRef]

17. Ali, A.; Sattar, M.; Riaz, T.; Khan, B.A.; Awais, M.; Militky, J.; Noman, M.T. Highly stretchable durable electro-thermal conductive yarns made by deposition of carbon nanotubes. *J. Text. Inst.* **2020**. [CrossRef]
18. Noman, M.T.; Amor, N.; Petru, M. Synthesis and applications of ZnO nanostructures (ZONSs): A review. *Crit. Rev. Solid State Mater. Sci.* **2021**. [CrossRef]
19. Zhang, Q.; Dandeneau, C.S.; Zhou, X.; Cao, G. ZnO nanostructures for dye-sensitized solar cells. *Adv. Mater.* **2009**, *21*, 4087–4108. [CrossRef]
20. Aditya, A.; Chattopadhyay, S.; Jha, D.; Gautam, H.K.; Maiti, S.; Ganguli, M. Zinc oxide nanoparticles dispersed in ionic liquids show high antimicrobial efficacy to skin-specific bacteria. *ACS Appl. Mater. Interfaces* **2018**, *10*, 15401–15411. [CrossRef] [PubMed]
21. Yang, T.; Hu, L.; Xiong, X.; Petrů, M.; Noman, M.T.; Mishra, R.; Militký, J. Sound Absorption Properties of Natural Fibers: A Review. *Sustainability* **2020**, *12*, 8477. [CrossRef]
22. Ali, A.; Nguyen, N.H.; Baheti, V.; Ashraf, M.; Militky, J.; Mansoor, T.; Noman, M.T.; Ahmad, S. Electrical conductivity and physiological comfort of silver coated cotton fabrics. *J. Text. Inst.* **2018**, *109*, 620–628. [CrossRef]
23. Noman, M.T.; Petrů, M. Functional properties of sonochemically synthesized zinc oxide nanoparticles and cotton composites. *Nanomaterials* **2020**, *10*, 1661. [CrossRef] [PubMed]
24. Noman, M.T.; Petru, M.; Amor, N.; Yang, T.; Mansoor, T. Thermophysiological comfort of sonochemically synthesized nano $TiO_2$ coated woven fabrics. *Sci. Rep.* **2020**, *10*, 17204. [CrossRef]
25. Behera, P.; Noman, M.T.; Petrů, M. Enhanced Mechanical Properties of Eucalyptus-Basalt-Based Hybrid-Reinforced Cement Composites. *Polymers* **2020**, *12*, 2837. [CrossRef] [PubMed]
26. Jamshaid, H.; Mishra, R.; Militky, J.; Pechociakova, M.; Noman, M.T. Mechanical, thermal and interfacial properties of green composites from basalt and hybrid woven fabrics. *Fibers Polym.* **2016**, *17*, 1675–1686. [CrossRef]
27. Shetti, N.P.; Malode, S.J.; Nayak, D.S.; Bagihalli, G.B.; Kalanur, S.S.; Malladi, R.S.; Reddy, C.V.; Aminabhavi, T.M.; Reddy, K.R. Fabrication of ZnO nanoparticles modified sensor for electrochemical oxidation of methdilazine. *Appl. Surf. Sci.* **2019**, *496*, 143656. [CrossRef]
28. Zhang, D.; Yang, Z.; Li, P.; Zhou, X. Ozone gas sensing properties of metal-organic frameworks-derived $In_2O_3$ hollow microtubes decorated with ZnO nanoparticles. *Sens. Actuators B Chem.* **2019**, *301*, 127081. [CrossRef]
29. Jamshaid, H.; Mishra, R.; Militký, J.; Noman, M.T. Interfacial performance and durability of textile reinforced concrete. *J. Text. Inst.* **2018**, *109*, 879–890. [CrossRef]
30. Mansoor, T.; Hes, L.; Bajzik, V.; Noman, M.T. Novel method on thermal resistance prediction and thermo-physiological comfort of socks in a wet state. *Text. Res. J.* **2020**, *90*, 1987–2006. [CrossRef]
31. Kim, D.; Yong, K. Boron doping induced charge transfer switching of a C3N4/ZnO photocatalyst from Z-scheme to type II to enhance photocatalytic hydrogen production. *Appl. Catal. B Environ.* **2021**, *282*, 119538. [CrossRef]
32. Xiao, B.; Huang, Q.; Chen, H.; Chen, X.; Long, G. A fractal model for capillary flow through a single tortuous capillary with roughened surfaces in fibrous porous media. *Fractals* **2021**, *29*, 2150017. [CrossRef]
33. Xiao, B.; Zhang, Y.; Wang, Y.; Jiang, G.; Liang, M.; Chen, X.; Long, G. A fractal model for Kozeny–Carman constant and dimensionless permeability of fibrous porous media with roughened surfaces. *Fractals* **2019**, *27*, 1950116. [CrossRef]
34. Maharjan, B.; Park, J.; Kaliannagounder, V.K.; Awasthi, G.P.; Joshi, M.K.; Park, C.H.; Kim, C.S. Regenerated cellulose nanofiber reinforced chitosan hydrogel scaffolds for bone tissue engineering. *Carbohydr. Polym.* **2021**, *251*, 117023. [CrossRef]
35. Zhang, Z.; Fang, Z.; Xiang, Y.; Liu, D.; Xie, Z.; Qu, D.; Sun, M.; Tang, H.; Li, J. Cellulose-based material in lithium-sulfur batteries: A Review. *Carbohydr. Polym.* **2020**, *255*, 117469. [CrossRef] [PubMed]
36. Sun, Y.; Chu, Y.; Wu, W.; Xiao, H. Nanocellulose-based Lightweight Porous Materials: A Review. *Carbohydr. Polym.* **2020**, *255*, 117489. [CrossRef] [PubMed]
37. Ulu, A.; Birhanlı, E.; Köytepe, S.; Ateş, B. Chitosan/polypropylene glycol hydrogel composite film designed with TiO2 nanoparticles: A promising scaffold of biomedical applications. *Int. J. Biol. Macromol.* **2020**, *163*, 529–540. [CrossRef] [PubMed]
38. Ahmed, M.; Hameed, B.; Hummadi, E. Review on recent progress in chitosan/chitin-carbonaceous material composites for the adsorption of water pollutants. *Carbohydr. Polym.* **2020**, *247*, 116690. [CrossRef] [PubMed]
39. Nasrollahzadeh, M.; Sajjadi, M.; Iravani, S.; Varma, R.S. Starch, cellulose, pectin, gum, alginate, chitin and chitosan derived (nano) materials for sustainable water treatment: A review. *Carbohydr. Polym.* **2020**, *251*, 116986. [CrossRef]
40. Theerthagiri, J.; Salla, S.; Senthil, R.; Nithyadharseni, P.; Madankumar, A.; Arunachalam, P.; Maiyalagan, T.; Kim, H.-S. A review on ZnO nanostructured materials: Energy, environmental and biological applications. *Nanotechnology* **2019**, *30*, 392001. [CrossRef]
41. Nada, A.A.; El Aref, A.T.; Sharaf, S.S. The synthesis and characterization of zinc-containing electrospun chitosan/gelatin derivatives with antibacterial properties. *Int. J. Biol. Macromol.* **2019**, *133*, 538–544. [CrossRef] [PubMed]
42. Noman, M.T.; Militky, J.; Wiener, J.; Saskova, J.; Ashraf, M.A.; Jamshaid, H.; Azeem, M. Sonochemical synthesis of highly crystalline photocatalyst for industrial applications. *Ultrasonics* **2018**, *83*, 203–213. [CrossRef]
43. Abramova, A.V.; Abramov, V.O.; Bayazitov, V.M.; Voitov, Y.; Straumal, E.A.; Lermontov, S.A.; Cherdyntseva, T.A.; Braeutigam, P.; Weiße, M.; Günther, K. A sol-gel method for applying nanosized antibacterial particles to the surface of textile materials in an ultrasonic field. *Ultrason. Sonochem.* **2020**, *60*, 104788. [CrossRef] [PubMed]
44. Barreto, G.P.; Morales, G.; Quintanilla, M.L.L. Microwave assisted synthesis of ZnO nanoparticles: Effect of precursor reagents, temperature, irradiation time, and additives on nano-ZnO morphology development. *J. Mater.* **2013**, *2013*, 478681. [CrossRef]

45. Noman, M.T.; Petru, M.; Amor, N.; Louda, P. Thermophysiological comfort of zinc oxide nanoparticles coated woven fabrics. *Sci. Rep.* **2020**, *10*, 21080. [CrossRef] [PubMed]
46. Król, A.; Pomastowski, P.; Rafińska, K.; Railean-Plugaru, V.; Buszewski, B. Zinc oxide nanoparticles: Synthesis, antiseptic activity and toxicity mechanism. *Adv. Colloid Interface Sci.* **2017**, *249*, 37–52. [CrossRef] [PubMed]
47. Pala, R.G.S.; Metiu, H. The structure and energy of oxygen vacancy formation in clean and doped, very thin films of ZnO. *J. Phys. Chem. C* **2007**, *111*, 12715–12722. [CrossRef]
48. Ong, C.B.; Ng, L.Y.; Mohammad, A.W. A review of ZnO nanoparticles as solar photocatalysts: Synthesis, mechanisms and applications. *Renew. Sustain. Energy Rev.* **2018**, *81*, 536–551. [CrossRef]
49. Noman, M.T.; Ashraf, M.A.; Ali, A. Synthesis and applications of nano-$TiO_2$: A review. *Environ. Sci. Pollut. Res* **2019**, *26*, 3262–3291. [CrossRef]
50. Hassan, N.; Hashim, M.; Bououdina, M. One-dimensional ZnO nanostructure growth prepared by thermal evaporation on different substrates: Ultraviolet emission as a function of size and dimensionality. *Ceram. Int.* **2013**, *39*, 7439–7444. [CrossRef]
51. Ju, D.; Xu, H.; Zhang, J.; Guo, J.; Cao, B. Direct hydrothermal growth of ZnO nanosheets on electrode for ethanol sensing. *Sens. Actuators B Chem.* **2014**, *201*, 444–451. [CrossRef]
52. Noman, M.T.; Wiener, J.; Saskova, J.; Ashraf, M.A.; Vikova, M.; Jamshaid, H.; Kejzlar, P. In-situ development of highly photocatalytic multifunctional nanocomposites by ultrasonic acoustic method. *Ultrason. Sonochem.* **2018**, *40*, 41–56. [CrossRef] [PubMed]
53. Yue, S.; Lu, J.; Zhang, J. Synthesis of three-dimensional ZnO superstructures by a one-pot solution process. *Mater. Chem. Phys.* **2009**, *117*, 4–8. [CrossRef]
54. Smijs, T.G.; Pavel, S. Titanium dioxide and zinc oxide nanoparticles in sunscreens: Focus on their safety and effectiveness. *Nanotech. Sci. Appl.* **2011**, *4*, 95. [CrossRef] [PubMed]
55. Rasmussen, J.W.; Martinez, E.; Louka, P.; Wingett, D.G. Zinc oxide nanoparticles for selective destruction of tumor cells and potential for drug delivery applications. *Expert. Opin. Drug. Deliv.* **2010**, *7*, 1063–1077. [CrossRef] [PubMed]
56. Mishra, P.K.; Mishra, H.; Ekielski, A.; Talegaonkar, S.; Vaidya, B. Zinc oxide nanoparticles: A promising nanomaterial for biomedical applications. *Drug Discov. Today* **2017**, *22*, 1825–1834. [CrossRef]
57. Zhang, Z.-Y.; Xiong, H.-M. Photoluminescent ZnO nanoparticles and their biological applications. *Materials* **2015**, *8*, 3101–3127. [CrossRef]
58. Jiang, J.; Pi, J.; Cai, J. The advancing of zinc oxide nanoparticles for biomedical applications. *Bioinorg. Chem. Appl.* **2018**, *2018*, 1062562. [CrossRef] [PubMed]
59. Ashraf, M.A.; Wiener, J.; Farooq, A.; Saskova, J.; Noman, M.T. Development of maghemite glass fibre nanocomposite for adsorptive removal of methylene blue. *Fibers Polym.* **2018**, *19*, 1735–1746. [CrossRef]
60. Noman, M.T.; Ashraf, M.A.; Jamshaid, H.; Ali, A. A novel green stabilization of $TiO_2$ nanoparticles onto cotton. *Fibers Polym.* **2018**, *19*, 2268–2277. [CrossRef]
61. Haque, F.Z.; Nandanwar, R.; Singh, P.; Dharavath, K.; Syed, F.F. Effect of Different Acids and Solvents on Optical Properties of $SiO_2$ Nanoparticles Prepared by the Sol-Gel Process. *Silicon* **2018**, *10*, 413–419. [CrossRef]
62. Khodadadi, A.; Farahmandjou, M.; Yaghoubi, M. Investigation on synthesis and characterization of Fe-doped $Al_2O_3$ nanocrystals by new sol–gel precursors. *Mater. Res. Exp* **2018**, *6*, 025029. [CrossRef]
63. Abebe, B.; Murthy, H.A.; Zerefa, E.; Adimasu, Y. PVA assisted ZnO based mesoporous ternary metal oxides nanomaterials: Synthesis, optimization, and evaluation of antibacterial activity. *Mater. Res. Exp* **2020**, *7*, 045011. [CrossRef]
64. Sui, R.; Charpentier, P. Synthesis of metal oxide nanostructures by direct sol–gel chemistry in supercritical fluids. *Chem. Rev.* **2012**, *112*, 3057–3082. [CrossRef]
65. Alias, S.; Ismail, A.; Mohamad, A. Effect of pH on ZnO nanoparticle properties synthesized by sol–gel centrifugation. *J. Alloys Compd.* **2010**, *499*, 231–237. [CrossRef]
66. Gupta, S.; Chang, C.; Anbalagan, A.K.; Lee, C.-H.; Tai, N.-H. Reduced graphene oxide/zinc oxide coated wearable electrically conductive cotton textile for high microwave absorption. *Compos. Sci. Technol.* **2020**, *188*, 107994. [CrossRef]
67. Gong, B.; Shi, T.; Liao, G.; Li, X.; Huang, J.; Zhou, T.; Tang, Z. UV irradiation assisted growth of ZnO nanowires on optical fiber surface. *Appl. Surf. Sci.* **2017**, *406*, 294–300. [CrossRef]
68. Koutavarapu, R.; Reddy, C.V.; Syed, K.; Reddy, K.R.; Shetti, N.P.; Aminabhavi, T.M.; Shim, J. Ultra-small zinc oxide nanosheets anchored onto sodium bismuth sulfide nanoribbons as solar-driven photocatalysts for removal of toxic pollutants and phtotoelectrocatalytic water oxidation. *Chemosphere* **2020**, *267*, 128559. [CrossRef] [PubMed]
69. Shetti, N.P.; Malode, S.J.; Ilager, D.; Raghava Reddy, K.; Shukla, S.S.; Aminabhavi, T.M. A novel electrochemical sensor for detection of molinate using ZnO nanoparticles loaded carbon electrode. *Electroanalysis* **2019**, *31*, 1040–1049. [CrossRef]
70. Noman, M.T.; Petru, M.; Militký, J.; Azeem, M.; Ashraf, M.A. One-Pot Sonochemical Synthesis of ZnO Nanoparticles for Photocatalytic Applications, Modelling and Optimization. *Materials* **2020**, *13*, 14. [CrossRef] [PubMed]
71. Ezeh, C.I.; Yang, X.; He, J.; Snape, C.; Cheng, X.M. Correlating ultrasonic impulse and addition of ZnO promoter with $CO_2$ conversion and methanol selectivity of $CuO/ZrO_2$ catalysts. *Ultrason. Sonochem.* **2018**, *42*, 48–56. [CrossRef]
72. Salmeri, M.; Ognibene, G.; Saitta, L.; Lombardo, C.; Genovese, C.; Barcellona, M.; D'Urso, A.; Spitaleri, L.; Blanco, I.; Cicala, G. Optimization of ZnO Nanorods Growth on Polyetheresulfone Electrospun Mats to Promote Antibacterial Properties. *Molecules* **2020**, *25*, 1696. [CrossRef] [PubMed]

73. Gnaneshwar, P.V.; Sudakaran, S.V.; Abisegapriyan, S.; Sherine, J.; Ramakrishna, S.; Rahim, M.H.A.; Yusoff, M.M.; Jose, R.; Venugopal, J.R. Ramification of zinc oxide doped hydroxyapatite biocomposites for the mineralization of osteoblasts. *Mater. Sci. Eng. C* **2019**, *96*, 337–346. [CrossRef] [PubMed]
74. Wang, M.; Zhang, M.; Pang, L.; Yang, C.; Zhang, Y.; Hu, J.; Wu, G. Fabrication of highly durable polysiloxane-zinc oxide (ZnO) coated polyethylene terephthalate (PET) fabric with improved ultraviolet resistance, hydrophobicity, and thermal resistance. *J. Colloid Interface Sci.* **2019**, *537*, 91–100. [CrossRef] [PubMed]
75. Priya, A.; Arumugam, M.; Arunachalam, P.; Al-Mayouf, A.M.; Madhavan, J.; Theerthagiri, J.; Choi, M.Y. Fabrication of visible-light active BiFeWO6/ZnO nanocomposites with enhanced photocatalytic activity. *Colloids Surf. Phys. Eng. Asp.* **2020**, *586*, 124294.
76. Jellal, I.; Nouneh, K.; Toura, H.; Boutamart, M.; Briche, S.; Naja, J.; Soucase, B.M.; Touhami, M.E. Enhanced photocatalytic activity of supported Cu-doped ZnO nanostructures prepared by SILAR method. *Opt. Mater.* **2021**, *111*, 110669. [CrossRef]
77. Sbardella, F.; Rivilla, I.; Bavasso, I.; Russo, P.; Vitiello, L.; Tirillò, J.; Sarasini, F. Zinc oxide nanostructures and stearic acid as surface modifiers for flax fabrics in polylactic acid biocomposites. *Int. J. Biol. Macromol.* **2021**, *177*, 495–504. [CrossRef] [PubMed]
78. Rabani, I.; Lee, S.-H.; Kim, H.-S.; Yoo, J.; Park, Y.-R.; Maqbool, T.; Bathula, C.; Jamil, Y.; Hussain, S.; Seo, Y.-S. Suppressed photocatalytic activity of ZnO based Core@ Shell and RCore@ Shell nanostructure incorporated in the cellulose nanofiber. *Chemosphere* **2021**, *269*, 129311. [CrossRef] [PubMed]
79. Raja, A.; Ashokkumar, S.; Marthandam, R.P.; Jayachandiran, J.; Khatiwada, C.P.; Kaviyarasu, K.; Raman, R.G.; Swaminathan, M. Eco-friendly preparation of zinc oxide nanoparticles using Tabernaemontana divaricata and its photocatalytic and antimicrobial activity. *J. Photochem. Photobiol. B Biol.* **2018**, *181*, 53–58. [CrossRef]
80. Khan, S.A.; Noreen, F.; Kanwal, S.; Iqbal, A.; Hussain, G. Green synthesis of ZnO and Cu-doped ZnO nanoparticles from leaf extracts of Abutilon indicum, Clerodendrum infortunatum, Clerodendrum inerme and investigation of their biological and photocatalytic activities. *Mater. Sci. Eng. C* **2018**, *82*, 46–59. [CrossRef] [PubMed]
81. Ghayempour, S.; Montazer, M. Ultrasound irradiation based in-situ synthesis of star-like Tragacanth gum/zinc oxide nanoparticles on cotton fabric. *Ultrason. Sonochem.* **2017**, *34*, 458–465. [CrossRef]
82. Chakrabarti, S.; Banerjee, P. Preparation and characterization of multifunctional cotton fabric by coating with sonochemically synthesized zinc oxide nanoparticle-flakes and a novel approach to monitor its self-cleaning property. *J. Text. Inst.* **2015**, *106*, 963–969. [CrossRef]
83. Noman, M.T.; Petru, M. Effect of Sonication and Nano $TiO_2$ on Thermophysiological Comfort Properties of Woven Fabrics. *ACS Omega* **2020**, *5*, 11481–11490. [CrossRef] [PubMed]
84. Nair, M.G.; Nirmala, M.; Rekha, K.; Anukaliani, A. Structural, optical, photo catalytic and antibacterial activity of ZnO and Co doped ZnO nanoparticles. *Mater. Lett.* **2011**, *65*, 1797–1800. [CrossRef]
85. Kaur, J.; Singhal, S. Facile synthesis of ZnO and transition metal doped ZnO nanoparticles for the photocatalytic degradation of Methyl Orange. *Ceram. Int.* **2014**, *40*, 7417–7424. [CrossRef]
86. Bukkitgar, S.D.; Shetti, N.P.; Kulkarni, R.M.; Reddy, K.R.; Shukla, S.S.; Saji, V.S.; Aminabhavi, T.M. Electro-catalytic behavior of Mg-doped ZnO nano-flakes for oxidation of anti-inflammatory drug. *J. Electrochem. Soc.* **2019**, *166*, B3072. [CrossRef]
87. Bukkitgar, S.; Shetti, N.; Kulkarni, R.; Nandibewoor, S. Electro-sensing base for mefenamic acid on a 5% barium-doped zinc oxide nanoparticle modified electrode and its analytical application. *RSC Adv.* **2015**, *5*, 104891–104899. [CrossRef]
88. Karthikeyan, C.; Arunachalam, P.; Ramachandran, K.; Al-Mayouf, A.M.; Karuppuchamy, S. Recent advances in semiconductor metal oxides with enhanced methods for solar photocatalytic applications. *J. Alloys Compd.* **2020**, *828*, 154281. [CrossRef]
89. Shouli, B.; Liangyuan, C.; Dianqing, L.; Wensheng, Y.; Pengcheng, Y.; Zhiyong, L.; Aifan, C.; Liu, C.C. Different morphologies of ZnO nanorods and their sensing property. *Sens. Actuators B Chem.* **2010**, *146*, 129–137. [CrossRef]
90. Al-Gaashani, R.; Radiman, S.; Daud, A.; Tabet, N.; Al-Douri, Y. XPS and optical studies of different morphologies of ZnO nanostructures prepared by microwave methods. *Ceram. Int.* **2013**, *39*, 2283–2292. [CrossRef]

polymers

MDPI

*Article*

# Geometric Analysis of Three-Dimensional Woven Fabric with in-Plane Auxetic Behavior

**Muhammad Zeeshan [1], Hong Hu [1,2,*] and Ehsan Etemadi [1,3]**

[1] School of Fashion and Textiles, The Hong Kong Polytechnic University, Hong Kong, China
[2] Research Institute for Intelligent Wearable Systems, The Hong Kong Polytechnic University, Hong Kong, China
[3] Department of Mechanical Engineering, Hakim Sabzevari University, Sabzevar 9617976487, Iran
* Correspondence: hu.hong@polyu.edu.hk

**Abstract:** Auxetic textiles are emerging as an enticing option for many advanced applications due to their unique deformation behavior under tensile loading. This study reports the geometrical analysis of three-dimensional (3D) auxetic woven structures based on semi-empirical equations. The 3D woven fabric was developed with a special geometrical arrangement of warp (multi-filament polyester), binding (polyester-wrapped polyurethane), and weft yarns (polyester-wrapped polyurethane) to achieve an auxetic effect. The auxetic geometry, the unit cell resembling a re-entrant hexagon, was modeled at the micro-level in terms of the yarn's parameters. The geometrical model was used to establish a relationship between the Poisson's ratio (PR) and the tensile strain when it was stretched along the warp direction. For validation of the model, the experimental results of the developed woven fabrics were correlated with the calculated results from the geometrical analysis. It was found that the calculated results were in good agreement with the experimental results. After experimental validation, the model was used to calculate and discuss critical parameters that affect the auxetic behavior of the structure. Thus, geometrical analysis is believed to be helpful in predicting the auxetic behavior of 3D woven fabrics with different structural parameters.

**Keywords:** auxetic woven fabrics; 3D textile structure; negative Poisson's ratio; geometrical analysis

**Citation:** Zeeshan, M.; Hu, H.; Etemadi, E. Geometric Analysis of Three-Dimensional Woven Fabric with in-Plane Auxetic Behavior. *Polymers* **2023**, *15*, 1326. https://doi.org/10.3390/polym15051326

Academic Editors: Michal Petrů and Muhammad Tayyab Noman

Received: 18 January 2023
Revised: 1 March 2023
Accepted: 2 March 2023
Published: 6 March 2023



## 1. Introduction

Auxetic textile is a sub-class of meta-materials that are differentiated from conventional materials due to their negative Poisson's ratio (NPR) [1]. So far, scientists have successfully developed fibers [2,3], polymers [4,5], yarns [6–8], and fabrics [9,10] with the uncommon property of an NPR. Textile fabrics with an NPR mean that they expand when they are subjected to tensile loading and shrink when compressive loading is applied. Due to this unusual response to the applied loading, auxetic fabrics show outstanding extensibility in both warp and weft directions, better drapability, and become more compact upon compression [1]. Therefore, auxetic fabrics are considered to be the first choice for applications in protective textiles, medical textiles, smart textiles, sportswear, and auxetic composites [11]. Researchers have conducted an immense amount of work to develop auxetic fabrics in every possible way. Therefore, auxetic fabrics could be realized in woven form two-dimensional (2D) auxetic woven fabrics [12–14], three-dimensional (3D) auxetic textiles [15–17], knitted form (warp-knitted auxetic fabrics [18,19], weft-knitted auxetic fabrics [20,21]), non-woven form [22,23], and auxetic laminated fabrics [24,25]. Mainly, there are two techniques for introducing auxetic property into textile fabrics. The first and more simple technique is to use auxetic yarns and a conventional weave pattern. This technique was first introduced by Miller et al. [6], and they used their invented double helix yarn (DHY) in weft direction of a plain-woven fabric. The reason for using the DHY in weft direction was to ensure the out-of-register position of each neighboring DHY strand to maximize the auxetic effect. When the fabric was subjected to tensile tension, the DHY strands overlap each other in the thickness direction, which causes an out-of-plane auxetic effect by increasing the thickness of the sample. In this study, the maximum achieved out-of-plane Poisson's ratio (PR) was −0.1. Similarly, the same research group also developed auxetic plain-woven fabrics using helical auxetic yarn (HAY) in warp direction and multi-filament polyester yarn in weft direction [26]. This study concluded that the auxetic effect is associated with the weft yarn's material and weave geometry. Recently, auxetic woven fabrics were produced using auxetic plied yarns (4 ply and 6 ply) and DHY [27]. For fabric production, auxetic yarns were used in the warp direction only, while conventional elastic yarn was used in the weft direction. Three weave designs (plain, twill, and satin) were selected, along with other design parameters of plied yarn, to evaluate their effect on the NPR of the woven fabrics. Although producing auxetic woven fabrics with this method is the easiest way, there are still many drawbacks that limit the practical application of such fabrics. Those drawbacks are: the NPR of auxetic yarn transfers partially to the fabric due to weaving constraints, the unstable auxetic effect, and the out-of-register orientation of auxetic yarns. Furthermore, the orientation of auxetic yarn inside the fabric structure is crucial and difficult, i.e., the yarns should be as straight as possible to obtain a better auxetic effect. Due to this limitation, this technique is only suitable for woven fabrics and cannot be applied to knitted fabrics.

The second technique of developing auxetic fabrics includes the use of non-auxetic yarns and the realization of special auxetic geometry. It is well known that the auxetic effect is associated with the geometrical arrangement of a structure's unit cells such as re-entrant hexagonal, rotating squares, etc. [28]. Therefore, the idea of developing auxetic fabrics progressed when scientists gave considerable attention to configuring conventional yarns in a special geometrical arrangement using knitting or weaving technologies. Up to now, various weft-knitted [20] and warp-knitted [18] auxetic fabrics have been developed using this technique. However, the low structural stability and low elastic recovery of the knitted auxetic fabrics make them impractical for many applications such as protective textiles, etc. Recently, a considerable amount of work has been reported on the development of 2D auxetic woven fabrics. For the first time, a uni-stretch auxetic woven fabric was developed using conventional weaving techniques and non-auxetic yarns [9]. Auxetic behavior was introduced to fabrics through foldable structures, re-entrant hexagonal geometry, and a rotating rectangle structure. The basic mechanism involved in the auxetic effect is the different shrinkage properties of tight weaves (plain) and loose weaves (twill or satin), along with the use of elastic and non-elastic yarns. These fabrics showed a zero PR or a very low NPR in one direction only. Therefore, another study was conducted by Zulifqar et al. [29] on developing bi-stretch auxetic woven fabrics. The methodology of the study was based on a similar principle of differential shrinkage, but elastic yarns were used in both the warp and weft directions along with non-elastic yarns. Among all the developed auxetic structures based on different geometries, the zig-zag folded stripes showed the highest NPR of −0.15. The developed 2D auxetic woven fabrics have the potential to be used for clothing, fashion garments, etc. However, these fabrics cannot be used for high-performance applications because they experience more longitudinal deformation under tension, a low NPR, and a sharp decrease in the NPR under increased strain.

Nowadays, 3D auxetic fabrics have been the central focus for many researchers because they can exhibit more extraordinary mechanical performances compared to those of 2D auxetic fabrics. Ge and Hu [17], for the first time, reported a novel 3D auxetic fabric developed by combining knitting and non-weaving techniques. After fabrication, the fabric structure was converted into a soft composite for stability. Uniquely, this fabric structure was designed to show the NPR effect under compression. Hence, it was recommended for indentation-resistant and impact-resistant applications. Similarly, 3D multi-layer auxetic woven structures having an out-of-plane NPR were developed using the differential modulus of the yarns [30]. The out-of-plane NPR effect is triggered by the Z-direction binding yarns when they are stretched longitudinally. Due to the limitation of the out-of-plane auxetic effect, the authors used their developed fabrics as a reinforcement in polymer composite. Furthermore, the authors claimed improved impact-resistant properties for the 3D auxetic woven fabric reinforced composites. Not long ago, a new milestone was achieved by developing 3D auxetic woven fabrics with in-plane NPR effect [31]. The development is considered to be very significant in view of broader application aspects because it addresses the drawbacks associated with the existing 2D and 3D auxetic textiles. Furthermore, the fabrics that exhibit in-plane auxetic behavior at a higher tensile tension are useful for releasing the contact pressure when they are used in applications such as automotive seat belts, garment belts, etc. Although significant interests have been shown on the design, fabrication, and characteristics of 3D auxetic woven structure, the important information on the fundamental geometrical analytics of structure is yet lacking.

In this study, a geometrical model of 3D woven fabric has been proposed at the micro-level (yarn level) to establish a relationship between the geometrical parameters (in terms of yarn parameters) and the tensile deformation. Since the geometrical analysis is carried out at the micro-level, the geometrical arrangement and the physical properties of yarns are very crucial to facilitate the geometrical model. Thus, a 3D auxetic woven fabric was fabricated to carefully observe the arrangements of the warp and weft yarns, forming auxetic geometry. After observing the geometry in a free state, the fabric was then subject to tensile deformation to examine the behavior of the yarns together with the geometrical deformation of the whole structure. Here, it was found that the structure extends longitudinally and axially under tensile deformation due to its auxetic nature, which causes tension and compression to the weft yarns. Thus, by recognizing the geometry and physical behavior of yarns, semi-empirical equations were drawn based on the geometrical arrangement of the warp and weft yarns, including the tension and compression effect of the yarns. The study shows that the extracted results from the established semi-empirical equations are in good agreement with the experimental results. Therefore, it could be said that the proposed geometrical model will be a useful tool to design, predict, and optimize the auxetic behavior of the 3D auxetic woven fabric.

## 2. Methodology

A comprehensive methodology was adopted to propose an effective geometrical model for the 3D auxetic woven structure. First, the 3D auxetic woven fabric structure was designed and developed with appropriate warp and weft yarns. Then, the fabric was subjected to a tensile loading to observe its deformation behavior at the micro-level and macro-level. Finally, a geometrical model was proposed by establishing semi-empirical equations based on the warp and weft yarn's geometry at the initial and deformed states.

### 2.1. Structure Design and Fabric Formation

A novel multi-layer 3D auxetic woven structure was previously designed by modifying the conventional 3D orthogonal through-the-thickness structure [31]. The auxetic effect was introduced by featuring an unusual lateral crimp to the warp yarns, which form a special geometry resembling a re-entrant hexagon, as shown in Figure 1a. The unusual lateral crimp of the warp yarns was induced by using two different weft yarn systems, i.e., a coarser, non-elastic yarn and a fine elastic yarn. A solid, coarser yarn was inserted with a 2/2 twill pattern as a binder (through the thickness), creating alternate empty spaces within the warp yarns. Meanwhile, multiple insertions of fine elastic yarn were made in stretched form, which shrank the structure in a relaxed state by forcing the warp yarns to crimp laterally, filling those empty spaces. As a result, an unusual lateral crimp, similar to a sinusoidal wave, was induced in the warp yarns. When this structure is stretched longitudinally, the warp yarns try to become straight, pushing the coarse binding yarns in a lateral direction. Thus, under tensile extension, the width of the structure also increases, and an auxetic effect is achieved.

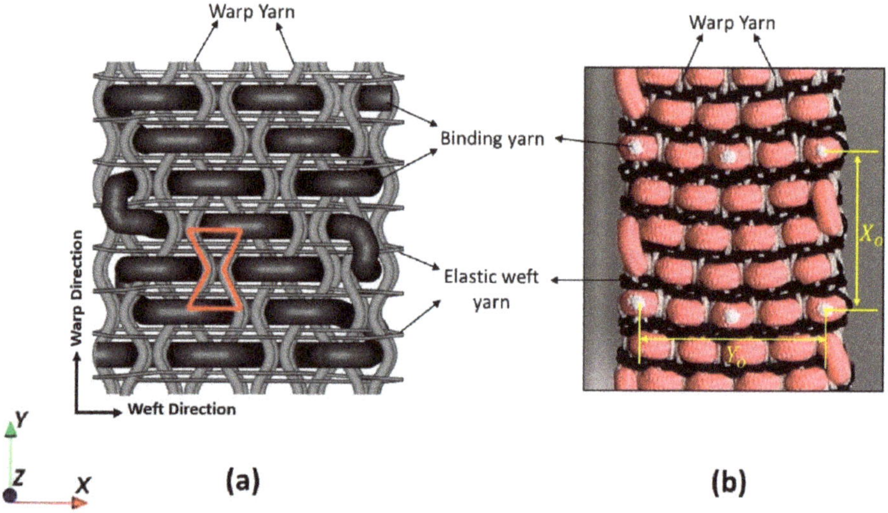


**Figure 1.** Three dimensional auxetic woven structure: (**a**) schematic representation; (**b**) real fabric.

For sample fabrication, the three types of polymer-based yarn were procured from Wai Hung Weaving Factory Limited with the desired properties. A single type of yarn made of multi-filament polyester with a 0.7 mm diameter was selected as a warp. Two types of yarn, namely coarse binding yarn and elastic yarn, made of polyester-wrapped polyurethane (PU) having diameters of 2.5 mm and 0.62 mm, respectively, were selected as the weft. In addition, other features of the warp yarn include stability and strength because it will bear tensile loading. Similarly, the coarse binding yarn was solid, but easier to bend to facilitate orthogonal shaping during weaving. A semi-automatic weaving machine with a maximum of 16 heald frames attached to the dobby shedding mechanism was used to produce the fabric sample. The required number of warp yarns were passed through 8 heald frames with a straight drawing-in draft to weave a two-layer 3D woven fabric. A special denting plan, i.e., two yarns per dent followed by an empty dent, was used. This denting sequence kept a gap between the warp yarns which facilitate the insertion of coarse binding yarn. Furthermore, elastic yarn was inserted in a stretch form to shrink the structure along the width direction when the sample was cut off the loom. The developed fabric sample is shown in Figure 1b.

It is worth mentioning that, in a previous study [31], the 3D auxetic woven structure was investigated with four highly influencing parameters: diameter of binding yarn, bending stiffness of binding yarn, repeats of elastic weft yarn, and stretch percentage of elastic weft yarn. Thereupon, it was concluded that the auxetic behavior of the 3D woven fabrics is primarily associated with the yarn's properties that limit the weaving process. For example, the higher diameter and stiffness of binding yarn were difficult to weave to achieve a favorable 3D auxetic structure. Therefore, it is preferred to select and develop a fabric sample with the optimized parameters having the maximum auxetic effect for validating the geometric model of the structure because it is considered that a sample with maximum auxetic behavior indicates the best formation of auxetic geometry. The physical properties of the three types of yarn used to develop the sample are given in Table 1. The bending stiffness of the binding yarn was calculated with the beam deflection method reported by Msalilwa et al. [32].

**Table 1.** Properties of the yarns used for sample preparation.

| Yarn | Material | Type | Diameter (mm) | Tensile Modulus (MPa) | Tensile Strength (N) | Bending Stiffness ($\times 10^{-6}$ Nm$^2$) |
|---|---|---|---|---|---|---|
| Warp yarn | Polyester multi-filament | Braided yarn | 0.70 | 153 | 106 | - |
| Binding yarn | Polyester-wrapped PU | Braided yarn | 2.28 | 8 | 203 | 0.66 |
| Elastic weft yarn | Polyester-wrapped PU | Braided yarn | 0.62 | 3 | 14 | - |

*2.2. Tensile Test and Measurement of Negative Poisson's Ratio*

To avoid possible slippage during the tensile test, aluminum tabs were mounted on both ends of each sample, covering the gripping area of the clamps. Epoxy EL2 and AT30 hardener were used to mount the aluminum tabs onto the specimen. The narrow-woven fabric samples of width 25 mm and length 150 mm were prepared according to the guidelines provided by tensile testing standard ASTM D5035-11 [33]. Further, the gauge length between the two jaws of the machine was kept at 75 mm. An Instron 5982 universal testing machine (UTM) with a maximum load capacity of 100 kN was used in this experiment. The samples were subjected to a strain-controlled tensile test, where the lower jaw was fixed, and the upper jaw moved at the rate of 30 mm per minute. The test was performed three times for each set of samples, and average values were reported.

The UTM can only provide raw data of the mechanical properties, therefore, a separate digital camera (Canon EOS 800D) was set up in front of the machine to record the change in dimensions of the fabric samples during the tensile deformation, as shown in Figure 2. Thus, each sample was marked with four dots (two longitudinally and two vertically) at a distance of 25 mm before clamping them onto the UTM. The video recording and the tensile test started simultaneously. First, the recorded videos were processed, and pictures were extracted at a rate of 1.5 s, which was equivalent to 1% tensile strain. Then, the extracted pictures were analyzed with screen ruler software to calculate the lateral and longitudinal deformation of the fabric. Finally, the PR for each sample was calculated using Equation (1).

$$\upsilon = -\frac{\varepsilon_y}{\varepsilon_x} = -\frac{(Y' - Y_o)/Y_o}{(X' - X_o)/X_o} = -\frac{(Y' - Y_o)X_o}{(X' - X_o)Y_o} \quad (1)$$

where $\varepsilon_x$ is strain in longitudinal direction and $\varepsilon_y$ is strain in the lateral direction; $Y_o$ and $X_o$ are the initial width and length as shown in Figure 1b; $Y'$ and $X'$ are width and length in the stretched state, respectively.

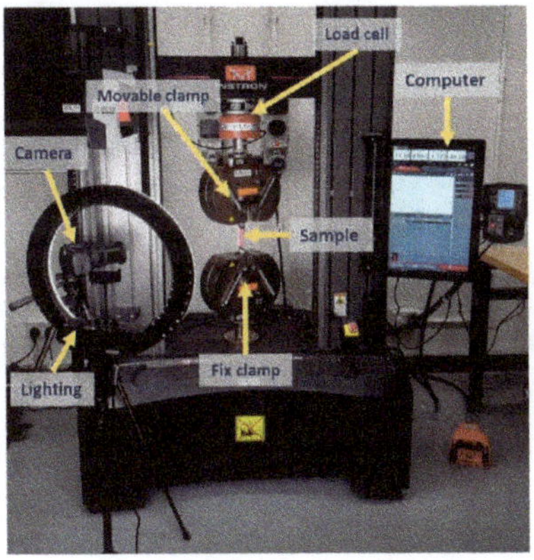


**Figure 2.** Testing setup on a universal testing machine for Poisson's ratio measurement.

*2.3. Yarns Analysis*

In a previous study [31], it was found that the auxetic effect of the 3D woven fabric is highly associated with the extension of elastic weft yarn and compression of coarse binding yarn during the tensile deformation. Therefore, the inclusion of these two characteristics of yarns are essential for developing a geometrical model.

The extension of elastic weft yarn was simply characterized by performing a tensile test on a single yarn using the Instron 5566 machine. The yarn was fixed between the machine's jaws, where the lower jaw wax fixed, and the upper jaw moved at a constant velocity of 30 mm/min. Similarly, the compression of coarse binding yarn was determined by performing a compression test. However, conventional flat compression could be applied here because the compression of the binding yarn was applied by the warp yarns. Hence, a customized assembly, as shown in Figure 3, was prepared to perform the test. The assembly was designed in such a way that the two pins, whose diameter was equal to that of the warp yarn, applied the compressive force on the binding yarn. So, the binding yarn was placed between the upper and lower parts of the assembly, perpendicular to the pins. Then, the assembly was placed between the two platens of the Instron 5566 machine. To perform the compression test, the upper platen was moved downward at a constant velocity of 1 mm/min, and the lower platen was fixed. A very sensitive load cell of capacity 50 N was used to obtain precise results.

**Figure 3.** Binding yarn compression test assembly.

*2.4. Geometrical Analysis*

The parametric terms used in this geometrical analysis are given below:

$a_o$—initial diameter of binding yarn.
$b_o$—diameter of warp yarn at the initial state and during the first stage of deformation.
$b_{ii}$—diameter of warp yarn during the second stage of deformation.
$l_o$—length of warp yarn between the centers of the two binding yarns at the initial state and during the first stage of deformation.
$l_{ii}$—length of warp yarn between the centers of the two binding yarns during the second stage of deformation.
$\theta_o$—initial inclination angle of warp yarn.
$\theta_i$—inclination angle of warp yarn in a deformed state.
$D_o$—sum of the radii of binding yarn and warp yarn at the initial state.
$D_i$—sum of the radii of binding yarn and warp yarn during the first stage of deformation.
$D_{ii}$—sum of the radii of binding yarn and warp yarn during the second stage of deformation.
$a_i$—height of binding yarn after compression.
$a_i^{'}$—width of binding yarn after compression.

A geometric model is proposed to predict the NPR effect of the 3D in-plane auxetic woven structure. The following assumptions are made based on the deformation behavior of the fabric during the tensile test:

1. Figure 4 represent the schematic diagram of the 3D auxetic woven structure while the detailed geometry of the unit cell of the structure is illustrated in Figure 5. It is assumed that all the repeating unit cells are similar in size and shape in the initial state and deformed symmetrically during extension.

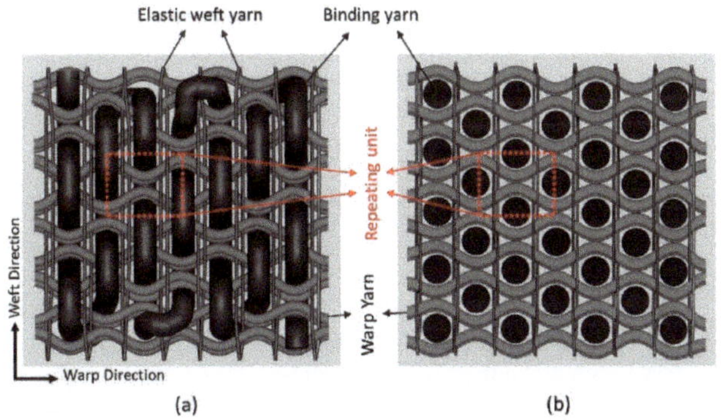


**Figure 4.** Schematic diagram of 3D auxetic woven structure: (**a**) top view; (**b**) top cross-sectional view [31].

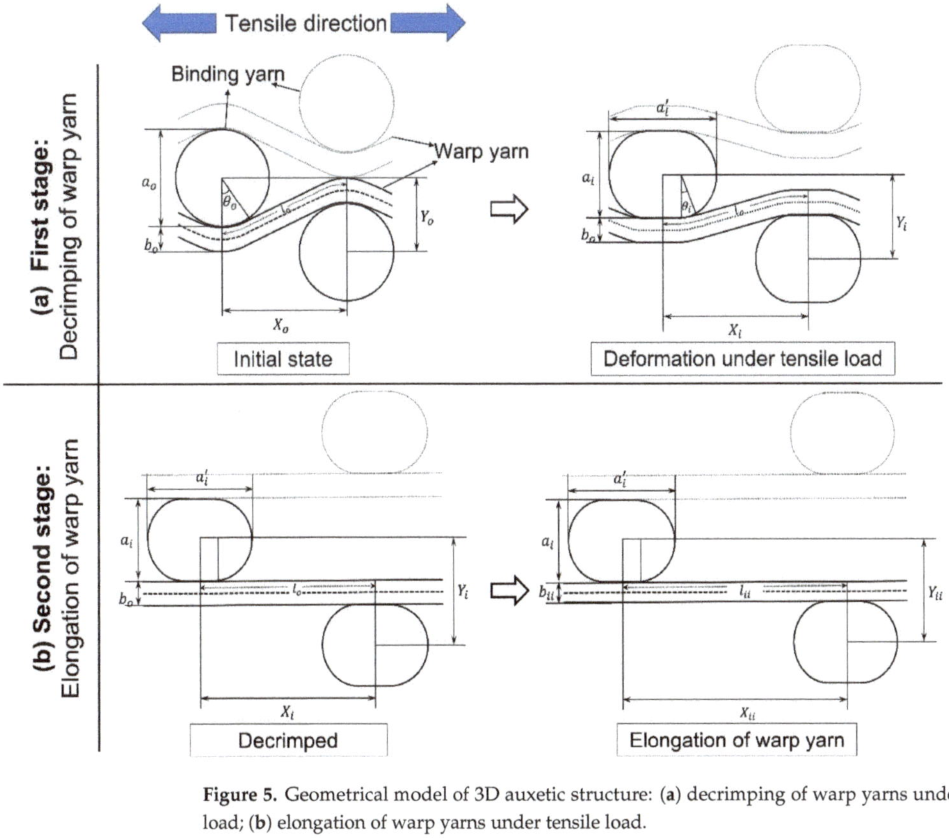


**Figure 5.** Geometrical model of 3D auxetic structure: (**a**) decrimping of warp yarns under tensile load; (**b**) elongation of warp yarns under tensile load.

2. In the geometrical model, the presentation of an elastic weft yarn has been omitted because the warp and binding yarn analysis is suitable enough to predict the auxetic behavior. However, the effect of the elastic weft yarn, i.e., causing compression to the binding yarn during tensile loading, was included.
3. The initial cross-section of the binding yarn is circular, which, during compression, changes to a race track shape due to the compression of the warp yarns caused by the restoring force of the elastic weft yarns. However, the cross-sectional area of the binding yarn remains constant all the time.
4. The warp yarns are crimped by the binding yarns at the initial state. It is assumed that the part of the warp yarn that is in contact with the binding yarn is circular, while the non-contact part is straight and tangent to the circular part.
5. There is no slippage at the contact points of binding and warp yarns during tensile deformation.
6. During tensile stretching, the structure deforms in two stages: decrimping (Figure 5a) and elongation (Figure 5b) of the warp yarns under tensile loading. Furthermore, it is also assumed that at the decrimping stage, the cross-section of binding yarn changes from a circular one to a race track one, while the diameter of warp yarn remains unchanged. However, in the elongation stage, there is no further change to the dimensions of the race-track-shaped binding yarn, while the diameter of the warp yarn changes due to elongation.
    i. First stage: In the first stage of deformation, the warp yarns start decrimping at a constant rate and become fully straight at a particular tensile strain. In this stage, the length and diameter of the warp yarn remain constant, as shown in Figure 5a. Based on the above assumptions, the geometrical model of the first stage can be re-illustrated in detail, as shown in Figure 6.

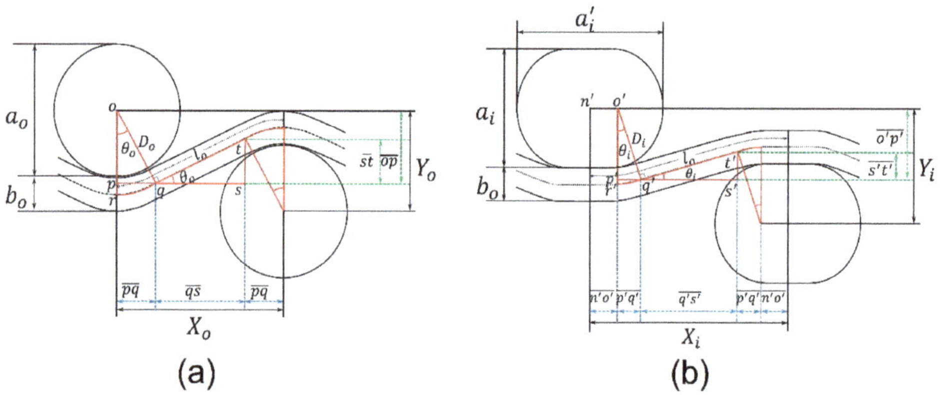


**Figure 6.** First stage of deformation of a unit-cell: (**a**) initial state; (**b**) deformed state.

From Figure 6, the initial distance ($X_o$) between the centers of two binding yarns is given by:

$$X_o = 2\overline{pq} + \overline{qs} \tag{2}$$

The length of $\overline{pq}$ and $\overline{qs}$ in terms of yarn's parameters can be calculated by solving right angle triangles $\Delta opq$ and $\Delta qst$, respectively.

$$\overline{pq} = D_o \sin\theta_o \tag{3}$$

$$\overline{qs} = \overline{qt} \cos\theta_o \tag{4}$$

Whereas:

$$\overline{qt} = l_o - 2\hat{qr} \tag{5}$$

$$\hat{qr} = D_o\theta_o \tag{6}$$

By solving Equations (4)–(6), the following relation can be obtained

$$\overline{qs} = (l_o - 2D_o\theta_o)\cos\theta_o \tag{7}$$

Substituting Equations (3) and (7) into Equation (2) gives the following relation

$$X_o = 2D_o\sin\theta_o + (l_o - 2D_o\theta_o)\cos\theta_o \tag{8}$$

Likewise, the initial height ($Y_o$) between the centers of two binding yarns is given as

$$Y_o = 2\overline{op} - \overline{st} \tag{9}$$

To solve $\Delta opq$ and $\Delta qst$, the following relation can be made for $\overline{op}$ and $\overline{st}$

$$\overline{op} = D_o\cos\theta_o \tag{10}$$

$$\overline{st} = \overline{qt}\sin\theta_o \tag{11}$$

Substituting Equations (5) and (6) into Equation (11) gives the following equation

$$\overline{st} = (l_o - 2D_o\theta_o)\sin\theta_o \tag{12}$$

Substituting Equations (10) and (12) into Equation (9) gives the following relation

$$Y_o = 2D_o\cos\theta_o - (l_o - 2D_o\theta_o)\sin\theta_o \tag{13}$$

This relation can be used for calculating the initial height ($Y_o$) using the inclination angle ($\theta_o$), however, $Y_o$ can also be calculated simply by using the following relation.

$$Y_o = \frac{a_o}{2} + b_o \tag{14}$$

Similarly, the distance between the centers of the two binding yarns at a deformed state ($X_i$), is given below.

$$X_i = 2\overline{p'q'} + \overline{q's'} + 2\overline{n'o'} \tag{15}$$

Whereas:

$$\overline{n'o'} = \left(a_i' - a_i\right)/2 \tag{16}$$

By solving the right triangles $\Delta o'p'q'$ and $\Delta q's't'$, the lengths $\overline{p'q'}$ and $\overline{q's'}$ can be expressed with Equations (17) and (18), respectively.

$$\overline{p'q'} = D_i\sin\theta_i \tag{17}$$

$$\overline{q's'} = \left[l_o - 2D_i\theta_i - \left(a_i' - a_i\right)\right]\cos\theta_i \tag{18}$$

Substituting Equations (16)–(18) into Equation (15) gives the following relation

$$X_i = 2D_i\sin\theta_i + \left[l_o - 2D_i\theta_i - \left(a_i' - a_i\right)\right]\cos\ \theta_i + \left(a_i' - a_i\right) \tag{19}$$

The height, $Y_i$, between the coarse binding yarns can express mathematically as,

$$Y_i = 2\overline{o'p'} - \overline{s't'} \tag{20}$$

From $\Delta o'p'q'$ and $\Delta q's't'$, $\overline{o'p'}$and $\overline{s't'}$ can be expressed in the following Equations

$$\overline{o'p'} = D_i\cos\theta_i \tag{21}$$

$$\overline{s't'} = \left(l_o - 2D_i\theta_i - \left(a_i' - a_i\right)\right)\sin\theta_i \tag{22}$$

Substituting Equations (21) and (22) into Equation (20) gives the following relation

$$Y_i = 2D_i\cos\theta_i - \left[l_o - 2D_i\theta_i - \left(a_i' - a_i\right)\right]\sin\theta_i \tag{23}$$

As both the lateral and longitudinal strains in the initial and deformed states are known, the PR ($\upsilon_i$) for the first stage of deformation can be calculated. By putting Equations (8), (13), (19), and (23) into Equation (1), the following relation can be obtained

$$\upsilon_i = -\frac{\left[2D_i\cos\theta_i - \left[l_o - 2D_i\theta_i - a_i' + a_i\right]\sin\theta_i - 2D_o\cos\theta_o + \left(l_o - 2D_o\theta_o\right)\sin\theta_o\right] \times \left[2D_o\sin\theta_o + \left(l_o - 2D_o\theta_o\right)\cos\theta_o\right]}{\left[2D_i\sin\theta_i + \left[l_o - 2D_i\theta_i - a_i' + a_i\right]\cos\theta_i + a_i' - a_i - 2D_o\sin\theta_o - \left(l_o - 2D_o\theta_o\right)\cos\theta_o\right] \times \left[2D_o\cos\theta_o - \left(l_o - 2D_o\theta_o\right)\sin\theta_o\right]} \tag{24}$$

ii. Second stage: As the warp yarns become decrimped and become fully straight in the first stage, they then elongate at a constant rate due to the applied tensile loading. In addition, the warp yarn's diameter will also decrease due to elongation, as shown in Figure 7.

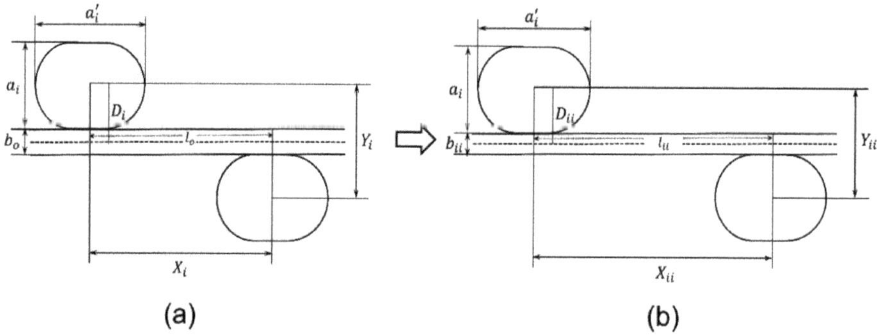


**Figure 7.** Second stage of deformation of a unit cell: (**a**) decrimped and (**b**) elongation of warp yarn.

It should be noted that the height and width of the race-track-shaped binding yarns will remain constant, while according to assumption 6, the length and diameter of the warp yarn will change during the second stage of deformation. Therefore, the sum of the radii of warp yarn and binding yarn $D_i$ will change to $D_{ii}$, which is given in Equation (25), and the length of warp yarn $l_o$ will change to $l_{ii}$, which is written in Equation (26).

$$D_{ii} = \frac{a_i}{2} + \frac{b_{ii}}{2} \tag{25}$$

$$l_{ii} = l_o + \Delta l \tag{26}$$

where $\Delta l$ is the change in the length of warp yarn, it can be calculated from the magnitude of tensile extension.

As the warp yarns are decrimped and straightened, the inclination angle ($\theta_i$) becomes "0". By putting $\theta_i = 0$ into Equations (19) and (23), the following relations can be obtained, respectively

$$X_{ii} = l_{ii} \tag{27}$$

$$Y_{ii} = 2D_{ii} \tag{28}$$

As the lateral and longitudinal strains at the second stage of deformation are known, the PR $\upsilon_{ii}$ for the second stage of deformation can be calculated by putting Equations (8), (13), (27), and (28) into Equation (1)

$$\upsilon_{ii} = -\frac{[2D_{ii} - 2D_o \cos\theta_o + (l_o - 2D_o\theta_o)\sin\theta_o] \times [2D_o \sin\theta_o + (l_o - 2D_o\theta_o)\cos\theta_o]}{[l_{ii} - 2D_o \sin\theta_o - (l_o - 2D_o\theta_o)\cos\theta_o] \times [2D_o \cos\theta_o - (l_o - 2D_o\theta_o)\sin\theta_o]}. \quad (29)$$

## 3. Results and Discussion

After successfully developing the geometrical model, the first step is to validate it through experimental results. Hence, the model can be used to predict, present, and analyze critical findings based on different structural parameters of the 3D auxetic woven fabric.

### *3.1. Experimental Observation*

#### 3.1.1. Determining the Poisson's Ratio during the First Stage of Deformation

To validate the geometrical model, the calculated tensile strain and PR results should be based on the same geometrical parameters as that of the developed 3D woven fabric sample. Considering that, all the parameters in the initial state are reasonably straightforward to obtain. For example, the radii of the coarse binding yarn ($\frac{a_o}{2}$) and warp yarn ($\frac{b_o}{2}$) can be taken from Table 1, the initial distance ($X_o$) between two coarse binding yarns can be determined by dividing the total length of a sample by the total repeats of coarse binding yarn, and the initial height ($Y_o$) can be calculated using Equation (14). When these parameters are known, the initial inclination angle ($\theta_o$) and length ($l_o$) of the warp yarn can be calculated using Equations (8) and (13).

In contrast, other parameters, particularly the compression effect of coarse binding yarn in a deformed state, are comparatively complicated to calculate. According to assumption (3), the cross-section of binding yarn changes to a race track shape during tensile deformation due to the compression of warp yarns caused by the restoring force of the elastic weft yarns. Therefore, the magnitude of force applied by elastic weft yarn should be known. During tensile deformation, the force applied by elastic weft yarn increases because of the lateral expansion of the structure. So, the first step is to calculate the amount of lateral deformation of the structure without assuming the coarse binding yarn's compression. For this purpose, Equations (8), (13), (19), and (23) can be used, while considering $a_i' = a_i = a_o$. When the percentage of lateral deformation of the structure is known, the magnitude of force applied by elastic weft yarn can be calculated from the force–strain curve of elastic weft yarn as shown in Figure 8. Eventually, the compression caused to the coarse binding yarn at a given force can be determined from the force–compressive extension curve of the coarse binding yarn, as shown in Figure 9.

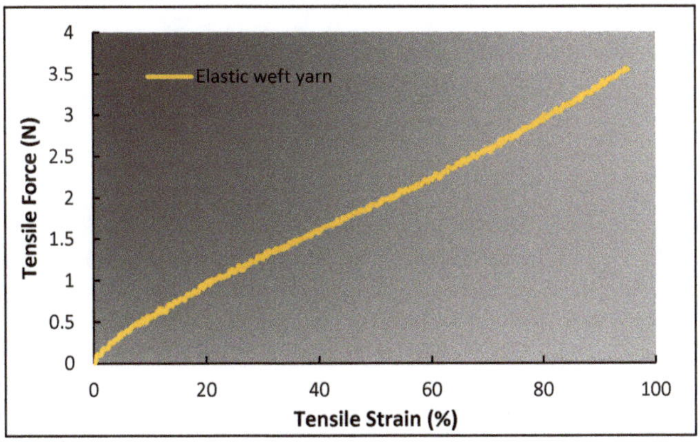


**Figure 8.** Tensile force–strain curve of elastic weft yarn.

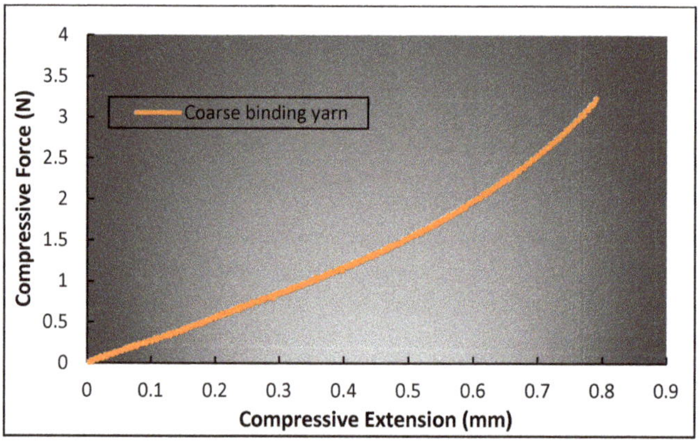


**Figure 9.** Compressive force–extension curve of coarse binding yarn.

After calculating the compression of coarse binding yarn in a deformed state, other parameters such as the final height ($a_i$) and width ($a_i'$) of race-track-shaped binding yarn can be calculated based on assumption 3. Now, using Equation (19), the inclination angle ($\theta_i$) can be calculated at a certain tensile deformation ($X_i$). Thus, all the variables are known, therefore, the PR ($\upsilon_i$) during the first stage of deformation can be determined using Equation (24). The calculated values of PR for the first stage of deformation are given in Table 2.

**Table 2.** Structural parameters and the calculated Poisson's ratio results of 3D auxetic woven fabric.

| State | | Tensile Strain (%) | $a_o$ or $a_i$ (mm) | $b_o$ or $b_{ii}$ (mm) | $D_o$ or $D_i$ or $D_{ii}$ (mm) | $X_o$ or $X_i$ or $X_{ii}$ (mm) | $Y_o$ or $Y$ (mm) | $\theta_o$ or $\theta_i$ (rad) | $l_o$ or $l_{ii}$ (mm) | PR |
|---|---|---|---|---|---|---|---|---|---|---|
| **Initial state** | | 0 | 2.28 | 0.70 | 1.490 | 3.000 | 1.840 | 0.460 | 3.242 | 0 |
| **Under tensile deformation** | First stage | 1 | 2.19 | 0.70 | 1.447 | 3.030 | 1.845 | 0.436 | 3.242 | −0.29 |
| | | 2 | 2.12 | 0.70 | 1.410 | 3.060 | 1.862 | 0.408 | 3.242 | −0.61 |
| | | 3 | 2.05 | 0.70 | 1.377 | 3.090 | 1.890 | 0.376 | 3.242 | −0.90 |
| | | 4 | 2.00 | 0.70 | 1.351 | 3.120 | 1.933 | 0.337 | 3.242 | −1.26 |
| | | 5 | 1.93 | 0.70 | 1.316 | 3.150 | 1.974 | 0.295 | 3.242 | −1.46 |
| | | 6 | 1.87 | 0.70 | 1.286 | 3.180 | 2.037 | 0.243 | 3.242 | −1.79 |
| | | 7 | 1.78 | 0.70 | 1.245 | 3.210 | 2.108 | 0.177 | 3.242 | −2.08 |
| | | 8 | 1.62 | 0.70 | 1.158 | 3.240 | 2.226 | 0.047 | 3.242 | −2.61 |
| | | **8.07** | **1.62** | **0.70** | **1.113** | **3.242** | **2.226** | **0** | **3.242** | **−2.59** |
| | Second stage | 9 | 1.62 | 0.62 | 1.110 | 3.270 | 2.221 | 0 | 3.27 | −2.33 |
| | | 10 | 1.62 | 0.61 | 1.108 | 3.300 | 2.216 | 0 | 3.3 | −2.07 |
| | | 11 | 1.62 | 0.60 | 1.105 | 3.330 | 2.211 | 0 | 3.33 | −1.86 |
| | | And so on | | | | | | | | |

### 3.1.2. Determining the Poisson's Ratio during the Second Stage of Deformation

The inclination angle ($\theta_i$) reaches 0 at a tensile strain of 8.07%, which indicates that the warp yarn is decrimped and becomes fully straight. At this point, the second stage of deformation starts, where no changes occur to the other parameters except to the length and diameter of the warp yarn. This specific point is determined as "critical strain," and the PR is called "critical Poisson's ratio". After the critical strain, the warp yarns elongate at a rate of tensile strain, which is a known parameter. However, the decrease in the diameter of the warp yarn ($b_{ii}$) at a particular tensile strain needs to be calculated. So, a customized tensile test was performed on the warp yarn, in which the cross head of UTM machine needed to be stopped after every 1% strain for 10 s to observe the diameter of the warp yarn using a thickness gauge. The variation trend of warp yarn's diameter ($b_{ii}$) as a function of tensile strain ($\varepsilon$) is shown in Figure 10. It can be observed that the change in $b_{ii}$ has a polynomial trend, therefore, according to a second-degree polynomial trend, the following equation can be established

$$b_{ii} = m_1\varepsilon_i^2 + m_2\varepsilon_i + m_3 \tag{30}$$

where $\varepsilon_i$ is the tensile strain, and $m_1$, $m_2$, and $m_3$ are the constants that can be determined from the experimental resutls of the warp yarn. Substituting Equation (30) in Equation (25) gives the following equation

$$D_{ii} = \frac{a_i}{2} + \frac{m_1\varepsilon^2 + m_2\varepsilon + m_3}{2} \tag{31}$$

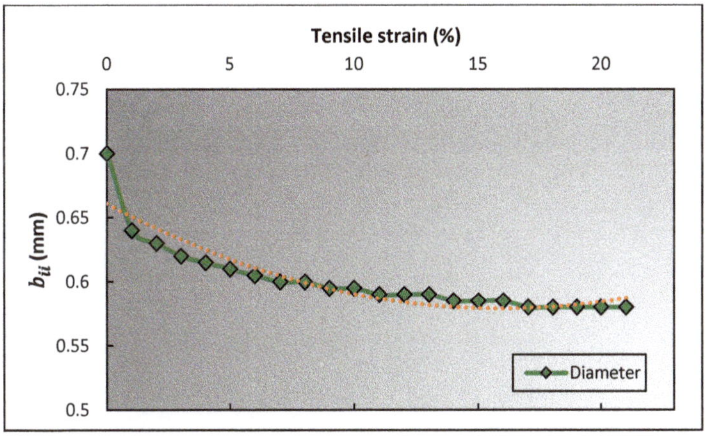


**Figure 10.** Trend of a diameter of warp yarn against tensile strain.

From Equation (31), $a_i$ is a known parameter, whereas the three other unknown constants were calculated experimentally from the relationship of warp yarn's diameter and the tensile strain, which are: $m_1 = 0.003$, $m_2 = -0.0102$, and $m_3 = 0.6608$.

When the sum of radii of the binding yarn and warp yarn $D_{ii}$ are known at a particular tensile extension, $X_{ii}$, then the PR at the second stage of deformation can be calculated using Equation (29). Finally, the yarn's parameters, the structural variables of the geometric model, and the calculated PR are given in Table 2.

It should be noted that if the tensile strain is less than the critical tensile strain, then $\upsilon_i$ (Equation (24)) will be used to calculate the PR, otherwise $\upsilon_{ii}$ Equation (29) will be used.

Figure 11 presents the PR curves plotted against the tensile strain of the developed 3D auxetic woven fabric and the calculated results obtained from the geometrical model. It can be seen that both the experimental and calculated PRs follow a similar trend. The agreement of calculated and experimental results confirms the reliability of the geometrical model. The PR first decreases, reaches a minimum at 8.07% tensile strain, and then follows a steadily increasing trend. From the analysis of the structure's geometry, it is clear that the lateral crimp of warp yarn plays an important role in the auxetic behavior under the tensile extension. The warp yarn starts decrimp under the tensile deformation, and the PR decreases. The effect of decrimping continues until the warp yarn becomes fully straight, which takes effect at 8.07% tensile strain. Hereafter, there is no decrimping of warp yarn that indicates there will be no lateral extension with further tensile extension. In addition, the diameter of warp yarns decreases due to longitudinal extension. As a result, the PR increased beyond 8.07% tensile strain.

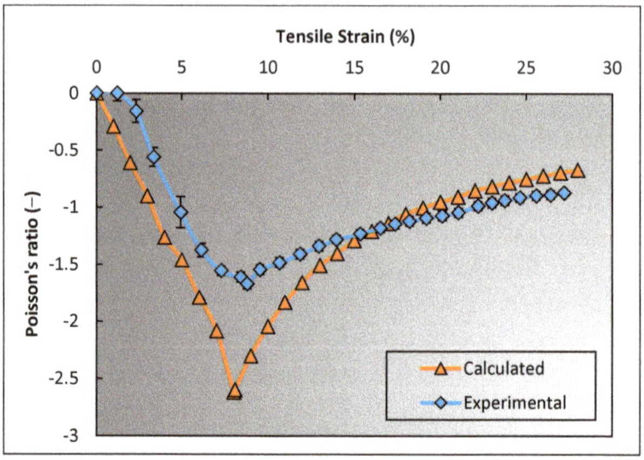


**Figure 11.** Comparison of calculated and experimental results of Poisson's ratio against tensile strain.

### 3.2. *Prediction of Auxetic Behavior*

#### 3.2.1. Effect of Binding Yarn Compression

A previous study [31] claimed that the coarse binding yarn becomes compressed during tensile extension. From the experimental assessment of the structure, it is quite clear that compression occurs due to the restoring force of elastic yarn. However, what could not be answered is how much the compression of binding yarn effects the auxetic behavior. In comparison, the geometrical model can calculate the effect of binding yarn compression on the auxetic behavior of 3D woven fabric. Figure 12 shows the PR versus tensile strain curves. The two curves, with compression (W-C) and without compression (W/O-C), are calculated using the geometrical model by keeping all the parameters constant, except for the compression effect of the binding yarn. For WO-C, it is supposed that the diameter of the binding yarn ($a_o$) does not change during the tensile strain, while for W-C, the diameter of the binding yarn decreases as the structure elongates in both the longitudinal and lateral directions. It is worth mentioning that the compression that occurred to the binding yarn is calculated experimentally, as explained in Section 3.1. From Figure 12, it can be observed that the auxetic behavior of W-C is 54% less than that of WO-C. This means that the binding yarn's compressive stiffness is critical and must be considered actively when one is developing 3D auxetic woven fabric.

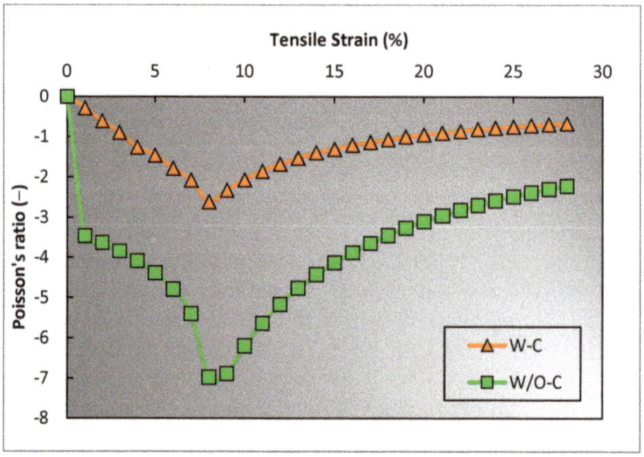


**Figure 12.** Effect of binding yarn's compression on auxetic behavior of 3D woven structure.

### 3.2.2. Effect of Binding Yarn Diameter

It is well known that [17] the diameters of two yarns (warp and binding yarns) significantly affect the NPR of 3D auxetic structures. Later on, it was proven again through an experimental study [31]. Since then, the effect of diameter of yarns on auxetic behavior has been understood, however, a scientific reason for the difference is still lacking. To provide a concrete reason, PR of the 3D woven structure was calculated using the geometrical model. Three different binding yarns diameter, i.e., 1.9 mm, 2.28 mm, and 2.75 mm, were used, while other parameters were kept constant. The PR–tensile strain curves of the 3D woven fabrics based on the different diameters of binding yarns are shown in Figure 13.

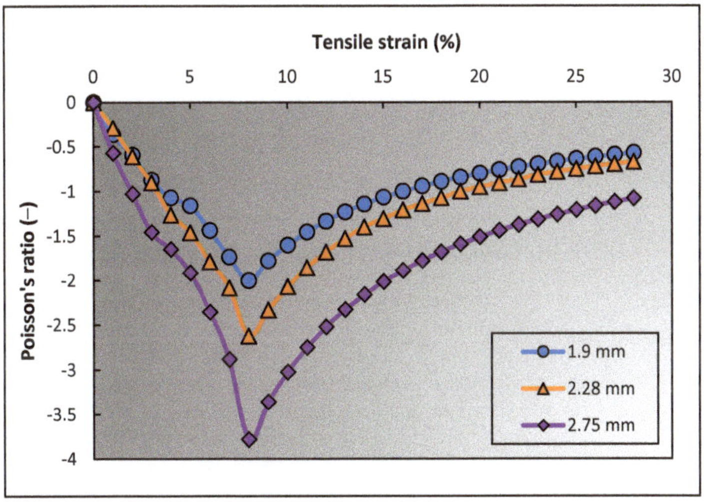


**Figure 13.** Effect of binding yarn diameter on auxetic behavior of 3D woven structure.

It can be found that the auxetic effect increases by increasing the diameter of the binding yarn, which validates the literature. For a solid reason, the numerical values generated by the geometrical model were analyzed. It is concluded that the initial lateral crimp percentage (%) of warp yarn decreases by increasing the diameter of the binding yarn. The crimp% can be calculated using Equation (32). For binding yarn diameters 1.9 mm, 2.28 mm, and 2.75 mm, the crimp% are 8.086, 8.011, and 7.949, respectively. Less lateral crimp in the warp yarn means the lateral expansion of auxetic structure will be triggered at a lower longitudinal strain, and the structure will reach its maximum lateral deformation under less strain. Since the longitudinal strain is inversely proportional to the NPR (Equation (1)), the auxetic behavior of the structure with the warp yarns being less laterally crimped is less notable, and vice versa.

$$Crimp\ \% = \frac{l_o - X_o}{X_o} \times 100 \quad (32)$$

### 3.2.3. Effect of Binding Yarn Spacing

Similar to the effect of binding yarn compression and binding yarn diameter, the spacing between the two binding yarns is also important to assess the auxetic behavior of the 3D auxetic woven structure. For this assessment, a similar methodological approach was applied, i.e., all the other parameters were kept constant, and the initial distance between the two binding yarns ($X_o$), also referred to as binding yarn spacing, was changed for three measurements. The three selected values for $X_o$ were 2.8 mm, 3 mm, and 3.2 mm. The PR values were then calculated using the developed geometrical model that are presented in Figure 14.

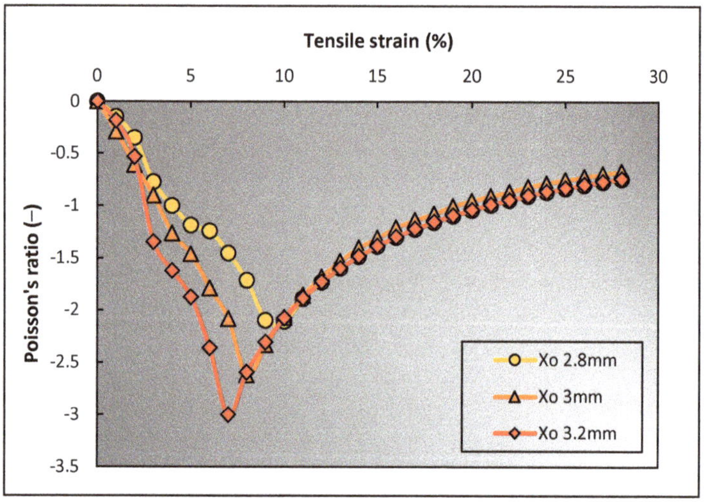


**Figure 14.** Effect of binding yarn spacing on auxetic behavior of 3D woven structure.

It can be explained that the binding yarn spacing has a direct relationship with the NPR effect, i.e., the NPR increases by increasing the yarn spacing ($X_o$), and vice versa. After evaluating the measured values of each parameter, it was found that the binding yarn spacing also effected the lateral crimping of the warp yarn. Higher yarn spacing caused less crimping of the warp yarn, producing a high NPR effect at a lower longitudinal strain value. The calculated lateral crimp values are 9.74 mm, 8.01 mm, and 6.97 mm when $X_o$ are 2.8 mm, 3 mm, and 3.2 mm , respectively.

## 4. Conclusions

The 3D woven fabric was designed and fabricated with in-plane auxetic behavior. The auxetic effect was visualized by the unusual lateral crimping of warp yarns that form a special geometry, resembling a re-entrant hexagon. Based on the geometrical arrangement of yarns in the structure, a micro-level geometrical model was proposed in terms of the yarn parameters to establish a relationship between PR and tensile strain through semi-empirical equations. The following conclusions can be drawn according to the study.

1. As the geometrical model is based on the micro-level analysis of the structure, therefore, it is useful to predict the auxetic behavior with given yarn parameters. Additionally, it can help to find specific reasons for the NPR effect in relation to a particular parameter.
2. Good agreements are observed between the experimental results and the results obtained from the established empirical equations based on the geometrical analysis.
3. It was found from the geometric analysis that the compression that is applied to binding yarns during lateral expansion significantly affects the NPR of the structure, i.e., the NPR decreases by 62.5% after the compression of the coarse binding yarn. Therefore, the compressive stiffness of binding yarn should be considered carefully.
4. The diameter of binding yarn and the spacing between the two binding yarns affect the lateral crimp percentage of the warp yarn. Both the diameter and spacing of binding yarns have an inverse relationship with the crimp percentage of the warp yarn. For example, the PR of the samples are −2.1 and −3.0 when the binding yarn spacing is 2.8 mm and 3.2 mm, respectively.
5. The lateral crimp of warp yarn has a strong relationship with the NPR of the structure. The NPR increases if the crimping degree is lower and decreases if the crimping degree is higher. For example, when the lateral crimping of warp yarn is 8.09%, the PR is −2.0, while the PR for a lateral crimping of 7.95% is −3.78.

The retainability of the auxetic effect is of great significance to its practical application. Therefore, the auxetic behavior of the fabrics under repetitive tensile forces can be evaluated experimentally. Furthermore, the current study uses coarse yarns to produce the fabric. The auxetic fabric can be used for broader applications if the fabric is developed with finer yarns.

**Author Contributions:** M.Z.: Methodology, Data curation, Writing—original draft, Writing—review and editing. H.H.: Visualization, Supervision, Methodology, Resources, Project administration, Writing—review and editing. E.E.: Methodology, Validation, Writing—review and editing. All authors have read and agreed to the published version of the manuscript.

**Funding:** This work was supported by the Research Institute for Intelligent Wearable Systems of The Hong Kong Polytechnic University in form of an internal project (No. P0039471).

**Acknowledgments:** The authors would like to acknowledge the support from the Technician, Ng Shui-wing, of the weaving workshop at The Hong Kong Polytechnic University.

**Conflicts of Interest:** The author(s) declared no potential conflicts of interest with respect to the research, authorship, and/or publication of this article.

## References

1. Liu, Y.; Hu, H. A Review on Auxetic Structures and Polymeric Materials. *Sci. Res. Essays* **2010**, *5*, 1052–1063.
2. Ravirala, N.; Alderson, K.L.; Davies, P.J.; Simkins, V.R.; Alderson, A. Negative Poisson's Ratio Polyester Fibers. *Text. Res. J.* **2006**, *76*, 540–546. [CrossRef]
3. Alderson, K.L.; Alderson, A.; Smart, G.; Simkins, V.R.; Davies, P.J. Auxetic Polypropylene Fibres:Part 1—Manufacture and Characterisation. *Plast. Rubber Compos.* **2002**, *31*, 344–349. [CrossRef]
4. Alderson, K.; Evans, K. The Fabrication of Microporous Polyethylene Having a Negative Poisson's Ratio. *Polymer (Guildf.)* **1992**, *33*, 4435–4438. [CrossRef]
5. Caddock, B.D.; Evans, K.E. Microporous Materials with Negative Poisson's Ratios. I. Microstructure and Mechanical Properties. *J. Phys. D Appl. Phys.* **1989**, *22*, 1877–1882. [CrossRef]
6. Miller, W.; Hook, P.B.; Smith, C.W.; Wang, X.; Evans, K.E. The Manufacture and Characterisation of a Novel, Low Modulus, Negative Poisson's Ratio Composite. *Compos. Sci. Technol.* **2009**, *69*, 651–655. [CrossRef]
7. Gao, Y.; Chen, X.; Studd, R. Experimental and Numerical Study of Helical Auxetic Yarns. *Text. Res. J.* **2021**, *91*, 1290–1301. [CrossRef]
8. Sloan, M.R.; Wright, J.R.; Evans, K.E. The Helical Auxetic Yarn—A Novel Structure for Composites and Textiles; Geometry, Manufacture and Mechanical Properties. *Mech. Mater.* **2011**, *43*, 476–486. [CrossRef]
9. Zulifqar, A.; Hua, T.; Hu, H. Development of Uni-Stretch Woven Fabrics with Zero and Negative Poisson's Ratio. *Text. Res. J.* **2018**, *88*, 2076–2092. [CrossRef]
10. Ali, M.; Zeeshan, M.; Qadir, M.B.; Ahmad, S.; Nawab, Y.; Anjum, A.S.; Riaz, R. Development and Comfort Characterization of 2D Woven Auxetic Fabric for Wearable and Medical Textile Applications. *J. Cloth. Text.* **2018**, *36*, 199–214. [CrossRef]
11. Wang, Z.; Hu, H. Auxetic Materials and Their Potential Applications in Textiles. *Text. Res. J.* **2014**, *84*, 1600–1611. [CrossRef]
12. Cao, H.; Zulifqar, A.; Hua, T.; Hu, H. Bi-Stretch Auxetic Woven Fabrics Based on Foldable Geometry. *Text. Res. J.* **2019**, *89*, 2694–2712. [CrossRef]
13. Ali, M.; Zeeshan, M.; Qadir, M.B.; Riaz, R.; Ahmad, S.; Nawab, Y.; Anjum, A.S. Development and Mechanical Characterization of Weave Design Based 2D Woven Auxetic Fabrics for Protective Textiles. *Fibers Polym.* **2018**, *19*, 2431–2438. [CrossRef]
14. Zeeshan, M.; Ali, M.; Riaz, R.; Anjum, A.S.; Nawab, Y.; Qadir, M.B.; Ahmad, S. Optimizing the Auxetic Geometry Parameters in Few Yarns Based Auxetic Woven Fabrics for Enhanced Mechanical Properties Using Grey Relational Analysis. *J. Nat. Fibers* **2022**, *19*, 4594–4605. [CrossRef]
15. Ahmed, H.I.; Umair, M.; Nawab, Y.; Hamdani, S.T.A. Development of 3D Auxetic Structures Using Para-Aramid and Ultra-High Molecular Weight Polyethylene Yarns. *J. Text. Inst.* **2021**, *112*, 1417–1427. [CrossRef]
16. Khan, M.I.; Akram, J.; Umair, M.; Hamdani, S.T.; Shaker, K.; Nawab, Y.; Zeeshan, M. Development of Composites, Reinforced by Novel 3D Woven Orthogonal Fabrics with Enhanced Auxeticity. *J. Ind. Text.* **2019**, *49*, 676–690. [CrossRef]
17. Ge, Z.; Hu, H. Innovative Three-Dimensional Fabric Structure with Negative Poisson's Ratio for Composite Reinforcement. *Text. Res. J.* **2013**, *83*, 543–550. [CrossRef]
18. Zhao, S.; Hu, H.; Kamrul, H.; Chang, Y.; Zhang, M. Development of Auxetic Warp Knitted Fabrics Based on Reentrant Geometry. *Text. Res. J.* **2020**, *90*, 344–356. [CrossRef]
19. Wang, Z.; Hu, H. 3D Auxetic Warp-Knitted Spacer Fabrics. *Phys. Status Solidi.* **2014**, *251*, 281–288. [CrossRef]
20. Glazzard, M.; Breedon, P. Weft-Knitted Auxetic Textile Design. *Phys. Status Solidi* **2014**, *251*, 267–272. [CrossRef]
21. Boakye, A.; Chang, Y.; Rafiu, K.R.; Ma, P. Design and Manufacture of Knitted Tubular Fabric with Auxetic Effect. *J. Text. Inst.* **2018**, *109*, 596–602. [CrossRef]

22. Bhullar, S.K.; Ko, J.; Cho, Y.; Jun, M.B.G. Fabrication and Characterization of Nonwoven Auxetic Polymer Stent. *Polym. Plast. Technol. Eng.* **2015**, *54*, 1553–1559. [CrossRef]
23. Verma, P.; Shofner, M.L.; Lin, A.; Wagner, K.B.; Griffin, A.C. Inducing Out-of-Plane Auxetic Behavior in Needle-Punched Nonwovens. *Phys. Status Solidi.* **2015**, *252*, 1455–1464. [CrossRef]
24. Yuping, C.; Liu, Y.; Hong, H. Deformation Behavior of Auxetic Laminated Fabrics with Rotating Square Geometry. *Text. Res. J.* **2022**, *92*, 4652–4665. [CrossRef]
25. Shukla, S.; Behera, B.K. Auxetic Fibrous Structures and Their Composites: A Review. *Compos. Struct.* **2022**, *290*, 115530. [CrossRef]
26. Wright, J.R.; Burns, M.K.; James, E.; Sloan, M.R.; Evans, K.E. On the Design and Characterisation of Low-Stiffness Auxetic Yarns and Fabrics. *Text. Res. J.* **2012**, *82*, 645–654. [CrossRef]
27. Ng, W.; Hu, H. Woven Fabrics Made of Auxetic Plied Yarns. *Polymers* **2018**, *10*, 226. [CrossRef]
28. Etemadi, E.; Gholikord, M.; Zeeshan, M.; Hu, H. Improved Mechanical Characteristics of New Auxetic Structures Based on Stretch-Dominated-Mechanism Deformation under Compressive and Tensile Loadings. *Thin-Walled Struct.* **2023**, *184*, 110491. [CrossRef]
29. Zulifqar, A.; Hu, H. Development of Bi-Stretch Auxetic Woven Fabrics Based on Re-Entrant Hexagonal Geometry. *Phys. Status Solidi. Basic Res.* **2019**, *256*, 1800172. [CrossRef]
30. Liaqat, M.; Samad, H.A.; Hamdani, S.T.A.; Nawab, Y. The Development of Novel Auxetic Woven Structure for Impact Applications. *J. Text. Inst.* **2017**, *108*, 1264–1270. [CrossRef]
31. Zeeshan, M.; Hu, H.; Zulifqar, A. Three-Dimensional Narrow Woven Fabric with in-Plane Auxetic Behavior. *Text. Res. J.* **2022**, *92*, 4695–4708. [CrossRef]
32. Msalilwa, L.R.; Kyosev, Y.; Rawal, A.; Kumar, U. Investigation of the Bending Rigidity of Double Braided Ropes. In *Recent Developments in Braiding and Narrow Weaving*; Kyosev, Y., Ed.; Springer International Publishing: Cham, Germany, 2016; pp. 47–57.
33. *ASTM D5035-11*; Standard Test Method for Breaking Force and Elongation of Textile Fabrics (Strip Method). ASTM International: West Conshohocken, PA, USA, 2019.

polymers MDPI

*Review*

# Recent Trends in Protective Textiles against Biological Threats: A Focus on Biological Warfare Agents

**Joana C. Antunes [1,2,*], Inês P. Moreira [1,2], Fernanda Gomes [3,4], Fernando Cunha [1,2], Mariana Henriques [3,4] and Raúl Fangueiro [1,2]**

1 Fibrenamics, Institute of Innovation on Fiber-based Materials and Composites, University of Minho, 4710-057 Guimarães, Portugal; ines.moreira@fibrenamics.com (I.P.M.); fernandocunha@det.uminho.pt (F.C.); rfangueiro@dem.uminho.pt (R.F.)
2 Centre for Textile Science and Technology (2C2T), University of Minho, 4710-057 Guimarães, Portugal
3 CEB, Centre of Biological Engineering, LIBRO-Laboratório de Investigação em Biofilmes Rosário Oliveira, University of Minho, 4710-057 Braga, Portugal; fernandaisabel@ceb.uminho.pt (F.G.); mcrh@deb.uminho.pt (M.H.)
4 LABBELS—Associate Laboratory, 4710-057 Braga, Portugal
* Correspondence: joanaantunes@fibrenamics.com

**Abstract:** The rising threats to worldwide security (affecting the military, first responders, and civilians) urge us to develop efficient and versatile technological solutions to protect human beings. Soldiers, medical personnel, firefighters, and law enforcement officers should be adequately protected, so that their exposure to biological warfare agents (BWAs) is minimized, and infectious microorganisms cannot be spread so easily. Current bioprotective military garments include multilayered fabrics integrating activated carbon as a sorptive agent and a separate filtrating layer for passive protection. However, secondary contaminants emerge following their accumulation within the carbon filler. The clothing becomes too heavy and warm to wear, not breathable even, preventing the wearer from working for extended hours. Hence, a strong need exists to select and/or create selectively permeable layered fibrous structures with bioactive agents that offer an efficient filtering capability and biocidal skills, ensuring lightweightness, comfort, and multifunctionality. This review aims to showcase the main possibilities and trends of bioprotective textiles, focusing on metal–organic frameworks (MOFs), inorganic nanoparticles (e.g., ZnO-based), and organic players such as chitosan (CS)-based small-scale particles and plant-derived compounds as bioactive agents. The textile itself should be further evaluated as the foundation for the barrier effect and in terms of comfort. The outputs of a thorough, standardized characterization should dictate the best elements for each approach.

**Keywords:** advanced protection; protective textiles; biological warfare agents; antimicrobial; metal–organic frameworks; zinc oxide nanoparticles; chitosan-based nanoparticles

**Citation:** Antunes, J.C.; Moreira, I.P.; Gomes, F.; Cunha, F.; Henriques, M.; Fangueiro, R. Recent Trends in Protective Textiles against Biological Threats: A Focus on Biological Warfare Agents. *Polymers* **2022**, *14*, 1599. https://doi.org/10.3390/polym14081599

Academic Editors: Muhammad Tayyab Noman and Michal Petrů

Received: 15 March 2022
Accepted: 11 April 2022
Published: 14 April 2022



## 1. Biological Warfare Agents (BWAs)

In their daily lives, the world population is exposed to several threats that put their wellbeing and health at risk. Chemicals and BWAs are some of these threats [1]. BWAs include bacteria, viruses, fungi, and biological toxins and are responsible for several diseases such as anthrax, plague, tularemia, botulism, smallpox, and viral hemorrhagic fever [2,3]. BWAs are higher-risk agents for use as biological weapons and present variable mortality rates that depend on the biological agent and the mode of transmission/route of exposure. Their use for this purpose can promote large-scale morbidity and mortality, affecting a large number of people [2,4,5]. The early detection of a biological attack, namely of the agent involved, is crucial to their effective management and resolution, so that lower mortality rates can be attained. According to several criteria such as the ease of transmission, the severity of morbidity and mortality, and the probability of use, BWAs were classified by Centers for Disease Control and Prevention (CDCs) into different categories, specifically: Category A (highest risk to the public and national security—high priority agents); B (second-highest priority agents); and C (third-highest priority agents—emerging threats for disease) [6]. Some of the most relevant BWAs that are most likely to be used, with high mortality rates and a high potential for a major public health impact, belong to category A and are listed below.

### 1.1. Bacteria

#### 1.1.1. Anthrax

*Bacillus anthracis*, a spore-forming Gram-positive rod bacterium, is one of the most popular biological weapons in bioterrorism. It causes anthrax. A relevant example of a *B. anthracis*-driven biological attack happened in the 21st century (2001) in the US via the postal system (letters containing spores). This attack resulted in 22 infected people, of whom 5 died. *B. anthracis* is considered an effective BWA due to its ability to be aerosolized, form spores, and be easily cultured, as well as its capacity to remain viable for a long period of time in the environment. It can persist in the spore state for years or even decades, with the spores being extremely resistant to heat, irradiation, desiccation, and disinfectant action [7]. This bacterium is in the top list of the Category A priority pathogens [8]. *B. anthracis* has a short incubation period, usually 48 h, but it may be up to 7 days [9]. Its symptoms include fever, nausea, vomiting, sweats, dyspnea, respiratory failure, and hemodynamic collapse [10]. Toxin production (exotoxins: lethal toxin and edema toxin) is one of its virulence factors, along with the presence of a capsule that helps *B. anthracis* to evade host immunity. The natural incidence of anthrax is rare, occurring via contact with contaminated soil, infected animals, and infected or contaminated animal products [10–12]. The global anthrax prevalence is around 28%. The incidence was decreased during the 20th century. According to the World Health Organization (WHO), the estimated anthrax annual incidence is between 2000 to 20,000 cases [10,11]. The mortality rate is very high, mainly in cases of gastrointestinal anthrax, where the average is 25–60%, though it can reach 100%. Cutaneous anthrax, the most common form of disease manifestation, is known to provoke death in less than 20% of cases [13]. Injectional anthrax, a more recent form of the disease, has a mortality rate of 35% despite medical treatment [12]. The inhalational form has the worst prognosis, with a fatality rate of 80% or higher [14]. Prompt treatment with antibiotics is curative and enhances the chances of a full recovery [15]. Cutaneous anthrax is easily treated, while inhalational anthrax can be fatal even in cases of adequate treatment. Antibiotic resistance, a global concern, is evidenced by *B. anthracis* in its interaction with penicillin, highlighting the need for effective treatment options avoiding the use of this antibiotic, as well as of related β-lactam antibiotics. Nowadays, a combination of antimicrobials is used in the treatment of anthrax [16]. The multidrug regimen includes at least one bactericidal agent (such as ciprofloxacin or doxycycline) along with a protein-synthesis inhibitor (such as linezolid or clindamycin) to suppress toxin production. An antitoxin product (such as raxibacumab, anthrax immunoglobulin) is also recommended in parallel to the multidrug regimen to neutralize *B. anthracis* toxins by inhibiting the binding of protective antigens and the translocation of toxins into cells [17]. There are also vaccines available for anthrax, but only for people from 18 to 65 years old and at increased risk of exposure. Thus, the vaccine is recommended only for a minority of cases, namely professionals who come into contact with animal hides and fur, and some members of the army. Anthrax vaccine adsorbed (AVA) and anthrax vaccine precipitated (AVP) are licensed anthrax vaccines whose immunological component is the protective antigen, the major constituent of anthrax toxins [18]. Anthrax vaccines show a protective efficacy of 93% against inhalational and cutaneous disease [19].

#### 1.1.2. Plague

Another bacterium listed in Category A of bioterrorism agents is *Yersinia pestis*, a Gram-negative bacterium of the family Enterobacteriaceae that causes plague, famously known as "the Black Death". It is associated with black scabs on skin sores. Although rare, plague caused by *Y. pestis* must be taken into consideration due to its possible intentional use as a bioterrorism weapon. The use of this biological agent as a biological weapon dates back to the Second World War [20,21]. Regardless, 75% of global plague cases have occurred in Madagascar, presenting an annual incidence of 200 to 700 suspected cases. Currently endemic, Madagascar endured an outbreak of plague in 2017, with a total of 2417 confirmed cases of plague and 209 patient deaths [22]. The mortality rates are indeed high, with pulmonary plague presenting a mortality rate of 40% and being fatal when untreated [23]. In parallel with *B. anthracis*, *Y. pestis* is one of the most virulent and deadliest BWAs, presenting mortality rates of 100% within 3 to 6 days postinfection [21,24,25]. *Y. pestis* is a nonmotile, non-spore-forming coccobacillus [20]. This bacterium has a short incubation period, usually 2 to 3 days, and symptoms include fever, headache, and general malaise. Plague can manifest in one of three clinical forms: bubonic plague, septicemic plague, and pulmonary plague, the latter being the most severe [26]. Plasminogen activator, Pla, is one virulence factor used by *Y. pestis* to overcome host immunity, since Pla adhesion and proteolytic ability have a crucial role in the manipulation of the fibrinolytic cascade and immune system [27]. Plague is a vector-borne illness transmitted by fleas from rodent reservoirs, but it can also be transmitted by direct contact or via aerosols (the inhalation of respiratory droplets). Fortunately, human cases are successfully treated with antibiotics (such as streptomycin, gentamicin, or ciprofloxacin). However, there are at least two cases of strains isolated in Madagascar (*Y. pestis* 16/95 and 17/95) exhibiting antibiotic resistance [20]. This poses an additional challenge for the control and management of the disease. Promising vaccine candidates are being created [28–30]. However, as of now, no licensed vaccine exists for plague. Once again, a quick diagnosis and treatment with antibiotics is crucial to a full recovery [23].

#### 1.1.3. Tularemia

Tularemia is another potential BWA [31,32]. In fact, tularemia is nowadays recognized as a reemerging disease due to the role of *Francisella tularensis* and its potential for misuse as a biological terrorism weapon [31,33]. This disease is caused by the Gram-negative coccobacillus-shaped bacterium *F. tularensis* [31,32]. *F. tularensis* is a pleomorphic, non-motile and non-spore-forming bacterium. This infectious bacterium is easily disseminated by aerosols, has a low infectious dose, and is associated with rapid and fatal disease. Tularemia can be spread by vectors, direct contact with water contamination, sick animals, and inhalation. *F. tularensis* virulence factors consist mainly in their envelope (capsule, outer membrane, lipopolysaccharide, periplasm, inner membrane, among others), an outer structure that confers protection from host immunity and promotes infection and disease [34]. The incubation period for this bacterium is typically short, 3 to 5 days on average, up to 2 weeks [35]. Tularemia symptoms are highly variable and depend on the route of infection [36]. However, the most common include fever, headache, chills, malaise, and a sore throat [34,37]. The worldwide incidence of tularemia is not known [31], but it is known that the incidence of cases of tularemia declined during the 20th century [38]. Currently, in the US, about 200 cases of tularemia per year are reported [39]. Tularemia is a disease characterized by high morbidity and mortality. In untreated cases, the mortality rate ranges from 30 to 60%, while with treatment the death rate is less than 2% [40]. After tularemia recovery, some sequelae might occur, such as residual scars, lung and kidney damage, and muscle loss [36].

The treatment of this disease consists of antimicrobial therapy, specifically, antibiotics (quinolones, tetracyclines, or aminoglycosides) [32]. Although *F. tularensis* showed antibiotic resistance to, for example, ampicillin, meropenem, daptomycin, clindamycin, and linezolid, and is only susceptible to a small range of antibiotics, so far it has responded well to the antibiotics usually used to treat tularemia (gentamicin, ciprofloxacin, levofloxacin, and doxycycline) [41]. No vaccine is yet available for the prevention of this disease. However, clinical assays have been developed in order to find a vaccine against tularemia, and a mutant strain (ΔpdpC) tested in animals (mice and monkeys) was demonstrated to be a good candidate for a live attenuated vaccine against *F. tularensis* [42].

### *1.2. Virus*

#### 1.2.1. Smallpox

Although declared eradicated in 1980, smallpox, caused by the variola virus, remains a major threat to humanity due to its possible use as bioweapon [43,44]. The variola virus is an orthopox virus, one of the largest viruses to infect humans, belonging to the Poxviridae family [45]. It has a high mortality rate, high stability in an aerosol state, high transmissibility and high contagiousness among humans, a significant impact, and a great need for special preparedness. It is one of the most fatal diseases to have ever existed, presenting mortality rates of up to 30% (variola virus variant) [46]. The smallpox virus has a long incubation period, usually 11 to 14 days, and early symptoms include a fever and nonspecific macular rashes [47]. The variola virus is transmitted via respiratory droplets, cutaneous lesions, infected body fluids, and fomites. Smallpox sequelae include permanent scarring, which may be extensive; blindness resulting from corneal scarring; the loss of lip, nose, and ear tissue; arthritis; and osteomyelitis [48]. Smallpox inhibitor of complement enzymes (SPICE) and chemokine-binding protein type II (CKBP II) are considered two virulence factors of the variola virus, helping it evade the human immune system [49].

The smallpox vaccine, discovered by Edward Jenner in the 18th century, was the first vaccine to be successfully developed, involving the use of the cowpox virus to prevent smallpox [50]. In the 20th century, the first-generation vaccine comprised a strain of vaccinia virus followed by a second-generation vaccine based on the use of clones of the vaccinia viral strains used in the first-generation vaccine [51]. However, due to the controversial and severe adverse reactions to these vaccines, a safe and effective third-generation vaccine is being considered. KVAC103, a highly attenuated vaccinia virus strain, was recently proposed as such a candidate [45]. Tecovirimat, a small molecule used to treat smallpox, was the first smallpox antiviral therapeutic approved by the US Food and Drug Administration, but the smallpox virus has demonstrated resistance to it [44]. The latter constitutes a current concern that highlights the urgent need for multitherapeutic and effective strategies to fight this disease [52].

#### 1.2.2. Viral Hemorrhagic Fever

Ebola

The Ebola virus, which is suitable to be used as a BWA, belongs to the filoviridae family and is one of the causative agents of viral hemorrhagic fever in humans. This virus was first discovered in 1976 in the Democratic Republic of Congo, where the first Ebola outbreak occurred [53]. Currently, Ebola outbreaks continue to be recurrent in Africa, and its increased incidence requires an early detection in order to avoid the risk of an epidemic [54]. Since its discovery, over 20 outbreaks have occurred. Ebola fever is a fatal disease presenting a mortality rate ranging from 25 to 90%, and it is easily transmitted by direct contact with infected individuals (body fluids) [55]. The Ebola virus is a filamentous virus with a characteristic twisted thread shape. It has an incubation period of 2–21 days, with symptoms including fever, malaise, headache, diarrhea, and vomiting, and it can evolve into multiorgan failure (lungs, heart, kidney, liver), shock, and death [56]. Recovery is possible, though some sequelae can occur after disease recovery, including joint and vision problems, tiredness, and headaches [57]. The main virulence factors of the Ebola virus include some proteins such as virion proteins 35 and 24 (interferon antagonists) and glycoprotein, which interfere with the activation of a dysfunctional immune response and facilitate the attachment to host-cell surface receptor molecules and viral entry, respectively [58]. Currently, Ebola vaccines are being developed, including five promissory candidates, of which Ervebo, Zabdeno/Mvabea, and cAd3-EBOZ are the most advanced, based on a viral vector or on a modified version of a harmless surrogate virus. Among these, two are licensed (Ervebo and Zabdeno/Mvabea). The CanSino and GamEvac vaccines are also licensed, but only for emergency use in China and Russia, respectively. Although vaccines are available for Ebola, several questions remain unclear regarding their durability, safety, interaction with other therapeutics and vaccines, stability, etc. Other issues are related to vaccine costs, the narrow range of action (protection against only one species of Ebola virus), and the likely occurrence of intraspecies mutations that can affect the effectiveness of the vaccine [59]. Vaccinations are routinely administered for the Ebola disease only for individuals at high risk of exposure, due to the limited vaccine quantities, their unpredictable nature, and the relative rarity of Ebola outbreaks (mostly occurring in the regions of Central and West Africa) [60,61].

Lassa Fever

The Lassa virus is the causative agent of Lassa fever in humans, and it is an enveloped, single-stranded, bisegmented, negative-strand RNA virus belonging to the arenavirus family. It is responsible for 2 million cases of Lassa fever and 5000–10,000 deaths annually [62]. Lassa fever is an often-fatal hemorrhagic disease, first discovered in 1969 in Nigeria. This infection occurs mainly in West Africa and Nigeria and poses significant epidemic threats due to its high mortality (21–69% [63]) and morbidity rate and its highly contagious nature [64,65]. Lassa fever can have a zoonotic origin or can be transmitted by direct contact (aerosols or fluid secretions) with infected individuals. Its incubation period is 1 to 3 weeks. Lassa fever is normally asymptomatic in the initial stage or can present nonspecific symptoms such as fever, headache, malaise, and general fatigue, which can lead to a delay in diagnosis and treatment [66]. The progress of the disease leads to multiorgan collapse and hemorrhagic fever [67]. A prompt diagnosis and treatment is crucial to full recovery and, in fact, cases of severe Lassa fever with complete recovery were recently reported [66]. The recurrent outbreaks of Lassa fever and the emergence of the Lassa virus as well as its epidemic potential have highlighted the need for research into vaccines and treatments. To date, no approved vaccine is available to prevent the disease, and the therapeutic choices are limited [67]. Ribavarin, a synthetic nucleoside, is the only antiviral option available for the treatment of Lassa fever [68–70]. Currently, other therapeutic strategies are being developed and evaluated in humans and animal models. Of these, favipiravir and a human monoclonal antibody cocktail (Inmazeb) have shown potential to be used in clinical settings [62]. In parallel, several vaccine candidates are being examined, the most promising of which is based on the recombinant vesicular stomatitis virus, reassortants expressing Lassa virus antigens, and a deoxyribonucleic acid platform [71]; however, to date, no vaccine has passed the preclinical stage and evidenced both safety and efficacy in humans [62,71,72]. The main target used for the design of antibody-based therapeutics and Lassa virus vaccines is the envelope glycoprotein complex. This protein displayed on the surface of the Lassa virus can be considered a virulence factor, since it is essential for the attachment and entry of the virus into human cells [73].

### 1.3. *Toxins*

Botulism

In the case of botulism, another concerning BWA, the causative agent is the highly potent biological toxin botulinum neurotoxin produced by neurotoxigenic clostridia such as *Clostridium botulinum*. This toxin is the main virulence factor of this bacterium [74].

*C. botulinum* is a Gram-positive bacillus, spore-forming, anaerobic bacterium [75]. Natural cases of botulism are rare. Still, this toxin is easily produced, stored, and disseminated and presents extreme toxicity (lethal dose (LD50) = 1–3 ng/kg of body mass [76]). As a bioweapon, botulinum neurotoxin could be spread in food sources and via aerosolization. Between 1920 and 2014, only 197 outbreaks were reported, of which 55% occurred in the US, with an average of 110 cases reported annually. Botulism is a serious paralytic disease [77]. The toxin acts by blocking the release of a neurotransmitter, acetylcholine, at the neuromuscular junction, interfering with the nervous impulse and causing muscle paralysis [78].

Symptoms usually appear within 12–72 h after contact with the toxin. If untreated, botulism can progress to cause paralysis in various parts of the body, including respiratory muscles, leading to patient death. Patients with botulism may have a slow recovery that lasts days or even years. A prompt diagnosis and treatment can lead to full recovery in 2 weeks. In fact, reduced mortality was observed with the early administration of antitoxins and high-quality supportive care [79]. However, some sequelae can occur, such as feeling tired, shortness of breath, and ongoing breathing problems for a long time. Antitoxin therapy is the first-line therapeutic strategy used to promote toxin neutralization and elimination from blood circulation, being more effective when administered early in the course of the disease. It consists of antibodies or antibody antigen-binding fragments, whose purpose is to block the neurotoxin produced by *C. botulinum* [80]. However, patients may additionally require mechanical ventilation and/or other supportive measures until total recovery from paralysis. The availability of antitoxins and improvements in supportive and intensive respiratory care have substantially reduced the mortality rate by up to 5–10% in humans [77,81,82]. Unfortunately, although current treatment modalities can help to mitigate the progression/symptoms and accelerate recovery, no true antidote exists following exposure to botulinum neurotoxin [76,77,83]. Fortunately, vaccines are being developed to confer appropriate immune responses following incubation with the BWA, either in the case of a biothreat emergency or infectious disease outbreak [84–86].

## 2. COVID-19

COVID-19 is an acute respiratory illness that ranks third in terms of fatal coronavirus diseases threatening public health, with this kind of virus having emerged as a threat to people in the 21st century [87,88]. COVID-19 is caused by SARS-CoV-2, a beta-coronavirus, which was first reported in 2019 in China. Since then, SARS-CoV-2 has quickly spread all over the world, resulting in a pandemic situation that was declared by the WHO as a Public Health Emergency of International Concern. Its high morbidity and mortality rate have resulted so far in over 120 million infections and 2.5 million deaths worldwide in 1 year [89,90]. Although it has not been classified as a BWA by CDCs, and the origin/cause of its emergence is controversial, it is considered a global threat to health and safety and is already regarded as the greatest threat of this century [87]. The extremely high transmission rate of SARS-CoV-2 was one of the factors that contributed to its rapid propagation [91,92]. The virus is primarily transmitted by respiratory droplets and aerosol and contact routes. The implementation of the use of face masks or coverings was one of the strategies used to prevent virus transmission during the pandemic [90]. Such biological threats, whether of natural or intentional origin, highlight the extreme importance of bioprotective materials as fundamental to minimizing the consequences of this kind of threat.

## 3. Antimicrobial Activity Test Methods

Microorganisms can be carried by textiles and even multiply themselves in this environment, which is the reason why this kind of substrate is regarded as a possible vector of infection and disease transmission in hospitals and communities [93]. On the other hand, textiles can be used as means of protection against the transmission of diseases, including biological and chemical threats. In reality, there is a growing body of research concerning the development and application of textiles for military use, aiming at providing protection in a wide range of hostile environments and with a rapid effect on bacteria, fungi, viruses, and even toxins.

Biological threats do not have simulants in the same way as CWAs; however, for bacteria, several standard strains are typically used to evaluate the biocidal capacity of proposed textiles [94]. These selected strains are easily handled in the laboratory using well-established assaying protocols and representative bacterial strains of each group, including *Staphylococcus aureus* (Gram-positive bacteria) and *Escherichia coli* (Gram-negative bacteria) [94–98]. Gram-positive and Gram-negative bacteria differ in their cell wall structure, and this difference affects their susceptibility to antimicrobials [99,100]. The cytosol of Gram-positive bacteria is encircled by a cytoplasmic membrane attached to a thick peptidoglycan layer, while the cell wall of Gram-negative bacteria contains two distinct lipid membranes, the cytoplasmic cell membrane and the outer membrane, with a thin layer of peptidoglycans in between [100–102]. In addition to the two aforementioned microorganisms, which are the most commonly used in this type of evaluation, *Candida albicans*, a unicellular fungus, is another regularly assessed species [103]. However, many others are also routinely used, of which *Pseudomonas aeruginosa* and *Klebsiella pneumoniae*, both Gram-negative bacteria, can be emphasized [95,103–109]. The viricidal potential is often examined using model viruses such as bacteriophage MS2 (a surrogate of the SARS-CoV-2 virus) and P22 (a surrogate of the *Salmonella* virus), even though rotavirus and severe-acute-respiratory-syndrome-associated coronavirus (SARS-CoV) have been used to test potentially protective textiles [94]. In short, test microorganisms should be selected according to the intended application of the textile [99].

Test standards for antimicrobial textiles usually consist of two types of testing method: qualitative (first-step screening of the antimicrobial activity of antimicrobial textiles) and quantitative [110]. Among the various standards available, AATCC 147, JIS L1902, AATCC 100, and ISO 20645 are the most relevant examples [100,110]. Qualitative methods (agar diffusion assay) are based on the measurement of the halo, a clear zone of inhibition around the sample. In quantitative methods, the evaluation of the antimicrobial activity is more efficient and is based on the measurement of the number of microorganisms (or colony-forming units) after 18–24 h of contact with the textile material [93]. The different standards differ in the inoculation method, sample size, inoculum concentration, culture medium, and buffer formulation, among other things [110].

## 4. Biological Protective Textiles

The development of protective clothing is crucial nowadays, as there are increased levels of harmful biological threats, both for military forces and civilians [111]. The main purpose of barrier textiles is to protect the user against external hazards such as BWAs while maintaining safety and comfort next to the skin [112]. Figure 1 illustrates, in a simple manner, the different types of conventional biological protection, namely an impermeable membrane (A), an air-permeable shell layer (B), a semipermeable shell layer (C), and a selectively permeable membrane (D). However, most of the available protective clothing systems rely on passive protection, acting as a full barrier against air, vapors, and liquids, as in hazardous materials (HAZMAT) suits (Figure 1A) [111]. Materials that are chemically or mechanically unresponsive to the environment must be engineered to meet performance specifications under worst-case-scenario conditions, often sacrificing performance for the sake of other parameters [113]. Air-permeable overgarments are most frequently composed of an activated-carbon layer to adsorb toxic vapors, designed to be worn over battledress duty uniforms (Figure 1B) [111,112]. Although activated-carbon adsorption material has protective properties, it is limited by a nonselective adsorption, poor protection performance against large toxic liquid droplets, and secondary pollution. Hence, current needs, new materials, and new technologies are acting together to promote the advances of permeable protective suits in pursuit of high performance, multifunctionality, lightweightness, and comfort [114]. The development of new protective clothing with different features that can adsorb hazardous agents is envisioned, which can be accomplished by using different fibrous materials and by following a specific design. Selectively permeable fabrics are important to improving the user's comfort by reducing the airflow through the fabric layers while keeping a high water-vapor permeability [114]. As an example, the integration of electrospun nanofiber membranes in textile fibrous structures produces a high aerosol filtration efficiency, good air permeability, low surface density, and low-pressure loss, thanks to the small but highly interconnected pores and large surface area of built nanofibers [1,114]. In addition, active protection appeared as a promising concept to detect and inactivate/degrade microorganisms and BWAs, while considering that materials capable of responding to their environment may achieve optimal performance under a much wider set of conditions [1,113]. This can be achieved either by using fibers such as the ones prepared via electrospinning or by functionalizing textiles with nanomaterials that possess those capabilities.

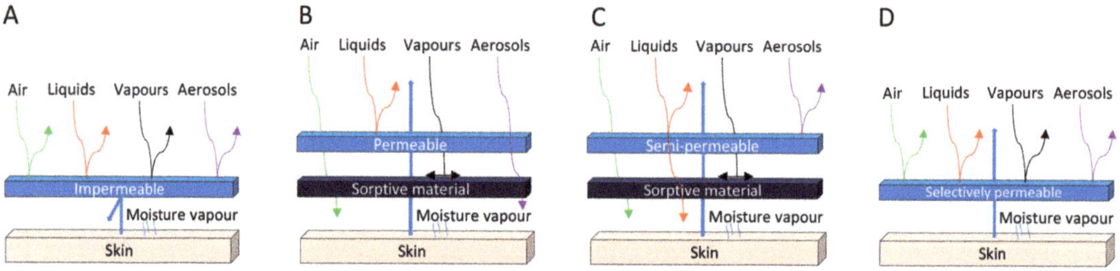


**Figure 1.** Schematic representation of the different types of conventional biological protection: (**A**) impermeable membrane; (**B**) air-permeable shell layer; (**C**) semipermeable shell layer; (**D**) selectively permeable membrane (adapted from [111,112]).

The development of biological protective clothing depends on a combination of different requirements, such as a barrier to liquids, water vapor permeability, and stretch properties. However, it also depends on parameters such as weight and comfort for the wearer, which will ultimately influence the level and durability of the protection. The type of biological threat also impacts this selection and constitutes one of the reasons why the requirements must be established beforehand [115].

This section will focus on the different materials and techniques, from the conventional to the innovative protection methods.

### 4.1. Fibrous Materials

Protective clothing can be achieved through the usage of several different fibrous materials, which are listed in this subsection with regard to the current solutions and the new developments.

#### 4.1.1. Conventional Protection

Commonly used materials for totally impermeable protective clothing are butyl and halogenated butyl rubber, neoprene, and other elastomers [115]. Even though they are effective in conferring a barrier against liquids, vapors, and aerosols, they impede moisture vapor from travelling from the user's body and skin to the environment. This is why fibrous materials are exploited in the development of protective clothing.

Conventionally, synthetic fibers such as polyester, polyethylene, polypropylene, polyamide, and polyurethane are used to fabricate protective clothing [116]. Natural fibers such as cotton, wool, and those regenerated from naturally available polymers can also be employed to provide not only protection (mostly thermal) but also comfort. These are advantageous for protective textiles in comparison to synthetic fibers due to their biocompatibility and low cost, among other things, but they normally require combination with high-performance fibers or post-treatment and finishing processes [1,117–119]. While collecting data on protective textiles, the dominance of cotton fiber is evident. This is mainly because of its natural comfort, appearance, and excellent performance, such as its alkali resistance, hydrophilicity, and moisture retention. Cotton fiber has, however, poor crease recovery, poor dye fixation, microbial growth, photo-yellowing, and poor color fastness properties that need to be improved [120]. Nevertheless, numerous strategies are being developed to overcome such limitations.

#### 4.1.2. Innovative Protection

Some specific fibers can be used in a way that provides sensing and responsive capabilities, making active protection possible. For instance, high performance fibers such as ceramic fibers, carbon fibers, stainless steel, and aluminum fibers can be employed [121]. However, most of these lack moisture management properties and are not durable, which is the reason why they are normally mixed with conventional fibers or interwoven in fabrics.

### *4.2. Fibrous Structures*

Different fibrous structures can be developed, and these are presented in this subsection with respect to conventional and active innovative protection.

#### 4.2.1. Conventional Protection

Completely impermeable suits can be achieved by film-laminated fabrics as a full hazardous barrier (Figure 1A). However, these do not meet the comfort requirements after a long operational time, as the water vapor permeability is high, which causes heat stress for the wearer.

Air-permeable fabrics are usually made of a woven shell fabric, an activated-carbon layer, and a liner fabric (Figure 1B) [115]. The activated-carbon layer is crucial for adsorbing toxic chemical vapors, since the outer layer is permeable not only to air, liquids, and aerosols, but also to vapors.

Another technique to improve the comfort of protective clothing is to use an impermeable material as a barrier for the outer part and a more breathable material for the inner part. To this end, semipermeable fabrics are designed (Figure 1C). In addition, a perm-selective membrane that allows the permeation of water vapor molecules but inhibits the passage of larger organic molecules (Figure 1D) can also be developed. Several materials, mostly polymers, have been used for these semipermeable or selectively permeable membranes (SPMs), such as poly(vinyl alcohol), cellulose acetate, cellulosic cotton, or poly(allylamine). The development of different membranes for protective textiles has been thoroughly reviewed, from the barrier films and breathing membranes to the future directions that advocate the use of selectively permeable barriers, which are schematically represented in Figure 2.

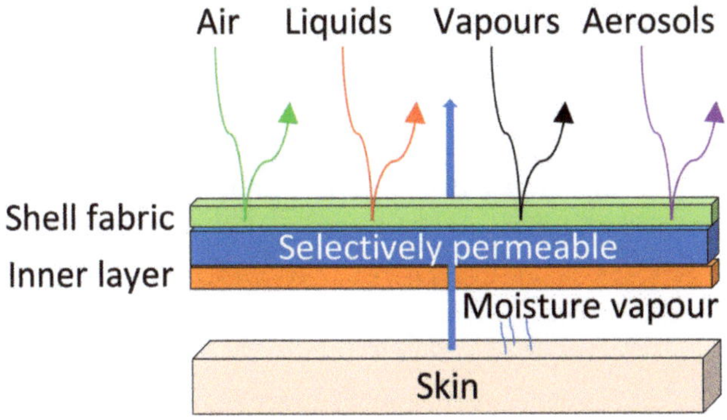


**Figure 2.** Detailed schematic drawing of a selectively permeable membrane (adapted from [111,112]).

Nonwoven fabrics made of a three-layered composite (spun-bonded, melt-blown, spun-bonded) are also a common option for biological protection [1,122]. However, the passage of BWAs through multilayered protective clothing is rather complex and thus must be thoroughly studied. The combination of different layers and barrier properties, in addition to the skin breathability and comfort, must be optimized. The gas/vapor transport by diffusion and convection should be studied and correlated with the vapor and liquid sorption of the protective fabrics in order to assess the degree of protection. The results highly depend on the properties of the materials used, such as yarns and fibers, but also on the fabric construction and clothing assembly [123]. Additionally, the intertwined interactions between some parameters are key, such as the fabric thickness, adsorption, and air permeability properties. Modeling work appears as a promising tool for the prediction and representation of air flow through designed fibrous arrangements and structures. With this, 2D structures can be developed and assessed in terms of their performance.

#### 4.2.2. Innovative Protection

There is an increasing interest in the development of active solutions for protection, with the ability to neutralize BWAs. Smart textiles have appeared and present a wide range of applications, including self-cleaning, phase-transition fabrics and protective clothing [124,125]. Nanotechnology has appeared as a promising solution to develop protective textiles with specific functionalities, such as UV protection, antimicrobial activity, and chemical resistance [112]. Particularly, the use of nanotechnology in chemical, biological, radiological, and nuclear (CBRN) protection clothing has arisen as an excellent possibility. The properties of nanoparticles, nanowires, nanotubes, nanostructures, and nanocomposites are distinctive from those of bulk materials.

An ultrahigh surface area and high surface concentrations are desired for the attachment of biocides and the destruction of adsorbents. This way, nanofibrous networks and consequent closely packed assembly has turned electrospinning into a highly attractive technique to produce membranes for biological protection [126]. A matt of nanofibers can be deposited, creating a randomly oriented fibrous assembly comparable to a nonwoven fabric, but this random assembly can also be collected and oriented into a yarn. The production of electrospun nanofiber-based membranes is promising for the achievement of a clothing system with a lighter weight. In addition, the small pores between fibers improve particulate retention, absorbing hazardous microorganisms. Electrospun polyurethane fibers have been shown to be effective in regard to their elasticity. Since biological agents penetrate fabric and skin in a slow manner, the decontamination of the surface is crucial and does not require immediate neutralization to make sure that the fabric and skin are not penetrated [111]. This, once again, points to the functionalization of fabrics as a promising solution. The combination of this with structured multilayered protective clothing can be highly advantageous for future developments.

### *4.3. Bioactive Agents*

The latest research has directed its efforts at the study of metal–organic frameworks (MOFs); quantum dots; and inorganic particles integrating silver (Ag), copper (Cu), zinc (Zn), and titanium (Ti) cations. Glimpses of the potential of natural polymer chitosan (CS) or derivatives as BWA-counteracting agents, applied as a coating layer or in the form of organic particles (loaded or not with plant-derived compounds such as plant extracts and essential oils (EOs)) can be perceived. Hydrogen-bonded organic frameworks (HOFs), which emerged recently, are also showing high potential to act as self-cleaning materials. The following sections will describe the aforementioned bioactive agents, unveiling the details of their biocidal potential, mechanisms of action, and known limitations.

#### 4.3.1. Metal Organic Frameworks (MOFs)

Zr is ubiquitous in nature, favoring research with Zr-based porous materials, namely zirconium dioxide ($ZrO_2$) or zirconia, which have outstanding optical and electrical features for the development of transparent optical devices, capacitors, fuel cells, and catalysts. Recently, a new class of Zr-based highly porous hybrid materials has emerged, consisting of inorganic metal-ion or metal-oxide clusters bridged by organic linkers, possessing tunable pore sizes, surface area, pore volumes, and responsiveness to visible light [127]. Zr-based MOFs are attracting tremendous attention from the scientific community and have started to become known for having the ability to degrade BWAs (research was first directed at CWAs) and thus having great potential as protective layers in suits or masks or in air purification systems (capturing toxic gases), since the metal-containing secondary building units function as Lewis acid sites for the catalytic hydrolysis of hazardous compounds [57,127–129].

The overall use of MOFs is, however, hindered by the intractable powdery or crystalline forms of the prepared catalysts, which additionally require complex instrumental settings for their processing [128,129]. Another limitation stems from the fact that, in order for them to act as antimicrobial agents, their structure needs to be robust; a release of metal ions (or active linkers) leads to the collapse of the structure. As a consequence, these structures may only be used as temporary microbicidal surfaces. Regardless, MOFs have been instrumental as light-induced disinfectants for pathogens [94]. Scarce, but solid, literature exists linking MOFs to military biological protection. Cheung and colleagues [128] screened an MOF derivative against both CWAs and BWAs. They introduced regenerable MOFs, using a N–chlorine biocide, into a textile via a porous UiO-66-$NH_2$ (a stable zirconium-based MOF with -$NH_2$ functional groups in its organic ligands) as the regenerable carrier. The active chlorine atoms were bonded to the amine-functionalized linker in the ordered framework to form chloramine groups by a simple immersion process in commercial bleaching solutions. The active-chlorine-loaded MOF/fiber composite (UiO-66-NH-Cl/PET; PET = polyethylene terephthalate) quickly killed both Gram-positive *S. aureus* and Gram-negative *E. coli* bacteria, as well as the SARS-CoV-2 virus, after a few minutes. The active N−Cl in the modified Zr-MOF coating was stable and regenerable, acting through the slow release of active chlorine through the pores of the MOF when in contact with the pathogens. The active chlorine could then be generated after water (from the surroundings) was used to hydrolyze the N−Cl to form HClO. The porosity of the MOF allowed the diffusion and slow release of the active chlorine, as the chlorine on the surface was consumed. N–chloramides had previously been deemed as self-decontaminating and regenerable against multiple CWAs, with the goal of using them within military textiles [130].

#### 4.3.2. Inorganic NPs

Nanoparticulate systems are colloidal-sized particles with diameters between 1 and 1000 nm. Their size offers a high surface/volume ratio and a correlation with the structural sizes of biological components: they are small enough to pass through biological barriers, internalize some target cells, and influence multiple cellular processes [131]. Inorganic NPs comprehend metallic, bimetallic, metal oxide, and magnetic elements in their structure [132], with metal oxide NPs containing silver (Ag), copper (Cu), zinc (Zn), or titanium (Ti) cations being the most studied in the fight against microbes such as those that constitute BWAs, alone or combined for synergistic activities. However, there are different types of inorganic NPs with particular characteristics and mechanisms of action against pathogens. In addition to their inherent physical structure, one of the main antimicrobial mechanisms exerted by inorganic NPs is reactive oxygen species (ROS) production [133]. When in direct contact with cells, NPs act through electrostatic attraction, ligand–receptor interactions, hydrophobic reactions, and van der Waals forces. Bioactive metallic ions are likewise released through the metal oxides that absorb the cell's peripheral layers, allowing them to interact with the functional groups of biomolecules, such as proteins and nucleic acids, extra- or intracellularly. This triggers cell metabolic and structural changes, generating homeostatic imbalances [134].

Silver (Ag) NPs wield bacteriocidal effects on both Gram-negative and Gram-positive bacteria at relatively low drug dosages, but side effects such as cytotoxicity in vitro and allergic responses in vivo may happen in the case of an overdose or prolonged use. Moreover, Ag NPs are prone to aggregate and have poor stability, even though stabilizers such as polyvinylpyrrolidone (PVP) or sodium dodecyl sulphate can be used to assist in shielding the corona of Ag NPs from disintegration, augmenting their diffusivity and contact with the microbes while decreasing their toxicity. Alternatively, Ag NPs can be anchored to the surface of some materials, thus relieving the weight of Ag NP-associated disadvantages [135]. In all cases, the main mechanism of action of Ag NPs against pathogens require the attachment and interaction of multiple NPs to the cell surface [53]. This induces the disruption of the microbe outer layer functions and the dissipation of the proton motive force. Small Ag NPs of a few nanometers may even alter the morphology of the cell wall, increasing their internalization and ultimately killing the cell [102]. Compared to Ag and gold (Au), copper (Cu) is cheaper and more attainable, biocompatible, and environmentally friendly. Cu NPs dissolve faster than other noble metals by outward ion release. Cu is an essential element to life, and it is a key regulator in several pathways that are essential for living. As such, Cu ion release can take part in some of these pathways. On the other hand, Cu NPs may accumulate in the body or release too many ions, causing long-term toxicity or contributing to the development of related diseases [134]. The work of Bhattacharjee et al. [104] disclosed that the application of either Ag NPs or Cu NPs enhanced the antimicrobial potency of the built structures for future use in protective clothing and medical textiles. Ag and Cu have broad intrinsic spectra of antimicrobial activity. The first biological barrier of microorganisms is traditionally negatively charged. Hence, these cationic NPs are able to disrupt cell membranes due to electrostatic attraction and form hydroxyl free radicals, resulting in lipid and protein oxidation. The results of the antibacterial activity underlined Ag NP-embedded samples as the most efficient bactericides. A plausible explanation could be the formation of an oxide layer on the Cu NPs, given that Cu NPs are highly susceptible to oxidation, when stored under ambient conditions. However, considering the toxicity of Ag NPs and the much lower cost of Cu, Cu NPs are becoming more attractive nowadays.

ZnO NPs are well-known for their low cost, availability, biocompatibility, biodegradability, and hexagonal prism shape, which allows an increase in surface roughness that ultimately enhances cell anchorage points. Their UV protection, photocatalytic activity, antimicrobial, self-cleaning, energy-harvesting, and biosafety features can confer multiple functionalities to their substrates: water resistance, antimicrobial action, UV blocking, flame retardancy, corrosion inhibition, and electrical conductivity [136]. Zn-doped NPs are indeed capable of endowing a fabric (e.g., cotton-derived) with superhydrophobic properties that facilitate cleaning [136], among other functionalities, including a microbicidal capacity [137]. Noorian and colleagues [138] showed excellent UV protection and significant antibacterial efficacy even after 20 washing cycles and 100 abrasion cycles following the in situ production of ZnO NPs, showcasing their potential for use in advanced protective textiles. The suggested mechanisms of action were again ROS formation, Zn-ion release, membrane dysfunction, and NP internalization, as taken from the literature. Nonmetal and metal doping may effectively change the active wavelength threshold of the absorbed light to the visible area [139], thus enhancing the antimicrobial characteristics in settings where UV light is absent. Doping metals such as Ag, Cu, Au, La, Sm, and Fe and nonmetals such as N, F, C, and S on the ZnO structure [139], or even carbon-based materials [140], enables the possibility of achieving such outcome. However, problems related to the stability, dispersion, and crystalline structure control of ZnO NPs in an aqueous medium seriously hinder the industrial application of this bioactive agent [141]. Moreover, although ZnO NPs offer significant safety and biocompatibility, several authors argue that their toxicity within biological systems should be better understood and controlled [142–144]. These toxic effects have so far been attributed to the high solubility of the particles, resulting in the cytotoxicity, oxidative stress, and mitochondrial dysfunction of mammalian cells [144].

Finally, the work on $TiO_2$ NPs has revealed good photochemical and chemical stability, hydrophobicity, biocompatibility, a low cost, and high photocatalytic and hydrophilic activity. These NPs are activated under UV-light irradiation and generate electron–hole pairs that dispense $Ti^{4+}$ to $Ti^{3+}$ cations and oxidize $O^{2-}$ anions to oxygen atoms. The ejection of oxygen atoms from the $TiO_2$ complexes produces oxygen vacancies that are occupied by water molecules, which in turn leave OH groups on the surface of $TiO_2$ NPs and make them hydrophilic. The generated electron–hole pairs induce bacterial growth inhibition and produce ROS. The addition of carbon-based materials such as graphite enlarges the activation range of $TiO_2$ nanoparticles to visible light and causes increased hydrophilic, photocatalytic, and antibacterial properties. To stabilize $TiO_2$ and $TiO_2$ composites, they can also be uniformly dispersed in polymeric substrates [145]. Görgülüer et al. [146] revealed that the photocatalytic activity of $TiO_2$ NPs was improved by the deposition of metal NPs (notably Ag NPs) on the $TiO_2$ surface, since the formation of a Schottky barrier at the metal–semiconductor interface resulted in the more efficient capture of photogenerated electron–hole pairs. Moreover, the surface plasmon absorption of Ag NPs can broaden the absorption spectrum in the visible region. Regardless, a lotus leaf effect on the tested assemblies [147] and antimicrobial activity [148,149] are generally present when $TiO_2$ NPs are added to the proposed substrates. However, their toxicity to human health and the ecosystem is also a considerable concern related to their extended use [150].

#### 4.3.3. Organic Small-Scale Particles

Organic small-scale particles comprise polymeric structures that are widely studied in the literature as drug delivery systems. Specifically, proteins, lipids, polysaccharides, nucleic acids, and other biomolecules are capable of being processed into small-scale particles, with increasingly significant research pinpointing their utility for drug delivery. These biomolecules can also be combined with inorganic nanomaterials to produce hybrid materials showcasing features from both types of material [151].

Some recent studies have explored CS-based small-scale particles loaded with plant-derived molecules to prevent or control infections while interspersed within fabrics to function as protective textiles. CS is widely recognized for its tuneable biocompatibility, bioactivity, chemical versatility, and ease of processing into a variety of structures, thus finding itself considered of high value for numerous applications [95,131,152–156]. Plant extracts or essential oils (widely used as folk medicine) are increasingly being studied as antimicrobial agents, as several natural drugs have already been approved for clinical use. Their modes of action comprise: the inhibition of cell wall synthesis, the permeabilization and disintegration of microbial peripheral layers, the restriction of microbial physiology, oxygen uptake and oxidative phosphorylation, efflux pump inhibition, the modulation of antibiotic susceptibility, biofilm inhibition, the hindrance of the microbial protein adhesion to the host's polysaccharide receptors, and the attenuation of pathogen virulence [131]. EOs in particular act through their inherent hydrophobicity, which enables them to accumulate in the cell membrane, disturbing its structure and functionality and causing an increase in their permeability to a point at which cell lysis and death is unavoidable [153]. Notwithstanding, their loading onto/into organic particles has also been the object of several studies, as a way of enhancing molecules' biostability and bioactivity, along with controlled release, thus holding the power to provide strong and durable effects. CS-based small-scale organic carriers have tremendous potential [131,153,156–158]. Recent efforts from the team of Bouaziz et al. [96] demonstrated that coacervated CS microcapsules, with cinnamon EO in their cores, could substantially inhibit the growth of the tested Gram-positive and Gram-negative bacterial strains. The antibacterial results were mainly due to the cinnamaldehyde (the major constituent of their cinnamon EO batch) after the oil release from the microcapsules and were not attributed to the CS itself or to the built architecture, even though the authors did not test unloaded particles. However, a fact is that the antimicrobial potency of CS alone is highly variable, depending on its cationic nature, when its amine groups are protonated (which traditionally occurs at $9.5 < pH < 6.5$, depending on the degree of acetylation). CS either accumulates at the cell surface, forming a polymer layer that prevents substance exchanges such as nutrient intake and metabolic disposal, or, as in the case of CS with a low Mw, reaches the intracellular compartments, adsorbing electronegative substances, disrupting the cells' equilibrium, and killing them [131]. If the environmental pH is above CS's pKa, the inhibitory effect is instead governed by hydrophobic interactions and the chelating capacity of divalent metal ions rather than the electrostatic interactions between its protonated amines and anionic bacterial outer-layer structures [153]. The known limitations are associated with batch-to-batch variability, stability in physiologically compatible media, burst release, washing durability, and poor mechanical properties [131].

Another study, defended by Wang et al. [159], created hydrogen-bonded organic frameworks (HOFs), which are supramolecular self-assembled π-conjugated structures of rigid and large functional tectons that demonstrated a significant enhancement in daylight-driven ROS generation capacity and ROS storage lifetime under dark conditions. After daylight stimulation for 2.5 min, the fluorinated HOF-101-F/fiber killed almost 95% of *E. coli*. The composite shows excellent sterilization efficiency under light irradiation and dark treatments for five cycles without decreasing its performance. HOF-101-F, after exposure to daylight for 30 min, could kill over 99.99% of *S. aureus*, *Klebsiella pneumoniae*, and *Mycobacterium marinum*. However, the applicability of HOFs is still in its infancy, with large conformationally flexible building blocks remaining a challenge because of rigid molecule approximations and limitations in the accuracy of force fields to rank diverse energy landscapes reliably, especially those where interpenetration is present [160].

Regardless, one truth is that the presence and efficiency of organic NPs within textiles is still sporadic.

#### 4.3.4. Carbon Nanodots

Carbon dots can be divided into carbon nanodots (CNDs), carbon quantum dots (CQDs), and graphene quantum dots (GQDs) [151]. CQDs are newly emerging quasi-spherical NPs with a particle size of 1–10 nm. These carbon-based materials have high temperature resistance, outstanding electrical/thermal conductivity, high plasticity, corrosion resistance, UV blocking, a high adsorption rate, good water solubility, excellent biocompatibility, low toxicity, and a high catalytic performance [107,161,162]. Some studies can be found describing their potential use as fluorescent probes against BWAs, in particular dipicolinic acid, a biomarker of *B. anthracis* [162,163] or of *E. coli* [162], as a result of their low environmental hazard, high selectivity, greater sensitivity, good biocompatibility, changeable fluorescent properties, and excitation-dependent multicolor emission behaviour. CQDs are composed of sp2 carbon atoms formed in planes, with each carbon atom being mainly connected to the three nearest neighbors with a distance of 120 degrees. The implantation of oxygen-, sulfur-, and nitrogen-containing functional groups can be introduced to the sides of graphite sheets to overcome the intersheet van der Waals forces that subsequently result in the enlargement of the interlayered spacing. However, despite this, the applicability of CQDs is still narrowly exploited in batteries, fuel cells, supercapacitors, and transistors, with sensing and bioimaging being indeed more actively explored [162]. One particular study reported the integration of carbon quantum dots clustered from the fluorescent aromatic compound named 4–(2,4–dichlorophenyl)–6–oxo–2–thioxohexahydropyrimidine–5–carbonitrile within a textile matrix for military protective garments [107]. CQDs were able to completely eradicate all the tested species: *S. aureus*, *E. coli*, and *C. albicans*. Even after 10 washing cycles, microbial inhibitions were substantially high. CQDs, specifically their lateral functional moieties, act by creating microbial oxidative stress intracellularly under visible light and in an aqueous medium. Oxidative stress can be defined as differences in the subcellular and tissue compartmentalization of ROS that contribute to stress responses, provoking altered cellular activities, cell proliferation, extracellular matrix synthesis, the production of matrix-degrading enzymes, and cell apoptosis. ROS comprise singlet oxygen, singlet sulfur, singlet nitrogen, and hydroxyl free radicals. In lethal doses, ROS directly guide nucleic acids to fragmentation; corrupt gene expression and protein synthesis values; incite lipid peroxidation, gradual cell wall destruction, and necrosis/apoptosis; and encourage microbial cell death [107,164]. Regardless, high toxicity due to the use of heavy metals in production, complex processing methodologies, and poor control over dot size have been related to this type of bioactive agent [165,166].

#### 4.3.5. Graphene and Derivatives

Graphene is a thick layer of sp2-hybridized carbon atoms arranged in a honeycomb-like crystal lattice [167]. Graphene has become one of the most studied carbon-based materials in recent years due to its excellent mechanical characteristics, high electrical conductivity properties, high Joule-heating capacity, high UV shielding, rapid heat dissipation, high hydrophobicity, high thermal stability, high antimicrobial activity, and high biocompatibility. However, it is limited by low fabrication rates and a high cost, in addition to a strong aggregation tendency and hydrophobicity, which leads to insolubility in aqueous media [104,168–170]. Consequently, graphene derivatives such as graphene oxide (GO) and reduced GO (rGO) have been produced. GO can be synthesized from graphite powder. It has several oxygen-containing functional groups, which turn it into a chemically versatile material. However, in some cases, these oxygen-based functional groups reduce its functionality. Thus, it is reduced using chemical, electrochemical, or thermal approaches, creating rGO. rGO shows properties similar to pristine graphene and also relatively good conductivity. It can be easily prepared in the desired amounts from cost-effective GO [151]. GO and rGO are able to form covalent or hydrogen bonds with textiles such as cotton or silk via their carboxyl, hydroxyl, and epoxide groups. The addition of rGO to cotton or silk, as shown by Bhattacharjee et al. [104], resulted in mild antibacterial activity, which seemed to derive from the scissoring action of its sharp creases/edges and the generation of oxidative stress in the pathogenic cells through electron transfer. rGO has been shown to react with lipids, DNA, and amino acids via electrostatic and $\pi-\pi$ stacking interactions. The reaction between rGO's oxygen and the cell wall polysaccharides of bacteria has also been reported [104,168]. However, it remains difficult to precisely control the compositions and sizes of graphene sheets, which heavily affects the performance of the derivatives [171].

Figure 3 illustrates the current trends in the use of different types of antimicrobial agents within protective textiles, for military use or otherwise. It presents approximate frequency counts of the use of these materials within the published literature of the last 5 years (database: Scopus). Inorganic NPs are the major contributors to these numbers, but natural approaches fall shortly behind. Of the former, Ag-based strategies are the most commonly explored. On the other hand, research using natural biocidal approaches is highly unfocused, even though many studies explore plant extracts, CS and derivatives, CS-based architectures, and plant-extract-loaded CS-based small-scale particles. However, it becomes clear that the quest to find suitable bioactive agents has been narrowing over the years, with bioactive agents such as ZnO or CuO NPs gaining more importance lately. Besides, MOFs as well as carbon-based materials such as CQDs and graphene derivatives emerged around 3 years ago for this type of application. In addition, natural approaches have appeared, emphasizing the potential of CS, particularly if processed in the form of small-scale particles carrying biomolecules such as plant-derived compounds.

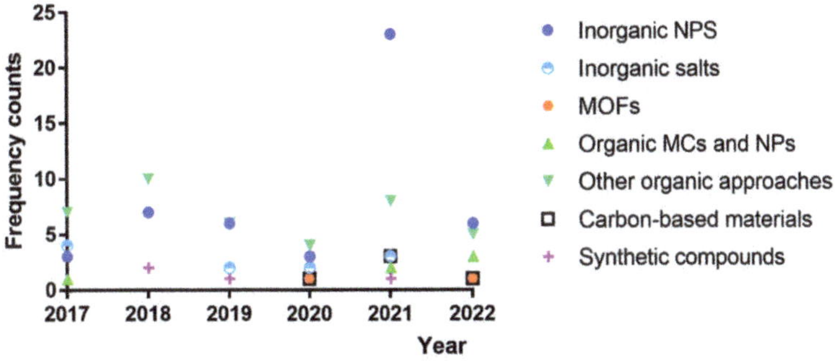


**Figure 3.** Approximate frequency counts of the usage of different categories of antimicrobial agents within protective textiles (intended for the military, first responders, or civilians) within published literature of the last 5 years (database: Scopus).

### 4.4. Textile Fabric Functionalization Methods

Textiles can carry microorganisms and also promote their survival, proliferation, and endurance. When a fabric is used for clothing, an infestation may create infections and constitute a biological threat. Antimicrobial functional finishes are therefore applied to textiles to protect the wearer and the fabric itself [172]. Various techniques exist to immobilize bioactive agents onto textile fibers, each one carrying its specifications, advantages, and limitations, with the fabric being previously treated and functionalized in order to improve the impregnation of the selected bioactive agents, as well as their durability within the textile. The dip-pad-cure method, the dip-and-dry method, the exhaustion method, the spray-dry method, the spray-cure method, the pad-batch method, and sol-gel and sonochemical coatings are a few relevant examples of the impregnation methods of bioactive agents [105,173]. However, coating and laminating procedures are increasingly important techniques for adding value to textiles, including coating approaches such as the lick-roll method; direct coating (knife on air, knife over table, knife over roller, knife over rubber blanket); foam coating; foam and crushed-foam coating; transfer coating; kiss-roll coating; rotary-screen printing; spray coating; calendar coating; hot-melt extrusion coating; and rotogravure [174].

Starting with MOFs, recent antimicrobial stars, some interesting studies have been performed. The work of Cheung and colleagues [128] stands out, as PET textiles had UiO-66-$NH_2$ MOFs grown in situ following chlorination with a hypochlorite bleach solution to obtain regenerable N–chlorine MOFs coating the textile. The same occurred elsewhere [106], but this time ZIF(Ni), ZIF-8(Zn), and ZIF-67(Co) were the MOFs synthesized into cotton fabrics. A silicate modification acted as a crosslinker between cotton on one side and ZIF-MOFs on the other, thereby increasing the number of MOFs adsorbed onto the fabrics. The fabrics were scoured for dirt removal or even bleached for discoloration [106,128], and sometimes functionalized to gain functional dopamine moieties [175] or the previously mentioned silicate modification [106] to reinforce binding with the bioactive agents through covalent bridges.

The same trend has been observed with inorganic NPs, with most of the NPs being grown in situ following textile incubation with metallic precursors. Despite their well-known handicaps, Ag NPs continue to be the most studied inorganic NPs in protective textiles, although often in combination with other microbicidal enhancers. Textile functionalization with the bioactive agents occurs mostly via the in situ formation of NPs [95,104,105,176,177]. As an example, El-Naggar and colleagues [105] showed that bleached and mercerized (an alkaline treatment to improve affinity towards subsequent chemical modifications) cotton fabric was rendered more hydrophilic through plasma treatment, then washed with a nonionic detergent to remove impurities and silanized to encourage metal–ligand binding with the Ag NPs. Silanization treatment forms silane groups that act as fiber–NP coupling agents, creating a siloxane bridge between the two components [178]. Finally, the treated fabric was immersed in a solution carrying metallic precursors, sonicated, padded, squeezed, and cured for thermal reduction to form Ag NPs. Görgülüer et al. [146] washed rayon fabric in an acetic acid solution and in a wet surfactant so that any chemical finishing, such as silicon, and softening on the fabric could be efficiently removed. Afterwards, the fabric was immersed in $TiO_2$ NPs; poly(dimethylsiloxane) (PDMS) to functionalize the later NPs with hydrophobic moieties; $AgNO_3$ and $NaBH_4$ as metallic precursor and reducing agent, respectively; and finally, tetrahydrofuran (THF) to assist in the production of compact and spherical Ag NPs. Samples were ready for characterization following a drying step. While using ZnO NPs to guarantee bacterial cell death in desized and bleached cotton fabrics, Noorian et al. [138] also washed the fabric in nonionic detergent, before performing oxidization by periodate and treatment with 4-aminobenzoic acid ligands (PABA). NPs were similarly built in situ after the immersion of the fabric in a ZnO precursor, ultrasonication, and chemical reduction.

The integration of CQDs into cotton fabric that had been scoured, bleached, and cationized with 3–chloro–2–hydroxypropyl trimethyl ammonium chloride ($C_6H_{15}Cl_2NO$), was indeed very simple [107]; it was achieved by dissolving previously prepared CQDs, impregnating the fabric with them while stirring, and drying. The addition of an rGO coating through a dip-dry process onto fabrics composed of cotton or silk [104] that had been previously washed with acetone and hot water and functionalized with a silane derivative allowed increased quantities of Ag and Cu NPs to be added subsequent to the composition, particularly with cotton, which is richer in hydroxyl groups than silk.

Botelho and team [95] washed PA taffeta and submitted it to plasma treatment. CS was then added through the dip-dry method, followed by the already prepared Ag NPs. Dip-pad-dry was the immobilization technique also selected by Verma et al. [120] to integrate dissolved CS, along with citric acid ($C_6H_8O_7$) to act as a linker to the enzymatically desized and scoured cotton fabric, with sodium hypophosphite ($NaPO_2H_2$) as the catalyst; this worked as a mordant to enhance the dyeability of the cotton. Samples were then padded, dried, and cured. A final step included a dyeing process with onion-skin dye. Some studies have additionally integrated plant-derived molecules into/onto CS-based small-scale particles [96,131,179]. Singh et al. [179] used the emulsification of gelatin and rosemary EO followed by ionic gelation between gelatin and CS to encapsulate the EO and produce a stable shell. Linen fabric was dipped in a microcapsule (MC) dispersion and low-temperature curable acrylic binder, padded, and dried. Verma and colleagues [96] encapsulated cinnamon EO within CS MCs produced by simple complexation with Tween 20. Dense taffeta cotton fabrics, which had been desized, bleached, and mercerized, were dipped into an MC dispersion and a binding agent (dimethyloldihydroxyethylene urea, DMDHEU), padded, dried, and cured; they were then autoclaved and stored. In another study, Wang and colleagues [159] explored HOFs that carried building units incorporating $CH_3$-, F-, or $NH_2$-groups on the ortho-position of the phenyl ring of the benzoic acid and were produced via a sol-gel method. These were spray-coated onto woven and knitted cotton fabric, as well as commercial chirurgical disposable face masks; dried; washed in acetone to remove unbound agent and solvent; and then dried again.

As mentioned above, multiple bioactive agents have been tested with textiles, alone or combined in order to obtain synergistic effects in the fight against pathogens. Many authors are also aware of the need to obtain durable bioactive effects, namely by retaining the bioactive compounds attached to fibers [128,146,159]. It is, however, noticeable that the past two years, during the COVID-19 pandemic situation, have been key for attempts to control the washing durability of finished fabrics, thereby responding to a major concern of the textile finishing industry [96]. Some authors have even followed standardized protocols to assess such features (the KS K ISO 6330 [104], IS: 3361-979 [120], AATCC-61 [180], or AATCC 2010 [107] standards), thus proving that the required bioactivity is present even after laundering activity. Table 1 summarizes the main, and representative, antimicrobial protective textiles designed for military purposes or for general use.

**Table 1.** Recent trends (2020–2022) in antimicrobial protective textiles designed for military purposes or for general use.

| Fabric | | Bioactive Agent | Impregnation Method | AM Testing | | | | Protective Textile | Ref. |
|---|---|---|---|---|---|---|---|---|---|
| Details | Cleaning and/or Pretreatment | | | Cell | Method | Main Results | Durability | | |
| Woven and knitted cotton fabric, plus commercial chirurgical disposable face masks | - | HOF-101-R (R=H, $CH_3$, F, $NH_2$), obtained by sol-gel method | Spray coating: HOF-101 tecton derivatives (1 mg/mL in DMF) were sprayed on various fiber materials (1 × 1 $cm^2$) for 10 s and dried (100 °C, 1 h). The procedure was repeated enough times until the sprayer was empty. Fibers were washed by acetone 3 times and dried (100 °C, 1 h). | *S. aureus*, *E. coli*, *K. pneumoniae*, and *M. marinum* | Shake-flask method, under simulated daylight and dark conditions | After illumination under simulated daylight for 2.5 min, the HOF-101-F/fiber killed 95% of *E. coli*. Following 12 h of solar irradiation and exposure to bacteria for 2 h, cell death was ≈46%. Performance maintained after light irradiation and dark treatments for 5 cycles. Over 99.99% of bacteria was eliminated after daylight treatment for 30 min. Antibacterial performance under complete dark conditions without preirradiation was much slower. | Washed in water without observable HOF loss. | Face masks | [159] |
| PET | Scoured in 3% NaOH solution (90 °C, 20 min), then washed with water | Regenerable N-chlorine, loaded into Zr-MOF UiO-66-$NH_2$ | In situ MOF synthesis: PET textile (20 cm × 20 cm), BDC-$NH_2$ (90 mmol, 16.2 g) and $ZrOCl_2 \cdot 8H_2O$ (60 mmol, 19.4 g) mixed in water (400 mL) and TFA (200 mL) in a sealed 1 L Schott bottle, sonicated for 0.5 h, placed at 100 °C for 6 h, cooled to RT, washed by water (2 × 500 mL) and acetone (3 × 500 mL), dried at RT, and activated at 110 °C for 24 h under dynamic vacuum. | *S. aureus*, *E. coli* and SARS-CoV-2 | Modified AATCC 100–2004 (with textile "sandwiched" using another identical sample for full contact), SEM of harvested bacteria, anti-SARS-CoV-2 virus test | Bacteria: 7-log reduction within 5 min. SARS-CoV-2: 5-log reduction within 15 min. | 23% loss in chlorine content after 40 days storage, sealed, under ambient conditions, still enabling total sterilization. | Cloth against BWAs and CWAs | [128] |

**Table 1.** *Cont.*

| Fabric | | Bioactive Agent | Impregnation Method | AM Testing | | | | Protective Textile | Ref. |
|---|---|---|---|---|---|---|---|---|---|
| Details | Cleaning and/or Pretreatment | | | Cell | Method | Main Results | Durability | | |
| 100% plain-woven cotton, 185 gm/m$^2$ | Scoured, bleached, then cationized with $C_6H_{15}Cl_2NO$ (50 °C, 2 h) | CQDs clustered from synthesized TM | Dip-dry: 0.25 g of prepared components (TM or CQDs) dissolved in 25 mL of $CHCl_3$. Fabric (0.25 g) impregnated in 0.25 g of TM or CQDs (1 h, continuous stirring), then air-dried. | *S. aureus*, *E. coli*, and *C. albicans* | Kirby–Bauer disk diffusion technique, MIC determination | 82%, 71%, and 62% growth inhibition, respectively, in 24 h. | 68%, 63%, and 67% growth inhibition, respectively, after 10 washing cycles. | Military clothing | [107] |
| Pristine CNWs fabricated from pulp and lyocell fibers | Drying (90 °C, 5 h) and hydrofobization with CI, plus UV-induced grafting of PTB | PHMG or NEO | Outer layer: grafting of antiviral/antibacterial agents by the ring-opening reaction of the PTB with -$NH_2$ of PHMG or NEO onto hydrophobic CI-functionalized CNWs. Middle layer: the same onto pristine CNWs. | *S. aureus*, *E. coli*, HcoV-229E virus, and SARS-CoV-2 virus | Colony count method and antiviral testing | Bacteria: >99.99%, 99.99 ± 0.01% growth inhibition rate after 10 min of incubation with CNWs-PTB-PHMG. Sars-Cov-2: 16.23 ± 1.69% survival after ~0.1 min with CNWs-PTB-NEO, 99.84% ± 0.14% after 30 min with CNWs-PTB-PHMG. | - | Face masks | [181] |
| 100% plain-weave cotton fabric: 80 ends/inch, 75 picks/inch, and 168 (g/m$^2$) | Scoured, bleached, and $C_8H_{11}NO_2$-modified (immersion in $C_8H_{11}NO_2$.HCl solution at pH 8.5, 24 h) | Ag NPs | Dip-dry: immersion in 10 mM $AgNO_3$ (continuous stirring, 30 °C, 8 h) and vacuum-drying (12 h, 40 °C). | *S. aureus* and *E. coli* | ASTM E2149-01 | Bacterial reduction of 86% for *S. aureus* and 93% for *E. coli* following 1 h of incubation, 100% after 24 h. | ~98% bacterial reduction after 20 washes. | Functional textiles | [177] |

**Table 1.** *Cont.*

| Fabric | | Bioactive Agent | Impregnation Method | AM Testing | | | | Protective Textile | Ref. |
|---|---|---|---|---|---|---|---|---|---|
| Details | Cleaning and/or Pretreatment | | | Cell | Method | Main Results | Durability | | |
| Woven viscose (120 $g/m^2$) | Fabric phosphorylation: immersion in DAPH at a molar ratio of 1:1; urea was also included as 3 equiv of DAHP, then rinse with water | ZPT | Dip-pad-dry: padding with 0.5 wt % aqueous solution of $N_2O_6Zn·6H_2O$ via the 2-dip-2-nip method. Then, water-soluble NaZPT was added at a molar ratio of 1:2 with respect to the metal precursor. Immersion in a ZPT ligand solution (2 h, 40 °C, orbital shaking at 120 rpm). Drying (80 °C, 10 min), curing (150 °C, 2 min), and rinsing with water. | *S. aureus*, *E. coli*, and *C. albicans* | Qualitative Kirby−Bauer disk diffusion method; quantitative AATCC-100, OD600, and bacteria survival (CFU) measurement methods; SEM and quantitative antifungal assay | Viscose-ZPT induced high ZoI (48 or 53 mm, respectively, against *S. aureus* or *E. coli*). | Viscose-ZPT induced high ZoI after 20 washes (38 or 43 mm, respectively, against *S. aureus* or *E. coli*). 96–97% growth inhibition (20 washes). | Protective clothing | [180] |
| 100% cotton or silk | Acetone and hot water (60 °C) washed; air-dried; soaking in coupling-agent solution (pH 4–5, $C_9H_{20}O_5Si$:water = 1:15) for 4 h at 60 °C; air-dried | rGO and Ag/Cu NPs | Immersion in 0.25 mg/mL rGO suspension (RT, 4 h), air drying (3 times), separately soaked in 0.05 M $AgNO_3$ and $CuSO_4·5H_2O$ solutions (2 h), air-dried, immersion in 2% wt/V $Na_2S_2O_4$ solution (chemical reduction, 4 h, 80 °C, 100 rpm), washed in water, dried (hotplate at 60 °C), and heat-treated in a vacuum oven (20 min, 175 °C). | *S. aureus*, *E. coli*, *P. aeruginosa*, and *C. albicans* | CFU counts | 69–99% (*S. aureus*), 92–100% (*E. coli*), and 97–100% (*P. aeruginosa*) growth inhibition, especially with Ag NPs after 24 h; 63–69% *C. albicans* growth inhibition with Cu NPs (50% with Ag NPs), namely using cotton. | 85−99% growth inhibition against Gram-negative bacteria; 62 to 90% against *S. aureus* after 10 washing cycles. | Protective clothing | [104] |

**Table 1.** *Cont.*

| Fabric | | Bioactive Agent | Impregnation Method | AM Testing | | | | Protective Textile | Ref. |
|---|---|---|---|---|---|---|---|---|---|
| Details | Cleaning and/or Pretreatment | | | Cell | Method | Main Results | Durability | | |
| Woven cotton fabric (areal mass density: 280 g/m$^2$; threads/cm: warp 48 ± 2; weft 37 ± 1; and CIE whiteness 80) | Desized, bleached, and mercerized | CS MCs, prepared by simple emulsion (with Tween 20) and loaded with cinnamon bark EO | Immersed in MCs (80 g/L) and the binding agent (40 g/L, DMDHEU), padded (wet pick up of 80%), dried (90 °C, 15 min), cured (150 °C, 5 min), autoclave-sterilized, and stored at RT. | *S. aureus* and *E. coli* | Diffusion assay method | 90% (*S. aureus*) and 97% (*E. coli*) growth inhibition. | 69% MC remaining after 5 washes, 12.5% after 10 washes. | Protective textiles | [96] |
| 100% cotton knitted fabric (194 g/m$^2$) with (1 x 1) interlock structure | Cleaned with acetone and water, mercerized | Ag NPs | Immersed into a solution of $C_6H_8O_6$ (5 min), dried (5 min, 80 °C); immersed into $AgNO_3$ solution (5 min), dried (5 min, 80 °C); 1–3 cycles. Encapsulation in a silicone binder solution in acetone at a ratio of 1:7 for 5 min (1 time), dried (10 min, 80 °C). | *S. aureus* and *E. coli* | AATCC 147, agar diffusion assay | Higher ZoI for 1-cycle samples after 24 h (0.531 mm with *S. aureus*, 0.25 mm with *E. coli*). | - | Protective textiles | [176] |
| Woven cotton fabric | Enzymatic desizing and scouring | CS and onion-skin dye | Dip-pad-dry: dip within CS (4%), $C_6H_8O_7$ (6%), and $NaH_2PO_2$ (5%) at 1:30 material:liquor ratio (pH 5, 90 °C, 45 min), pad (P = 2 kg/cm, expression of 70–75%), dry (100 °C, 5 min), and cure (140 °C, 4 min). Dyeing with onion-skin dye (exhaustion method): 6% dye, pH 5.5, 90 °C, 75 min, 1:30 material:liquor ratio. | *S. aureus* and *E. coli* | AATCC Test Method100, shake-flask | *S. aureus* (98.03%) and *E. coli* (97.20%) growth reduction after 24 h. | Reduction in *S. aureus* growth from 96.84 to 80.14% and *E. Coli* from 93.20 to 80.74% after 5–20 washing cycles. | Protective textiles | [120] |

**Table 1.** *Cont.*

| Fabric | | Bioactive Agent | Impregnation Method | AM Testing | | | | Protective Textile | Ref. |
|---|---|---|---|---|---|---|---|---|---|
| Details | Cleaning and/or Pretreatment | | | Cell | Method | Main Results | Durability | | |
| Rayon fabric | Acetic acid (3 g/L) and TEGO® wet surfactant (2 g/L) (Evonik) solution in DW (pH 3.5, 20 min), oven-drying | $TiO_2$, Ag NPs | Dip-dry: immersion in coating mixture (60 mL of 5% $TiO_2$ NPs + 9.7 mL PDMS + 8 mL of 1 M $AgNO_3$ + 10 mL 0.017 M $NaBH_4$ + 30 mL THF) 10 min, drying (70 °C, 4 h). | *S. aureus* and *E. coli* | Agar diffusion assay | ZoI of 14.44 mm (*S. aureus*) and 13.12 mm (*E. coli*) after 24 h. | Water contact angle remained nearly constant (152.3°) after 20 laundering cycles. | Multifunctional textiles | [146] |
| Polyamide taffeta (52 warp and 32 weft yarns, 100 $g/m^2$) | Washing, plasma treatment (RT, atmospheric pressure, width of 50 cm, gap distance of 3 mm, 10 kV, 40 Hz, 5 times, both sides) | Ag NPs, CS | Dip-dry: dip in each solution (5 min, RT) and dry (50 °C, 20 min). | *S. aureus* and *P. aeruginosa* | ASTM-E2149-01, shake-flask | *S. aureus* (80%) and *P. aeruginosa* (60%) growth reduction after 2 h. | - | Face masks | [95] |
| Bleached and mercerized cotton fabric | $O_2$ plasma treatment (13.56 MHz, 3 min, 400 W, 200 $cm^3$/min, 0.003 mbar); washing with nonionic detergent ($C_{32}H_{66}O_9$, 10 mmol); sonication (30 min); air-drying and washing with water; dipping in acetone solution of $C_9H_{22}O_3SSi$ (1%, 24 h); curing (75 °C, 30 min); rinsing with water | Ag NPs | In situ synthesis of Ag NPs: dip in 0.1–4 wt % $CH_3AgNO_2$, sonication (15 min), padding, squeezing, and curing (130 °C, 5 min). | *S. aureus*, *E. coli*, and *C. albicans* | Agar diffusion assay | Clear and large ZoI after 24–48 h. | - | Multifunctional textiles | [105] |
| Plain cotton fabric (135 $g/m^2$) | Immersion in 4 mg/mL $C_8H_{11}NO_2$. HCl (pH 8.5) | ZIF-8 | Immersion in $Zn(NO_3)_2.6H_2O$ (0.893 g, 15 mL) solution + $C_4H_6N_2$ (0.985 g, 15 mL) solution, autoclaving (100 °C, 12 h), washing, and drying (60 °C). | *E. coli* | Disc diffusion method | Defined ZoI after 24 h. | - | Multifunctional textiles | [175] |

**Table 1.** *Cont.*

| Fabric | | Bioactive Agent | Impregnation Method | AM Testing | | | | Protective Textile | Ref. |
|---|---|---|---|---|---|---|---|---|---|
| Details | Cleaning and/or Pretreatment | | | Cell | Method | Main Results | Durability | | |
| Cotton fabrics (shibeka, honeycomb, and crepe) | Bleached | CS or Ag NPs | Dip-dry: immersion in CS solution (10 min), squeezing for 100% wet pickup (constant pressure), drying (80 °C, 4 min), and curing (140 °C, 2 min); immersion in Ag NP dispersion (100–300 ppm), squeezing for 100% wet pickup (constant pressure), drying (80 °C, 3 min), and curing (140 °C, 2 min). | *S. aureus*, *P. aeruginosa*, *C. albicans*, and *A. niger* | Disc diffusion method | 20 or 13 (*S. aureus*), 15 or 11 (*P. aeruginosa*), 13 or 21 (*C. albicans*), and 12 or 11 mm (*A. niger*) with 6% CS (Crepe) or 300 ppm Ag NPs (Shebika), respectively, after 24 h. | - | Protective textiles | [108] |
| Desized and bleached cotton fabric (100% cellulose, 117.5 g/m$^2$) | Washed (30 min, 50 °C, nonionic detergent Adrasil HP P-836, 1 g/L, 1:60 L:G), water-rinsed, dried at RT; periodate oxidation in phosphate buffer (pH 8, L:G 1:50, dark), addition of $NaIO_4$ (5 g/L, 30 min, ultrasonication at 20 kHz, 750 W at 70% efficiency), water-washed, dried at RT; PABA treatment (10 g/L, 2 h) using acetate buffer solution (pH 5.5, ultrasonication), water-washed, dried at RT | ZnO NPs | In situ synthesis of ZnO NPs: immersion in 1 mM $ZnCl_2$ solution (30 min) and ultrasonication (pH 10 for 30 min by adding 4 g/L NaOH). Ultrasonication (extra 30 min, 60 °C), water washing, and drying (120 min, 110 °C). | *S. aureus* and *E. coli* | AATCC 100-2004, 24 h | 99.9% (*S. aureus*) and 99.4% (*E. coli*) growth inhibition. | 93.7% or 95.3% (*S. aureus*) and 93.4% or 95.4% (*E. coli*) after abrasion or washing process, respectively. | Protective textiles | [138] |

**Table 1.** *Cont.*

| Fabric | | Bioactive Agent | Impregnation Method | AM Testing | | | | Protective Textile | Ref. |
|---|---|---|---|---|---|---|---|---|---|
| Details | Cleaning and/or Pretreatment | | | Cell | Method | Main Results | Durability | | |
| Scoured and bleached plain-woven 100% cotton fabrics (165 gm/m$^2$) | Silicate modification: immersion in 100 mL of 5% NaOH (50 °C, 5 h, stirring), addition of 6 mL $C_3H_5ClO$ (5 h reaction), water and anhydrous ethanol washing, drying (60 °C); silicate mixture synthesized by dropwise addition of $SiC_8H_{20}O_4$ (12 mL) and methanol (80 mL) to a flask with 30 mL of ammonia and 320 mL of methanol; stirring 3 h, curing (110 °C, 1 h) | ZIF(Ni), ZIF-8(Zn), and ZIF-67(Co) MOFs | In situ synthesis of MOFs: immersion, separately, in 50 mL of methanol with metal salts (0.736 g of $Ni(NO_3)_2$, 0.758 g of $Zn(NO_3)_2$, and 0.733 g of $Co(NO_3)_2$), stirring 1 h at RT; pour three solutions individually from $C_4H_6N_2$ (1.623 g in 50 mL of methanol) above the three mixtures, stir 8 h; ethanol-wash and dried (vacuum, 60 °C, 12 h). | *S. aureus*, *B. cereus*, *E. coli*, and *C. albicans* | Kirby–Bauer disk diffusion method, overnight | ZoI: 25 (*S. aureus*), 23 (*B. cereus*), 15 (*E. coli*), 22 (*C. albicans*) for cotton–silicate–ZIF(Ni). | ZoI: 19 (*S. aureus*), 18 (*B. cereus*), 12 (*E. coli*), 18 (*C. albicans*) for cotton–silicate–ZIF(Ni) after 5 washing cycles. | Protective textiles | [106] |
| Inner layer: polystyrene fiber 3-ply twisted yarns (tex: 0.058, 0.115, or 0.230); outer layer: 3-ply twisted single yarns with PCMs, including use of functional fibers Resistex® Silver | Washed with 2.5 g/L nonionic detergent Felosan RG-N, 2.0 g/L $Na_2CO_3$, 3.0 g/L water softener CalgonVR Power (60 °C, 60 min), rinsed with 1 g/L acetic acid solution, centrifuged, air-dried | Silver | None | *S. aureus*, *E. coli*, and *K. pneumoniae* | EN ISO 20645 | Low bacterial growth. | - | Multifunctional socks | [109] |

PET: poly(ethylene terephthalate);TFA: trifluoroacetic acid; RT: room temperature; AATCC (American Association of Textile Chemists and Colorists); TM: 4–(2,4–dichlorophenyl)–6–oxo–2–thioxohexahydropyrimidine–5–carbonitrile; MIC: minimum inhibitory concentration; CNWs: cellulose nonwovens; CI: cyclohexyl isocyanate; UV: ultraviolet; PTB: poly(thiiran-2-yl methyl methacrylate-2-(4-benzoyl phenoxy)ethyl methacrylate; PHMG: polyhexamethyleneguanidine; NEO: neomycin sulfate; DAPH: diammonium hydrogen phosphate; ZPT: zinc pyrithione; ZoI: zone of inhibition; CFU: colony-forming units; DMDHEU: dimethyloldihydroxyethylene urea; ZIF-8: zeolite imidazole skeleton-8; PABA: 4-aminobenzoic acid ligand; L:G: liquor-to-fabric ratio; PCMs: phase-change materials.

## 5. Conclusions

The SARS-CoV-2 pandemic, which has generated a global health and economic crisis, has shown us that we need to be better prepared for the next global threat, which may be caused by pollutants, chemical toxins, or biohazards [94,182]. The urgency of obtaining effective solutions to degrade BWAs such as anthrax [7,8] has been increasing in response to a recent risk increment associated with the possible use of biological weapons. Consequently, it is essential to develop personally protective systems that can actively protect their user, ideally without compromising his/her comfort, which is highly pertinent, for instance, while working in war zones for long periods of time [1,183]. Active protection is preferred when compared to passive protection, since it allows the total degradation of hazards and does not require a post-decontamination process [1]. We need to develop protective textiles in which infectious pathogens cannot survive, proliferate, and persevere so easily [94]. The damage inflicted by these harmful agents can be avoided by taking appropriate preventive measures [184]. The development of active fibrous structures with MOFs, inorganic agents (e.g., ZnO NPs), carbon-based materials (such as CQDs and graphene or its derivatives), and/or organic players such as chitosan (CS)-based layers or small-scale particles (loaded or not with plant-derived compounds) as bioactive agents is paving the way in the manufacture of protective textiles such as army suits, general protective clothing, or face masks that can efficiently counteract the survival of these pathogens. The decision as to the best bioactive agents strongly depends on the specific application and requirements, but the advantage of inorganic NPs seems clear. The research studies presented and interlinked here reinforce that ZnO NPs are one of the most promising materials for the development of high-performance textile products and should therefore be intensively investigated in the future, as is also argued elsewhere [185]. Strategies should be applied to counteract their current limitations. Bioactive features should be thoroughly examined and controlled via standardized protocols.

The addition of such elements into selectively permeable barrier textiles would fill a gap that currently exists, for instance, in charcoal-based protective suits that are designed to solely confer passive protection, and it would likely not add significant extra weight to the composition [1,115,186,187]. Aspects such as fabric composition and construction and clothing assembly should be paid more attention to, as they can substantially contribute to the required barrier effect and comfort. Moreover, charcoal-based protective suits and similar items, such as the majority of face masks that are currently employed, are limited to a single use. Hence, contemporary challenges include the development of circular and multifunctional protective textiles with durable effects, regenerable bioactive agents, and recyclable/degradable materials [188,189]. The use of natural compounds can be a great strategy and an excellent alternative to the use of synthetic ones, due to their high abundance in nature, low cost, and biodegradability. The use of simple and greener methods is also preferred [1]. Overall, this area is presently a hot topic in both the scientific and industrial communities, being an object of intense research, yet it is unfortunately still highly dispersed. It thus seems to be imperative to apply all the efforts to successfully innovate and create scientific and technological breakthroughs, while rigorously defining all the requirements for a fully functional protective textile, performing all the needed standardized protocols to adequately evaluate each hypothesis, and allowing the results to speak for themselves regarding the definition of the best elements and/or combinations to use, so that substantial improvements in the field of antimicrobial protective textiles (namely against BWAs) can be achieved. On the verge of contact with dangerous pathogens, we seek products that actually work, making this entire pursuit worthwhile.

**Author Contributions:** Conceptualization, J.C.A., I.P.M. and R.F.; methodology, J.C.A. and I.P.M.; validation, R.F.; formal analysis, J.C.A.; investigation, J.C.A.; writing—original draft preparation, J.C.A., I.P.M. and F.G.; writing—review and editing, J.C.A. and I.P.M.; supervision, M.H. and R.F.; project administration, R.F.; funding acquisition, R.F. and F.C. All authors have read and agreed to the published version of the manuscript.

**Funding:** The authors acknowledge the Portuguese Foundation for Science and Technology (FCT), the FEDER funds by means of the Portugal 2020 Competitive Factors Operational Program (POCI), and the Portuguese Government (OE) for funding the project PluriProtech—"Desenvolvimentos de soluções multicamada para proteção ativa contra ameaças NBQR", ref. POCI-01-0247-FEDER-047012. The authors also acknowledge the strategic funding of UID/CTM/00264/2020 of 2C2T and UIDB/04469/2020 of CEB, given by FCT.

**Conflicts of Interest:** The authors declare no conflict of interest.

## References

1. Araújo, J.C.; Fangueiro, R.; Ferreira, D.P. Protective multifunctional fibrous systems based on natural fibers and metal oxide nanoparticles. *Polymers* **2021**, *13*, 2654. [CrossRef] [PubMed]
2. Hayoun, M.A.; King, K.C. *Biologic Warfare Agent Toxicity*; StatPearls Publishing LLC.: Treasure Island, FL, USA, 2022.
3. Rathish, B.; Pillay, R.; Wilson, A.; Pillay, V.V. *Comprehensive Review of Bioterrorism*; StatPearls Publishing LLC.: Treasure Island, FL, USA, 2022.
4. Galatas, I. The misuse and malicious uses of the new biotechnologies. *Ann. Des Mines Réalités Ind.* **2017**, *2017*, 103–108. [CrossRef]
5. O'Brien, C.; Varty, K.; Ignaszak, A. The electrochemical detection of bioterrorism agents: A review of the detection, diagnostics, and implementation of sensors in biosafety programs for Class A bioweapons. *Microsyst. Nanoeng.* **2021**, *7*, 16. [CrossRef] [PubMed]
6. Berger, T.; Eisenkraft, A.; Bar-Haim, E.; Kassirer, M.; Aran, A.A.; Fogel, I. Toxins as biological weapons for terror-characteristics, challenges and medical countermeasures: A mini-review. *Disaster Mil. Med.* **2016**, *2*, 7. [CrossRef]
7. WHO. Anthrax in Humans and Animals. Available online: https://www.ncbi.nlm.nih.gov/books/NBK310486/ (accessed on 3 March 2022).
8. Banerjee, D.; Chakraborty, B.; Chakraborty, B. Anthrax: Where Margins are Merging between Emerging Threats and Bioterrorism. *Indian J. Dermatol.* **2017**, *62*, 456–458. [CrossRef]
9. Plotkin, S.; Grabenstein, J.D. Countering Anthrax: Vaccines and Immunoglobulins. *Clin. Infect. Dis.* **2008**, *46*, 129–136. [CrossRef]
10. Simonsen, K.A.; Chatterjee, K. *Anthrax*; StatPearls Publishing LLC.: Treasure Island, FL, USA, 2022.
11. Kamal, S.M.; Rashid, A.K.; Bakar, M.A.; Ahad, M.A. Anthrax: An update. *Asian Pac. J. Trop. Biomed.* **2011**, *1*, 496–501. [CrossRef]
12. Zasada, A.A. Injectional anthrax in human: A new face of the old disease. *Adv. Clin. Exp. Med.* **2018**, *27*, 553–558. [CrossRef]
13. Chambers, J.; Yarrarapu, S.N.S.; Mathai, J.K. *Anthrax Infection*; StatPearls Publishing LLC.: Treasure Island, FL, USA, 2022.
14. Johari, M.R. Anthrax—Biological Threat in the 21st Century. *Malays. J. Med. Sci.* **2002**, *9*, 1–2.
15. CDC. Treatment of Anthrax Infection. Available online: https://www.cdc.gov/anthrax/treatment/index.html (accessed on 3 March 2022).
16. Heine, H.S.; Shadomy, S.V.; Boyer, A.E.; Chuvala, L.; Riggins, R.; Kesterson, A.; Myrick, J.; Craig, J.; Candela, M.G.; Barr, J.R.; et al. Evaluation of Combination Drug Therapy for Treatment of Antibiotic-Resistant Inhalation Anthrax in a Murine Model. *Antimicrob. Agents Chemother.* **2017**, *61*, e00788-17. [CrossRef]
17. Kummerfeldt, C.E. Raxibacumab: Potential role in the treatment of inhalational anthrax. *Infect. Drug. Resist.* **2014**, *7*, 101–109. [CrossRef]
18. Cybulski Jr, R.J.; Sanz, P.; O'Brien, A.D. Anthrax vaccination strategies. *Mol. Asp. Med.* **2009**, *30*, 490–502. [CrossRef]
19. CDC. Anthrax VIS. Available online: https://www.cdc.gov/vaccines/hcp/vis/vis-statements/anthrax.html (accessed on 3 March 2022).
20. Ditchburn, J.-L.; Hodgkins, R. Yersinia pestis, a problem of the past and a re-emerging threat. *Biosaf. Health* **2019**, *1*, 65–70. [CrossRef]
21. Tao, P.; Mahalingam, M.; Zhu, J.; Moayeri, M.; Sha, J.; Lawrence, W.S.; Leppla, S.H.; Chopra, A.K.; Rao, V.B. A Bacteriophage T4 Nanoparticle-Based Dual Vaccine against Anthrax and Plague. *mBio* **2018**, *9*, e01926-18. [CrossRef]
22. Nguyen, V.K.; Parra-Rojas, C.; Hernandez-Vargas, E.A. The 2017 plague outbreak in Madagascar: Data descriptions and epidemic modelling. *Epidemics* **2018**, *25*, 20–25. [CrossRef]
23. Randremanana, R.; Andrianaivoarimanana, V.; Nikolay, B.; Ramasindrazana, B.; Paireau, J.; ten Bosch, Q.A.; Rakotondramanga, J.M.; Rahajandraibe, S.; Rahelinirina, S.; Rakotomanana, F.; et al. Epidemiological characteristics of an urban plague epidemic in Madagascar, August–November 2017: An outbreak report. *Lancet Infect. Dis.* **2019**, *19*, 537–545. [CrossRef]
24. Gibbs, M.E.; Lountos, G.T.; Gumpena, R.; Waugh, D.S. Crystal structure of UDP-glucose pyrophosphorylase from Yersinia pestis, a potential therapeutic target against plague. *Acta Crystallogr. F Struct. Biol. Commun.* **2019**, *75*, 608–615. [CrossRef]
25. Sun, W.; Singh, A.K. Plague vaccine: Recent progress and prospects. *npj Vaccines* **2019**, *4*, 11. [CrossRef]
26. Sebbane, F.; Lemaître, N. Antibiotic Therapy of Plague: A Review. *Biomolecules* **2021**, *11*, 724. [CrossRef]

27. Sebbane, F.; Uversky, V.N.; Anisimov, A.P. Yersinia pestis plasminogen activator. *Biomolecules* **2020**, *10*, 1554. [CrossRef]
28. Kilgore, P.B.; Sha, J.; Andersson, J.A.; Motin, V.L.; Chopra, A.K. A new generation needle- and adjuvant-free trivalent plague vaccine utilizing adenovirus-5 nanoparticle platform. *npj Vaccines* **2021**, *6*, 21. [CrossRef] [PubMed]
29. Kilgore, P.B.; Sha, J.; Hendrix, E.K.; Motin, V.L.; Chopra, A.K. Combinatorial Viral Vector-Based and Live Attenuated Vaccines without an Adjuvant to Generate Broader Immune Responses to Effectively Combat Pneumonic Plague. *mBio* **2021**, *12*, e03223-21. [CrossRef] [PubMed]
30. Rosenzweig, J.A.; Hendrix, E.K.; Chopra, A.K. Plague vaccines: New developments in an ongoing search. *Appl. Microbiol. Biotechnol.* **2021**, *105*, 4931–4941. [CrossRef] [PubMed]
31. Markova, A.; Hympanova, M.; Matula, M.; Prchal, L.; Sleha, R.; Benkova, M.; Pulkrabkova, L.; Soukup, O.; Krocova, Z.; Jun, D.; et al. Synthesis and decontamination effect on chemical and biological agents of benzoxonium-like salts. *Toxics* **2021**, *9*, 222. [CrossRef]
32. Yeni, D.K.; Büyük, F.; Ashraf, A.; Shah, M.S.D. Tularemia: A re-emerging tick-borne infectious disease. *Folia Microbiol.* **2021**, *66*, 1–14. [CrossRef]
33. Wawszczak, M.; Banaszczak, B.; Rastawicki, W. Tularaemia—A diagnostic challenge. *Ann. Agric. Environ. Med.* **2021**, *29*, 12–21. [CrossRef]
34. Rowe, H.M.; Huntley, J.F. From the Outside-In: The Francisella tularensis Envelope and Virulence. *Front. Cell Infect. Microbiol.* **2015**, *5*, 94. [CrossRef]
35. Maurin, M. Francisella tularensis, Tularemia and Serological Diagnosis. *Front. Cell Infect. Microbiol.* **2020**, *10*, 646. [CrossRef]
36. Snowden, J.; Simonsen, K.A. *Tularemia*; StatPearls Publishing LLC.: Treasure Island, FL, USA, 2022.
37. Dennis, D.T.; Inglesby, T.V.; Henderson, D.A.; Bartlett, J.G.; Ascher, M.S.; Eitzen, E.; Fine, A.D.; Friedlander, A.M.; Hauer, J.; Layton, M.; et al. Tularemia as a biological weapon: Medical and public health management. *JAMA* **2001**, *285*, 2763–2773. [CrossRef]
38. Ellis, J.; Oyston, P.C.; Green, M.; Titball, R.W. Tularemia. *Clin Microbiol. Rev.* **2002**, *15*, 631–646. [CrossRef]
39. WDH. Tularemia. Available online: https://health.wyo.gov/publichealth/infectious-disease-epidemiology-unit/disease/tularemia/ (accessed on 3 March 2022).
40. MDH. Tularemia Fact Sheet. Available online: https://www.health.state.mn.us/diseases/tularemia/tularemiafs.html (accessed on 3 March 2022).
41. Caspar, Y.; Hennebique, A.; Maurin, M. Antibiotic susceptibility of Francisella tularensis subsp. holarctica strains isolated from tularaemia patients in France between 2006 and 2016. *J. Antimicrob. Chemother.* **2017**, *73*, 687–691. [CrossRef]
42. Tian, D.; Uda, A.; Ami, Y.; Hotta, A.; Park, E.-s.; Nagata, N.; Iwata-Yoshikawa, N.; Yamada, A.; Hirayama, K.; Miura, K.; et al. Protective effects of the Francisella tularensis ΔpdpC mutant against its virulent parental strain SCHU P9 in Cynomolgus macaques. *Sci. Rep.* **2019**, *9*, 9193. [CrossRef]
43. Delaune, D.; Iseni, F. Drug Development against Smallpox: Present and Future. *Antimicrob Agents Chemother* **2020**, *64*, e01683-19. [CrossRef]
44. Hutson, C.L.; Kondas, A.V.; Ritter, J.M.; Reed, Z.; Ostergaard, S.D.; Morgan, C.N.; Gallardo-Romero, N.; Tansey, C.; Mauldin, M.R.; Salzer, J.S.; et al. Teaching a new mouse old tricks: Humanized mice as an infection model for Variola virus. *PLOS Pathogens* **2021**, *17*, e1009633. [CrossRef]
45. Lim, H.; In, H.J.; Kim, Y.-J.; Jang, S.; Lee, Y.H.; Kim, S.H.; Lee, S.H.; Park, J.H.; Yang, H.J.; Yoo, J.-S.; et al. Development of an attenuated smallpox vaccine candidate: The KVAC103 strain. *Vaccine* **2021**, *39*, 5214–5223. [CrossRef]
46. Thèves, C.; Biagini, P.; Crubézy, E. The rediscovery of smallpox. *Clin. Microbiol. Infect.* **2014**, *20*, 210–218. [CrossRef]
47. Khan, S.; Rafique, I. Threats of bioterrorism in public health, Epidemiological clue, Detection and Safety pre-cautions for outbreaks. *Open J. Bac.* **2019**, *3*, 11–15. [CrossRef]
48. Tegnell, A.; Wahren, B.; Elgh, F. Smallpox—eradicated, but a growing terror threat. *Clin. Microbiol. Infect.* **2002**, *8*, 504–509. [CrossRef]
49. Dunlop, L.R.; Oehlberg, K.A.; Reid, J.J.; Avci, D.; Rosengard, A.M. Variola virus immune evasion proteins. *Microbes Infect.* **2003**, *5*, 1049–1056. [CrossRef]
50. Smith, K. Edward Jenner and the Small Pox Vaccine. *Front. Immunol.* **2011**, *2*, 21. [CrossRef]
51. Duarte, P.M.; Santana, V.T.P.d. Disinfection measures and control of SARS-CoV-2 transmission. *Global Biosecur.* **2020**, *1*. [CrossRef]
52. Van Schaik, E.J.; Chen, C.; Mertens, K.; Weber, M.M.; Samuel, J.E. Molecular pathogenesis of the obligate intracellular bacterium Coxiella burnetii. *Nat. Rev. Microbiol.* **2013**, *11*, 561–573. [CrossRef] [PubMed]
53. Lamunu, M.; Olu, O.O.; Bangura, J.; Yoti, Z.; Samba, T.T.; Kargbo, D.K.; Dafae, F.M.; Raja, M.A.; Sempira, N.; Ivan, M.L.; et al. Epidemiology of Ebola Virus Disease in the Western Area Region of Sierra Leone, 2014–2015. *Public Health Front.* **2017**, *5*, 33. [CrossRef]
54. Patel, P.R.; Shah, S.U. *Ebola Virus*; StatPearls Publishing LLC.: Treasure Island, FL, USA, 2022.
55. Fan, Y.; Moon, J.J. Particulate delivery systems for vaccination against bioterrorism agents and emerging infectious pathogens. *WIREs Nanomed. Nanobiotechnol.* **2017**, *9*, e1403. [CrossRef]
56. Dixon, M.G.; Schafer, I.J. Ebola viral disease outbreak–West Africa, 2014. *MMWR Morb. Mortal. Wkly. Rep.* **2014**, *63*, 548–551. [PubMed]
57. CDC. Ebola Treatment. Available online: https://www.cdc.gov/vhf/ebola/treatment/index.html (accessed on 3 March 2022).

58. Yamaoka, S.; Ebihara, H. Pathogenicity and Virulence of Ebolaviruses with Species- and Variant-specificity. *Virulence* **2021**, *12*, 885–901. [CrossRef] [PubMed]
59. Woolsey, C.; Geisbert, T.W. Current state of Ebola virus vaccines: A snapshot. *PLOS Pathog.* **2021**, *17*, e1010078. [CrossRef]
60. Rathjen, N.A.; Shahbodaghi, S.D. Bioterrorism. *Am. Fam. Physician* **2021**, *104*, 376–385.
61. WHO. Ebola virus disease: Vaccines. Available online: https://www.who.int/news-room/questions-and-answers/item/ebola-vaccines (accessed on 3 March 2022).
62. Hansen, F.; Jarvis, M.A.; Feldmann, H.; Rosenke, K. Lassa Virus Treatment Options. *Microorganisms* **2021**, *9*, 772. [CrossRef]
63. Strampe, J.; Asogun, D.A.; Speranza, E.; Pahlmann, M.; Soucy, A.; Bockholt, S.; Pallasch, E.; Becker-Ziaja, B.; Duraffour, S.; Bhadelia, N.; et al. Factors associated with progression to death in patients with Lassa fever in Nigeria: An observational study. *Lancet Infect. Dis.* **2021**, *21*, 876–886. [CrossRef]
64. Alli, A.; Ortiz, J.F.; Fabara, S.P.; Patel, A.; Halan, T. Management of Lassa Fever: A Current Update. *Cureus* **2021**, *13*, e14797. [CrossRef]
65. Happi, A.N.; Happi, C.T.; Schoepp, R.J. Lassa fever diagnostics: Past, present, and future. *Curr. Opin. Virol.* **2019**, *37*, 132–138. [CrossRef]
66. Onuh, J.A.; Uloko, A.E. Favourable Outcome of Severe Lassa Fever Following Early Diagnosis and Treatment: A Case Report. *West. Afr. J. Med.* **2021**, *38*, 395–397.
67. Wang, M.; Li, R.; Li, Y.; Yu, C.; Chi, X.; Wu, S.; Liu, S.; Xu, J.; Chen, W. Construction and Immunological Evaluation of an Adenoviral Vector-Based Vaccine Candidate for Lassa Fever. *Viruses* **2021**, *13*, 484. [CrossRef]
68. Lingas, G.; Rosenke, K.; Safronetz, D.; Guedj, J. Lassa viral dynamics in non-human primates treated with favipiravir or ribavirin. *PLoS Comput. Biol.* **2021**, *17*, e1008535. [CrossRef]
69. Merson, L.; Bourner, J.; Jalloh, S.; Erber, A.; Salam, A.P.; Flahault, A.; Olliaro, P.L. Clinical characterization of Lassa fever: A systematic review of clinical reports and research to inform clinical trial design. *PLoS Negl. Trop. Dis.* **2021**, *15*, e0009788. [CrossRef]
70. Salam, A.P.; Cheng, V.; Edwards, T.; Olliaro, P.; Sterne, J.; Horby, P. Time to reconsider the role of ribavirin in Lassa fever. *PLoS Negl. Trop. Dis.* **2021**, *15*, e0009522. [CrossRef]
71. Salami, K.; Gouglas, D.; Schmaljohn, C.; Saville, M.; Tornieporth, N. A review of Lassa fever vaccine candidates. *Curr. Opin. Virol.* **2019**, *37*, 105–111. [CrossRef]
72. Warner, B.M.; Safronetz, D.; Stein, D.R. Current research for a vaccine against Lassa hemorrhagic fever virus. *Drug. Des. Devel. Ther.* **2018**, *12*, 2519–2527. [CrossRef]
73. Müller, H.; Fehling, S.K.; Dorna, J.; Urbanowicz, R.A.; Oestereich, L.; Krebs, Y.; Kolesnikova, L.; Schauflinger, M.; Krähling, V.; Magassouba, N.F.; et al. Adjuvant formulated virus-like particles expressing native-like forms of the Lassa virus envelope surface glycoprotein are immunogenic and induce antibodies with broadly neutralizing activity. *npj Vaccines* **2020**, *5*, 71. [CrossRef]
74. Sebaihia, M.; Peck, M.W.; Minton, N.P.; Thomson, N.R.; Holden, M.T.G.; Mitchell, W.J.; Carter, A.T.; Bentley, S.D.; Mason, D.R.; Crossman, L.; et al. Genome sequence of a proteolytic (Group I) Clostridium botulinum strain Hall A and comparative analysis of the clostridial genomes. *Genome Res.* **2007**, *17*, 1082–1092. [CrossRef] [PubMed]
75. Fredrick, C.M.; Lin, G.; Johnson, E.A. Regulation of botulinum neurotoxin synthesis and toxin complex formation by arginine and glucose in Clostridium botulinum ATCC 3502. *Appl. Environ. Microbiol.* **2017**, *83*, e00642-17. [CrossRef] [PubMed]
76. Jeffery, I.A.; Karim, S. *Botulism*; StatPearls Publishing LLC.: Treasure Island, FL, USA, 2022.
77. Thirunavukkarasu, N.; Johnson, E.; Pillai, S.; Hodge, D.; Stanker, L.; Wentz, T.; Singh, B.; Venkateswaran, K.; McNutt, P.; Adler, M.; et al. Botulinum Neurotoxin Detection Methods for Public Health Response and Surveillance. *Front. Bioeng. Biotechnol.* **2018**, *6*, 80. [CrossRef] [PubMed]
78. Pero, R.; Laneri, S.; Fico, G. Botulinum Toxin Adverse Events. In *Botulinum Toxin, Serdev, N., Ed.*; IntechOpen: London, UK, 2018.
79. O'Horo, J.C.; Harper, E.P.; El Rafei, A.; Ali, R.; Desimone, D.C.; Sakusic, A.; Abu Saleh, O.M.; Marcelin, J.R.; Tan, E.M.; Rao, A.K.; et al. Efficacy of Antitoxin Therapy in Treating Patients with Foodborne Botulism: A Systematic Review and Meta-analysis of Cases, 1923-2016. *Clin. Infect. Dis.* **2017**, *66*, S43–S56. [CrossRef]
80. Ni, S.A.; Brady, M.F. *Botulism Antitoxin*; StatPearls Publishing LLC.: Treasure Island, FL, USA, 2022.
81. Clark, D.P.; Pazdernik, N.J. Biological Warfare: Infectious Disease and Bioterrorism. *Biotechnology* **2016**, 687–719. [CrossRef]
82. Lúquez, C.; Edwards, L.; Griffin, C.; Sobel, J. Foodborne Botulism Outbreaks in the United States, 2001–2017. *Front. Microbiol.* **2021**, *12*, 1982. [CrossRef]
83. Dhaked, R.K.; Singh, M.K.; Singh, P.; Gupta, P. Botulinum toxin: Bioweapon & magic drug. *Indian J. Med. Res.* **2010**, *132*, 489–503.
84. Gan, C.; Luo, W.; Yu, Y.; Jiao, Z.; Li, S.; Su, D.; Feng, J.; Zhao, X.; Qiu, Y.; Hu, L.; et al. Intratracheal inoculation of AHc vaccine induces protection against aerosolized botulinum neurotoxin A challenge in mice. *npj Vaccines* **2021**, *6*, 87. [CrossRef]
85. Kim, N.Y.; Son, W.R.; Lee, M.H.; Choi, H.S.; Choi, J.Y.; Song, Y.J.; Yu, C.H.; Song, D.H.; Hur, G.H.; Jeong, S.T.; et al. A multipathogen DNA vaccine elicits protective immune responses against two class A bioterrorism agents, anthrax and botulism. *Appl. Microbiol. Biotechnol.* **2022**, *106*, 1531–1542. [CrossRef]
86. Li, Z.; Lu, J.; Tan, X.; Wang, R.; Xu, Q.; Yu, Y.; Yang, Z. Functional EL-HN Fragment as a Potent Candidate Vaccine for the Prevention of Botulinum Neurotoxin Serotype E. *Toxins* **2022**, *14*, 135. [CrossRef]

87. Karcıoğlu, O.; Yüksel, A.; Baha, A.; Banu Er, A.; Esendağlı, D.; Gülhan, P.Y.; Karaoğlanoğlu, S.; Özçelik, M.; Şerifoğlu, İ.; Yıldız, E.; et al. COVID-19: The biggest threat of the 21st century: In respectful memory of the warriors all over the world. *Turk. Thorac. J.* **2020**, *21*, 409–418. [CrossRef]
88. Siddique, F.; Abbas, R.Z.; Mansoor, M.K.; Alghamdi, E.S.; Saeed, M.; Ayaz, M.M.; Rahman, M.; Mahmood, M.S.; Iqbal, A.; Manzoor, M.; et al. An Insight Into COVID-19: A 21st Century Disaster and Its Relation to Immunocompetence and Food Antioxidants. *Front. Vet. Sci.* **2021**, *7*, 1168. [CrossRef]
89. Chidambaram, V.; Tun, N.L.; Haque, W.Z.; Gilbert Majella, M.; Kumar Sivakumar, R.; Kumar, A.; Hsu, A.T.W.; Ishak, I.A.; Nur, A.A.; Ayeh, S.K.; et al. Factors associated with disease severity and mortality among patients with COVID-19: A systematic review and meta-analysis. *PLoS ONE* **2020**, *15*, e0241541. [CrossRef]
90. Zhou, L.; Ayeh, S.K.; Chidambaram, V.; Karakousis, P.C. Modes of transmission of SARS-CoV-2 and evidence for preventive behavioral interventions. *BMC Infect. Dis.* **2021**, *21*, 496. [CrossRef]
91. Ali Al Shehri, S.; Al-Sulaiman, A.M.; Azmi, S.; Alshehri, S.S. Bio-safety and bio-security: A major global concern for ongoing COVID-19 pandemic. *Saudi J. Biol. Sci.* **2022**, *29*, 132–139. [CrossRef]
92. WHO. Report on the WHO-China Joint Mission on Coronavirus Disease 2019 (COVID-19). Available online: https://www.who.int/docs/default-source/coronaviruse/who-china-joint-mission-on-COVID-19-final-report.pdf (accessed on 3 April 2022).
93. Freney, J.; Renaud, F.N.R. Textiles and microbes. In *NATO Science for Peace and Security Series B: Physics and Biophysics*; Springer: Dordrecht, The Netherlands, 2012; pp. 53–81. [CrossRef]
94. Jabbour, C.R.; Parker, L.A.; Hutter, E.M.; Weckhuysen, B.M. Chemical targets to deactivate biological and chemical toxins using surfaces and fabrics. *Nat. Rev. Chem.* **2021**, *5*, 370–387. [CrossRef]
95. Botelho, C.M.; Fernandes, M.M.; Souza, J.M.; Dias, N.; Sousa, A.M.; Teixeira, J.A.; Fangueiro, R.; Zille, A. New textile for personal protective equipment—Plasma chitosan/silver nanoparticles nylon fabric. *Fibers* **2021**, *9*, 3. [CrossRef]
96. Bouaziz, A.; Dridi, D.; Gargoubi, S.; Zouari, A.; Majdoub, H.; Boudokhane, C.; Bartegi, A. Study on the grafting of chitosan-essential oil microcapsules onto cellulosic fibers to obtain bio functional material. *Coatings* **2021**, *11*, 637. [CrossRef]
97. Davis, C.P.; Wagle, N.; Anderson, M.D.; Warren, M.M. Bacterial and fungal killing by iontophoresis with long-lived electrodes. *Antimicrob. Agents Chemother.* **1991**, *35*, 2131–2134. [CrossRef]
98. Savaloni, H.; Haydari-Nasab, F.; Abbas-Rohollahi, A. Antibacterial effect, structural characterization, and some applications of silver chiral nano-flower sculptured thin films. *J. Theor. Appl. Phys.* **2015**, *9*, 193–200. [CrossRef]
99. Haase, H.; Jordan, L.; Keitel, L.; Keil, C.; Mahltig, B. Comparison of methods for determining the effectiveness of antibacterial functionalized textiles. *PLoS ONE* **2017**, *12*, e0188304. [CrossRef] [PubMed]
100. Pinho, E.; Magalhães, L.; Henriques, M.; Oliveira, R. Antimicrobial activity assessment of textiles: Standard methods comparison. *Ann. Microbiol.* **2011**, *61*, 493–498. [CrossRef]
101. Román, L.E.; Gomez, E.D.; Solís, J.L.; Gómez, M.M. Antibacterial Cotton Fabric Functionalized with Copper Oxide Nanoparticles. *Molecules* **2020**, *25*, 5802. [CrossRef] [PubMed]
102. Tavares, T.D.; Antunes, J.C.; Padrão, J.; Ribeiro, A.I.; Zille, A.; Amorim, M.T.P.; Ferreira, F.; Felgueiras, H.P. Activity of specialized biomolecules against gram-positive and gram-negative bacteria. *Antibiotics* **2020**, *9*, 314. [CrossRef]
103. Kafafy, H.; Shahin, A.A.; Mashaly, H.M.; Helmy, H.M.; Zaher, A. Treatment of cotton and wool fabrics with different nanoparticles for multifunctional properties. *Egypt. J. Chem.* **2021**, *64*, 5257–5269. [CrossRef]
104. Bhattacharjee, S.; Joshi, R.; Yasir, M.; Adhikari, A.; Chughtai, A.A.; Heslop, D.; Bull, R.; Willcox, M.; Macintyre, C.R. Graphene-And Nanoparticle-Embedded Antimicrobial and Biocompatible Cotton/Silk Fabrics for Protective Clothing. *ACS Appl. Bio. Mat.* **2021**, *4*, 6175–6185. [CrossRef]
105. El-Naggar, M.E.; Khattab, T.A.; Abdelrahman, M.S.; Aldalbahi, A.; Hatshan, M.R. Development of antimicrobial, UV blocked and photocatalytic self-cleanable cotton fibers decorated with silver nanoparticles using silver carbamate and plasma activation. *Cellulose* **2021**, *28*, 1105–1121. [CrossRef]
106. Emam, H.E.; Darwesh, O.M.; Abdelhameed, R.M. Protective cotton textiles via amalgamation of cross-linked zeolitic imidazole frameworks. *Ind. Eng. Chem. Res.* **2020**, *59*, 10931–10944. [CrossRef]
107. Emam, H.E.; El-Shahat, M.; Hasanin, M.S.; Ahmed, H.B. Potential military cotton textiles composed of carbon quantum dots clustered from 4–(2,4–dichlorophenyl)–6–oxo–2–thioxohexahydropyrimidine–5–carbonitrile. *Cellulose* **2021**, *28*, 9991–10011. [CrossRef]
108. Ramadan, M.A.; Taha, G.M.; El- Mohr, W.Z.E.A. Antimicrobial and uv protection finishing of polysaccharide -based textiles using biopolymer and agnps. *Egypt. J. Chem.* **2020**, *63*, 2707–2716. [CrossRef]
109. Stygienė, L.; Varnaitė-Žuravliova, S.; Abraitienė, A.; Sankauskaitė, A.; Skurkytė-Papievienė, V.; Krauledas, S.; Mažeika, V. Development, investigation and evaluation of smart multifunctional socks. *J. Ind. Text.* **2020**, 1528083720970166. [CrossRef]
110. Song, X.; Padrão, J.; Ribeiro, A.I.; Zille, A. 16—Testing, characterization and regulations of antimicrobial textiles. In *Antimicrobial Textiles from Natural Resources*; Mondal, M.I.H., Ed.; Woodhead Publishing: Cambridgeshire, UK, 2021; pp. 485–511.
111. Schreuder-Gibson, H.L.; Truong, Q.; Walker, J.E.; Owens, J.R.; Wander, J.D.; Jones Jr, W.E. Chemical and biological protection and detection in fabrics for protective clothing. *MRS Bull.* **2003**, *28*, 574–578. [CrossRef]
112. Bhuiyan, M.A.R.; Wang, L.; Shaid, A.; Shanks, R.A.; Ding, J. Advances and applications of chemical protective clothing system. *J. Ind. Text.* **2019**, *49*, 97–138. [CrossRef]

113. Lundberg, D.J.; Brooks, A.M.; Strano, M.S. Design Rules for Chemostrictive Materials as Selective Molecular Barriers. *Adv. Eng. Mat.* **2022**, *24*, 2101112. [CrossRef]
114. Zhao, X.; Liu, B. Permeable Protective Suit: Status Quo and Latest Research Progress. *Cailiao Daobao/Mater Rev* **2018**, *32*, 3083–3089. [CrossRef]
115. Truong, Q.; Wilusz, E. 13—Advances in chemical and biological protective clothing. In *Smart Textiles for Protection*; Chapman, R.A., Ed.; Woodhead Publishing: Cambridgeshire, UK, 2013; pp. 364–377.
116. Paul, R.; Mao, N. Textile Materials for Protective Textiles. In *High Performance Technical Textiles*; Paul, R., Ed.; John Wiley and Sons Ltd.: Hoboken, NJ, USA, 2019.
117. Cruz, J.; Fangueiro, R. Surface Modification of Natural Fibers: A Review. *Procedia Eng.* **2016**, *155*, 285–288. [CrossRef]
118. Ferreira, D.P.; Costa, S.M.; Felgueiras, H.P.; Fangueiro, R. Smart and sustainable materials for military applications based on natural fibres and silver nanoparticles. In *Key Engineering Materials*; Trans Tech Publications Ltd.: Schwyz, Switzerland, 2019; Volume 812, pp. 66–74. [CrossRef]
119. Pereira, J.F.; Ferreira, D.P.; Pinho, E.; Fangueiro, R. Chemical and biological warfare protection and self-decontaminating flax fabrics based on CaO nanoparticles. In *Key Engineering Materials*; Trans Tech Publications Ltd.: Schwyz, Switzerland, 2019; Volume 812, pp. 75–83. [CrossRef]
120. Verma, M.; Gahlot, N.; Singh, S.S.J.; Rose, N.M. UV protection and antibacterial treatment of cellulosic fibre (cotton) using chitosan and onion skin dye. *Carbohydr. Polym.* **2021**, *257*, 117612. [CrossRef]
121. Hu, J.; Jahid, M.A.; Harish Kumar, N.; Harun, V. Fundamentals of the Fibrous Materials. In *Handbook of Fibrous Materials*; Wiley Online Library: Hoboken, NJ, USA, 2020; pp. 1–36.
122. Pais, V.; Mota, C.; Bessa, J.; Dias, J.G.; Cunha, F.; Fangueiro, R. Study of the filtration performance of multilayer and multiscale fibrous structures. *Materials* **2021**, *14*, 7147. [CrossRef]
123. Stylios, G.K. Protective clothing against chemical and biological agents. *Int. J. Cloth. Sci. Technol.* **2007**, *19*, 19–20. [CrossRef]
124. Gugliuzza, A.; Drioli, E. A review on membrane engineering for innovation in wearable fabrics and protective textiles. *J. Membr. Sci.* **2013**, *446*, 350–375. [CrossRef]
125. Sharma, N.; Nair, A.; Gupta, B.; Kulshrestha, S.; Goel, R.; Chawla, R. Chapter 6—Chemical, biological, radiological, and nuclear textiles: Current scenario and way forward. In *Advances in Functional and Protective Textiles*; Ul-Islam, S., Butola, B.S., Eds.; Woodhead Publishing: Cambridge, UK, 2020; pp. 117–140.
126. Hearle, J.W.S. 5—Fibres and fabrics for protective textiles. In *Textiles for Protection*; Scott, R.A., Ed.; Woodhead Publishing: Cambridge, UK, 2005; pp. 117–150.
127. Jung, H.; Kim, M.K.; Jang, S. Liquid-repellent textile surfaces using zirconium (Zr)-based porous materials and a polyhedral oligomeric silsesquioxane coating. *J. Colloid Interface Sci.* **2020**, *563*, 363–369. [CrossRef] [PubMed]
128. Cheung, Y.H.; Ma, K.; Van Leeuwen, H.C.; Wasson, M.C.; Wang, X.; Idrees, K.B.; Gong, W.; Cao, R.; Mahle, J.J.; Islamoglu, T.; et al. Immobilized Regenerable Active Chlorine within a Zirconium-Based MOF Textile Composite to Eliminate Biological and Chemical Threats. *J. Am. Chem. Soc.* **2021**, *143*, 16777–16785. [CrossRef] [PubMed]
129. Lee, J.; Kim, E.Y.; Chang, B.J.; Han, M.; Lee, P.S.; Moon, S.Y. Mixed-matrix membrane reactors for the destruction of toxic chemicals. *J. Membr. Sci.* **2020**, *605*, 118112. [CrossRef]
130. Salter, B.; Owens, J.; Hayn, R.; McDonald, R.; Shannon, E. N-chloramide modified Nomex® as a regenerable self-decontaminating material for protection against chemical warfare agents. *J. Mater. Sci.* **2009**, *44*, 2069–2078. [CrossRef]
131. Antunes, J.C.; Domingues, J.M.; Miranda, C.S.; Silva, A.F.G.; Homem, N.C.; Amorim, M.T.P.; Felgueiras, H.P. Bioactivity of chitosan-based particles loaded with plant-derived extracts for biomedical applications: Emphasis on antimicrobial fiber-based systems. *Mar. Drugs* **2021**, *19*, 359. [CrossRef]
132. Balderrama-González, A.S.; Piñón-Castillo, H.A.; Ramírez-Valdespino, C.A.; Landeros-Martínez, L.L.; Orrantia-Borunda, E.; Esparza-Ponce, H.E. Antimicrobial resistance and inorganic nanoparticles. *Int. J. Mol. Sci.* **2021**, *22*, 2890. [CrossRef]
133. Khorsandi, K.; Hosseinzadeh, R.; Sadat Esfahani, H.; Keyvani-Ghamsari, S.; Ur Rahman, S. Nanomaterials as drug delivery systems with antibacterial properties: Current trends and future priorities. *Expert Rev. Anti-Infect. Ther.* **2021**, *19*, 1299–1323. [CrossRef]
134. Ermini, M.L.; Voliani, V. Antimicrobial Nano-Agents: The Copper Age. *ACS Nano.* **2021**, *15*, 6008–6029. [CrossRef]
135. Xu, Q.; Hu, X.; Wang, Y. Alternatives to Conventional Antibiotic Therapy: Potential Therapeutic Strategies of Combating Antimicrobial-Resistance and Biofilm-Related Infections. *Mol. Biotechnol.* **2021**, *63*, 1103–1124. [CrossRef]
136. Boticas, I.; Dias, D.; Ferreira, D.; Magalhães, P.; Silva, R.; Fangueiro, R. Superhydrophobic cotton fabrics based on ZnO nanoparticles functionalization. *SN Appl. Sci.* **2019**, *1*, 1376. [CrossRef]
137. Costa, S.M.; Ferreira, D.P.; Ferreira, A.; Vaz, F.; Fangueiro, R. Multifunctional flax fibres based on the combined effect of silver and zinc oxide (Ag/zno) nanostructures. *Nanomaterials* **2018**, *8*, 1069. [CrossRef]
138. Noorian, S.A.; Hemmatinejad, N.; Navarro, J.A.R. Ligand modified cellulose fabrics as support of zinc oxide nanoparticles for UV protection and antimicrobial activities. *Int. J. Biol. Macromol.* **2020**, *154*, 1215–1226. [CrossRef]
139. Mirzaeifard, Z.; Shariatinia, Z.; Jourshabani, M.; Rezaei Darvishi, S.M. ZnO Photocatalyst Revisited: Effective Photocatalytic Degradation of Emerging Contaminants Using S-Doped ZnO Nanoparticles under Visible Light Radiation. *Ind. Eng. Chem. Res.* **2020**, *59*, 15894–15911. [CrossRef]

140. Ferreira, W.H.; Silva, L.G.A.; Pereira, B.C.S.; Gouvêa, R.F.; Andrade, C.T. Adsorption and visible-light photocatalytic performance of a graphene derivative for methylene blue degradation. *Environ. Nanotechnol. Monit. Manag.* **2020**, *14*, 100373. [CrossRef]
141. Shaba, E.Y.; Jacob, J.O.; Tijani, J.O.; Suleiman, M.A.T. A critical review of synthesis parameters affecting the properties of zinc oxide nanoparticle and its application in wastewater treatment. *Appl. Water Sci.* **2021**, *11*, 48. [CrossRef]
142. Anjum, S.; Hashim, M.; Malik, S.A.; Khan, M.; Lorenzo, J.M.; Abbasi, B.H.; Hano, C. Recent advances in zinc oxide nanoparticles (Zno nps) for cancer diagnosis, target drug delivery, and treatment. *Cancers* **2021**, *13*, 4570. [CrossRef]
143. EC. Zinc Oxide (Nano Form). Available online: https://ec.europa.eu/health/scientific_committees/opinions_layman/zinc-oxide/en/index.htm (accessed on 3 April 2022).
144. Elshama, S.S.; Abdallah, M.E.; Abdel-Karim, R.I. Zinc Oxide Nanoparticles: Therapeutic Benefits and Toxicological Hazards. *Open Nanomed. Nanotechnol. J.* **2018**, *5*, 16–22. [CrossRef]
145. Nosrati, R.; Olad, A.; Maryami, F. The use of graphite/TiO2 nanocomposite additive for preparation of polyacrylic based visible-light induced antibacterial and self-cleaning coating. *Res. Chem. Intermed.* **2018**, *44*, 6219–6237. [CrossRef]
146. Görgülüer, H.; Çakıroğlu, B.; Özacar, M. Ag NPs deposited TiO2 coating material for superhydrophobic, antimicrobial and self-cleaning surface fabrication on fabric. *J. Coat. Technol. Res.* **2021**, *18*, 569–579. [CrossRef]
147. Pais, V.; Navarro, M.; Guise, C.; Martins, R.; Fangueiro, R. Hydrophobic performance of electrospun fibers functionalized with TiO2 nanoparticles. *Text Res. J.* **2021**, 00405175211010669. [CrossRef]
148. Costa, S.M.; Pacheco, L.; Antunes, W.; Vieira, R.; Bem, N.; Teixeira, P.; Fangueiro, R.; Ferreira, D.P. Antibacterial and biodegradable electrospun filtering membranes for facemasks: An attempt to reduce disposable masks use. *Appl. Sci.* **2022**, *12*, 67. [CrossRef]
149. Khashan, K.S.; Sulaiman, G.M.; Abdulameer, F.A.; Albukhaty, S.; Ibrahem, M.A.; Al-Muhimeed, T.; Alobaid, A.A. Antibacterial activity of tio2 nanoparticles prepared by one-step laser ablation in liquid. *Appl. Sci.* **2021**, *11*, 4623. [CrossRef]
150. Rashid, M.M.; Tavčer, P.F.; Tomšič, B. Influence of titanium dioxide nanoparticles on human health and the environment. *Nanomaterials* **2021**, *11*, 2354. [CrossRef] [PubMed]
151. Sajid, M. Nanomaterials: Types, properties, recent advances, and toxicity concerns. *Curr. Op. Environ. Sci. Health* **2022**, *25*, 100319. [CrossRef]
152. Antunes, J.C.; Pereira, C.L.; Molinos, M.; Ferreira-Da-Silva, F.; Dessi, M.; Gloria, A.; Ambrosio, L.; Gonçalves, R.M.; Barbosa, M.A. Layer-by-layer self-assembly of chitosan and poly(γ-glutamic acid) into polyelectrolyte complexes. *Biomacromolecules* **2011**, *12*, 4183–4195. [CrossRef] [PubMed]
153. Antunes, J.C.; Tavares, T.D.; Teixeira, M.A.; Teixeira, M.O.; Homem, N.C.; Amorim, M.T.P.; Felgueiras, H.P. Eugenol-containing essential oils loaded onto chitosan/polyvinyl alcohol blended films and their ability to eradicate staphylococcus aureus or pseudomonas aeruginosa from infected microenvironments. *Pharmaceutics* **2021**, *13*, 195. [CrossRef]
154. Henriques, P.C.; Costa, L.M.; Seabra, C.L.; Antunes, B.; Silva-Carvalho, R.; Junqueira-Neto, S.; Maia, A.F.; Oliveira, P.; Magalhães, A.; Reis, C.A.; et al. Orally administrated chitosan microspheres bind Helicobacter pylori and decrease gastric infection in mice. *Acta Biomater.* **2020**, *114*, 206–220. [CrossRef]
155. Patel, D.K.; Ganguly, K.; Hexiu, J.; Dutta, S.D.; Patil, T.V.; Lim, K.T. Functionalized chitosan/spherical nanocellulose-based hydrogel with superior antibacterial efficiency for wound healing. *Carbohydr. Polym.* **2022**, *284*, 119202. [CrossRef]
156. Ribeiro, A.S.; Costa, S.M.; Ferreira, D.P.; Calhelha, R.C.; Barros, L.; Stojković, D.; Soković, M.; Ferreira, I.C.F.R.; Fangueiro, R. Chitosan/nanocellulose electrospun fibers with enhanced antibacterial and antifungal activity for wound dressing applications. *React. Funct. Polym.* **2021**, *159*, 104808. [CrossRef]
157. Felgueiras, H.P.; Homem, N.C.; Teixeira, M.A.; Ribeiro, A.R.M.; Antunes, J.C.; Amorim, M.T.P. Physical, thermal, and antibacterial effects of active essential oils with potential for biomedical applications loaded onto cellulose acetate/polycaprolactone wet-spun microfibers. *Biomolecules* **2020**, *10*, 1129. [CrossRef]
158. Mouro, C.; Gomes, A.P.; Ahonen, M.; Fangueiro, R.; Gouveia, I.C. Chelidonium majus l. Incorporated emulsion electrospun pcl/pva_pec nanofibrous meshes for antibacterial wound dressing applications. *Nanomaterials* **2021**, *11*, 1785. [CrossRef]
159. Wang, Y.; Ma, K.; Bai, J.; Xu, T.; Han, W.; Wang, C.; Chen, Z.; Kirlikovali, K.O.; Li, P.; Xiao, J.; et al. Chemically Engineered Porous Molecular Coatings as Reactive Oxygen Species Generators and Reservoirs for Long-Lasting Self-Cleaning Textiles. *Angew. Chem. Int.* **2022**, *61*, e202115956. [CrossRef]
160. Li, P.; Ryder, M.R.; Stoddart, J.F. Hydrogen-Bonded Organic Frameworks: A Rising Class of Porous Molecular Materials. *Acc. Mater. Res.* **2020**, *1*, 77–87. [CrossRef]
161. Laghari, S.H.; Memon, N.; Khuhawer, M.Y.; Jahangir, T.M. Fluorescent Carbon Dots and their Applications in Sensing of Small Organic Molecules. *Curr. Anal. Chem.* **2022**, *18*, 145–162. [CrossRef]
162. Pang, L.F.; Wu, H.; Wei, M.X.; Guo, X.F.; Wang, H. Cu(II)-assisted orange/green dual-emissive carbon dots for the detection and imaging of anthrax biomarker. *Spectrochim. Acta A Mol. Biomol. Spectrosc.* **2021**, *244*, 118872. [CrossRef]
163. Zhou, Z.; Wang, Z.; Tang, Y.; Zheng, Y.; Wang, Q. Optical detection of anthrax biomarkers in an aqueous medium: The combination of carbon quantum dots and europium ions within alginate hydrogels. *J. Mater. Sci.* **2019**, *54*, 2526–2534. [CrossRef]
164. Li, H.; Zhou, X.; Huang, Y.; Liao, B.; Cheng, L.; Ren, B. Reactive Oxygen Species in Pathogen Clearance: The Killing Mechanisms, the Adaption Response, and the Side Effects. *Front. Microbiol.* **2021**, *11*, 3610. [CrossRef]
165. Farshbaf, M.; Davaran, S.; Rahimi, F.; Annabi, N.; Salehi, R.; Akbarzadeh, A. Carbon quantum dots: Recent progresses on synthesis, surface modification and applications. *Artif. Cells Nanomed. Biotechnol.* **2018**, *46*, 1331–1348. [CrossRef]
166. Lim, S.Y.; Shen, W.; Gao, Z. Carbon quantum dots and their applications. *Chem. Soc. Rev.* **2015**, *44*, 362–381. [CrossRef]

167. Bhuyan, M.S.A.; Uddin, M.N.; Islam, M.M.; Bipasha, F.A.; Hossain, S.S. Synthesis of graphene. *Int. Nano. Lett.* **2016**, *6*, 65–83. [CrossRef]
168. Díez-Pascual, A.M. State of the art in the antibacterial and antiviral applications of carbon-based polymeric nanocomposites. *Int. J. Mol. Sci.* **2021**, *22*, 511. [CrossRef] [PubMed]
169. Francavilla, P.; Ferreira, D.P.; Araújo, J.C.; Fangueiro, R. Smart fibrous structures produced by electrospinning using the combined effect of pcl/graphene nanoplatelets. *Appl. Sci.* **2021**, *11*, 1124. [CrossRef]
170. Pereira, P.; Ferreira, D.P.; Araújo, J.C.; Ferreira, A.; Fangueiro, R. The potential of graphene nanoplatelets in the development of smart and multifunctional ecocomposites. *Polymers* **2020**, *12*, 2189. [CrossRef] [PubMed]
171. Razaq, A.; Bibi, F.; Zheng, X.; Papadakis, R.; Jafri, S.H.M.; Li, H. Review on Graphene-, Graphene Oxide-, Reduced Graphene Oxide-Based Flexible Composites: From Fabrication to Applications. *Materials* **2022**, *15*, 1012. [CrossRef] [PubMed]
172. Al-Balakocy, N.G.; Shalaby, S.E. Imparting antimicrobial properties to polyester and polyamide fibers-state of the art. *J. Text. Assoc.* **2017**, *78*, 179–201.
173. Natarajan, G.; Rajan, T.P.; Das, S. Application of Sustainable Textile Finishing Using Natural Biomolecules. *J. Nat. Fibers* **2020**, 1–18. [CrossRef]
174. Basuk, M.; Kherdekar, G. A synopsis on Coating and lamination in textiles: Process and applications. *Colourage* **2018**, *65*, 43–55.
175. Ran, J.; Chen, H.; Bi, S.; Guo, Q.; Deng, Z.; Cai, G.; Cheng, D.; Tang, X.; Wang, X. One-step in-situ growth of zeolitic imidazole frameworks-8 on cotton fabrics for photocatalysis and antimicrobial activity. *Cellulose* **2020**, *27*, 10447–10459. [CrossRef]
176. Islam, M.T.; Mamun, M.A.A.; Hasan, M.T.; Shahariar, H. Scalable coating process of AgNPs-silicone on cotton fabric for developing hydrophobic and antimicrobial properties. *J. Coat. Technol. Res.* **2021**, *18*, 887–898. [CrossRef]
177. Tania, I.S.; Ali, M.; Azam, M.S. Mussel-Inspired Deposition of Ag Nanoparticles on Dopamine-Modified Cotton Fabric and Analysis of its Functional, Mechanical and Dyeing Properties. *J. Inorg. Organomet. Polym. Mater.* **2021**, *31*, 4065–4076. [CrossRef]
178. Tavares, T.D.; Antunes, J.C.; Ferreira, F.; Felgueiras, H.P. Biofunctionalization of natural fiber-reinforced biocomposites for biomedical applications. *Biomolecules* **2020**, *10*, 148. [CrossRef]
179. Singh, N.; Sheikh, J. Novel Chitosan-Gelatin microcapsules containing rosemary essential oil for the preparation of bioactive and protective linen. *Ind. Crops Prod.* **2022**, *178*, 114549. [CrossRef]
180. Kumari, N.; Bhattacharya, S.N.; Das, S.; Datt, S.; Singh, T.; Jassal, M.; Agrawal, A.K. In Situ Functionalization of Cellulose with Zinc Pyrithione for Antimicrobial Applications. *ACS Appl. Mater. Interfaces* **2021**, *13*, 47382–47393. [CrossRef]
181. Deng, C.; Seidi, F.; Yong, Q.; Jin, X.; Li, C.; Zhang, X.; Han, J.; Liu, Y.; Huang, Y.; Wang, Y.; et al. Antiviral/antibacterial biodegradable cellulose nonwovens as environmentally friendly and bioprotective materials with potential to minimize microplastic pollution. *J. Hazard Mater.* **2022**, *424*, 127391. [CrossRef]
182. Domingues, J.M.; Teixeira, M.O.; Teixeira, M.A.; Freitas, D.; da Silva, S.F.; Tohidi, S.D.; Fernandes, R.D.V.; Padrão, J.; Zille, A.; Silva, C.; et al. Inhibition of Escherichia Virus MS2, Surrogate of SARS-CoV-2, via Essential Oils-Loaded Electrospun Fibrous Mats: Increasing the Multifunctionality of Antivirus Protection Masks. *Pharmaceutics* **2022**, *14*, 303. [CrossRef]
183. Sundarrajan, S.; Chandrasekaran, A.R.; Ramakrishna, S. An update on nanomaterials-based textiles for protection and decontamination. *J. Am. Ceram. Soc.* **2010**, *93*, 3955–3975. [CrossRef]
184. Kiani, S.S.; Farooq, A.; Ahmad, M.; Irfan, N.; Nawaz, M.; Irshad, M.A. Impregnation on activated carbon for removal of chemical warfare agents (CWAs) and radioactive content. *Environ. Sci. Pollut. Res.* **2021**, *28*, 60477–60494. [CrossRef]
185. Verbič, A.; Gorjanc, M.; Simončič, B. Zinc Oxide for Functional Textile Coatings: Recent Advances. *Coatings* **2019**, *9*, 550. [CrossRef]
186. Ealia, S.A.M.; Saravanakumar, M.P. A review on the classification, characterisation, synthesis of nanoparticles and their application. In *IOP Conference Series: Materials Science and Engineering*; IOP Publishing: Bristol, UK, 2017; Volume 263, p. 032019.
187. Wang, S.; Pomerantz, N.L.; Dai, Z.; Xie, W.; Anderson, E.E.; Miller, T.; Khan, S.A.; Parsons, G.N. Polymer of intrinsic microporosity (PIM) based fibrous mat: Combining particle filtration and rapid catalytic hydrolysis of chemical warfare agent simulants into a highly sorptive, breathable, and mechanically robust fiber matrix. *Mater. Today Adv.* **2020**, *8*, 100085. [CrossRef]
188. Botta, V. Durable, Repairable and Mainstream: How Ecodesign Can Make Our Textiles Circular. Available online: https://ecostandard.org/wp-content/uploads/2021/04/ECOS-REPORT-HOW-ECODESIGN-CAN-MAKE-OUR-TEXTILES-CIRCULAR.pdf (accessed on 3 March 2022).
189. Valdez-Salas, B.; Beltran-Partida, E.; Cheng, N.; Salvador-Carlos, J.; Valdez-Salas, E.A.; Curiel-Alvarez, M.; Ibarra-Wiley, R. Promotion of surgical masks antimicrobial activity by disinfection and impregnation with disinfectant silver nanoparticles. *Int. J. Nanomed.* **2021**, *16*, 2689–2702. [CrossRef]

polymers

MDPI

*Article*

# Mechanical and Thermal Behaviours of Weft-Knitted Spacer Fabric Structure with Inlays for Insole Applications

**Nga-Wun Li [1], Kit-Lun Yick [1,2,*], Annie Yu [3] and Sen Ning [2]**

[1] Laboratory for Artificial Intelligence in Design, Hong Kong, China; dorisli@aidlab.hk
[2] Institute of Textiles and Clothing, The Hong Kong Polytechnic University, Hong Kong, China; sen-karolyn.ning@connect.polyu.hk
[3] Department of Advanced Fibro Science, Kyoto Institute of Technology, Kyoto 606-8585, Japan; annieyu@kit.ac.jp
* Correspondence: tcyick@polyu.edu.hk; Tel.: +852-27666551

**Abstract:** Insoles provide resistance to ground reaction forces and comfort during walking. In this study, a novel weft-knitted spacer fabric structure with inlays for insoles is proposed which not only absorbs shock and resists pressure, but also allows heat dissipation for enhanced thermal comfort. The results show that the inlay density and spacer yarn increase compression resistance and reduce impact forces. The increased spacer yarn density provides better air permeability but reduces thermal resistance, while a lower inlay density with a random orientation reduces the evaporative resistance. The proposed structure has significantly positive implications for insole applications.

**Keywords:** compression; thermal comfort; weft-knitted spacer fabric; inlay knitting; cushioning insole; silicone inlay

**Citation:** Li, N.-W.; Yick, K.-L.; Yu, A.; Ning, S. Mechanical and Thermal Behaviours of Weft-Knitted Spacer Fabric Structure with Inlays for Insole Applications. *Polymers* **2022**, *14*, 619. https://doi.org/10.3390/polym14030619

Academic Editor: Muhammad Tayyab Noman

Received: 14 January 2022
Accepted: 1 February 2022
Published: 5 February 2022



## 1. Introduction

Insoles act as an important cushioning feature at the interface between the foot and footwear to manipulate the distribution of plantar pressure for foot protection and wear comfort. However, inappropriate insole designs and materials can lead to excessive moisture, heat and localized pressure points on the foot, which could then result in ulceration, arthritis or tendon dysfunctions. The growth of microorganisms in the enclosed environment of the inner shoe can also cause athlete's foot or undesirable odours [1]. The literature on footwear research has mainly focused on the structural design and fabrication materials of insoles to relieve plantar pressure, and thus reduce the risk of ulceration [2,3], falls [4,5], and growth of microbes [1] as well as ensure the sustainability of the insole [6]. Recently, spacer fabrics have been used as an alternative cushioning material because this material offers a wide range of good properties, and has good compression behaviour, high density and adequate thickness for different applications, including insoles [7], helmet liners [8], wound dressings [9], sports bras [10], etc. The structure of spacer fabric is like a sandwich, with two separate layers of fabric on the outside and a connective layer between them [11]. In weft knitted spacer fabric, the connective layer is developed by using spacer filaments to form tuck stitches by using the front and back needle beds to create support for the two surface layers of fabric [12]. This sandwich structure provides spacer fabric with excellent air and moisture permeabilities, cushioning performance and pressure distribution [8,10,13–15].

Previous studies have found that the compression properties are significantly affected by the spacer yarns in the connective layer. The spacer fabric is stiffer in compression when the spacer filaments stand perpendicular to the layer of surface fabric [14]. On the other hand, spacer fabric has better compression resistance and energy absorption when there is a smaller span distance between the spacer yarns, coarser monofilaments, and higher spacer yarn density [15]. Hamedi et al. developed weft knitted spacer fabric to construct insole materials for diabetic patients by using a shape memory alloy, nickel-titanium, as the spacer fabric monofilaments [16]. However, spacer fabrics deform easily under high compression stress due to body weight and lose their cushioning properties after prolonged use. Also, the use of metal wire in knitting requires special equipment and is time consuming. Yu et al. proposed the use of inlays to reinforce the spacer fabric structure and found that the Young's modulus of the inlaid silicone tubes is correlated with fabric compression [17]. Inlay knitting enables materials such as foam rods that usually cannot be knitted as part of the loops of the ground structure to be securely incorporated into the knitted fabric [18]. Its application can then be more widely used in the structural design of compression garments, buoyant swimwear, and sporting goods [19,20].

As for footwear and insole applications, increases in foot temperature incurred by the insoles during daily activities may result in a high level of foot discomfort from heat and perspiration. By taking the prolonged use and hygiene of insoles into consideration, the temperature and humidity transfer performance of insole fabrication material is regarded as the most important considerations for footwear comfort [21]. Transmission of moisture facilitates liquid to be evaporated through fabric to the environment which results in a cooling sensation for the wearer [22,23] and prevents a damp feeling which leads to discomfort. Nevertheless, few studies have focused on the thermal comfort of spacer fabric with inlays. While a preliminary study has shown that inlaid spacer fabric has superior air permeability [24], the impacts of the inlay density and orientation in the spacer fabric structure on the mechanical and thermal comfort properties have been largely absent from discussion. A more in-depth understanding of the effects of the inlay structure of spacer fabric and knitting parameters on mechanical and thermal behaviours is therefore paramount. The fabric structural parameters that affect the compression behaviour, impact force reduction, and thermal and evaporative resistance are systematically analysed. The findings of this study can greatly contribute to foot protection and orthotic treatment by providing insights into suitable insole materials.

## 2. Materials and Methods

### 2.1. Preparation of Inlaid Foam

Silicone foam rods with a diameter of 2.25 mm were used as the inlay as they are flexible enough to be laid into the fabric during the knitting process. Prior to knitting, the inlaid material was wrapped by using a knitted net to reduce the surface friction during insertion. The net was knitted by using a multifunction twisting machine for fancy yarn (SFM32-04, Kunshan Shun Feng Textile Co., LTD., Kunshan, China) with 4 needles. The yarn used was 2 ends of DRYARN® 140D 100% polypropylene yarn (Aquafil S.P.A., Via Linfano, Arco, Italy). The foam rods were then manually inserted into the knitted net. A microscopic view of the knitted nets and inserted silicone foam rods is provided in Figure 1.

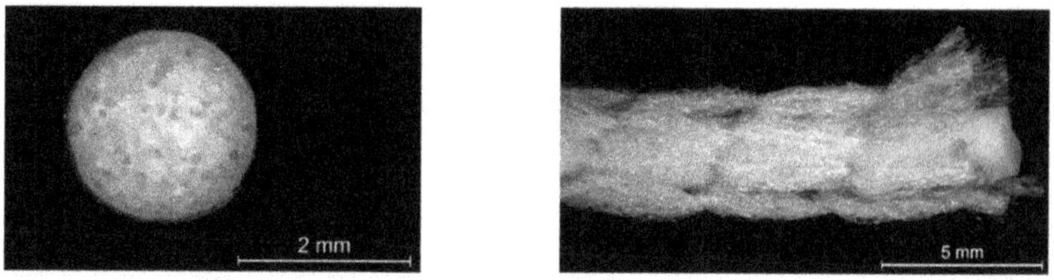


**(a)** **(b)**

**Figure 1.** Microscopic view of (**a**) cross section of silicone foam rod, and (**b**) side view of foam rod wrapped by the net.

### 2.2. Inlaid Spacer Fabric Samples

Ten spacer fabric samples with different inlay densities, inlay orientation and spacer yarn density were fabricated, and two knitted fabrics without inlays were used as the control. All of the fabric samples were knitted on a 14-gauge V-bed flat knitting machine (SVR123SP, SHIMA SEIKI, Wakayama, Japan) equipped with yarn feeder tips (4.5D) which are suitable for inserting materials with a diameter below 4 mm. Two ends of the 1/27NM 63% cupro 37% polyester and one end of 107D high power lycra were used as the yarn for the surface layer in a single jersey structure. The connective layer was constructed by using one end of 100% 220D polyester monofilament. All the samples were prepared by using the same knitting tension and parameters. Two of the same samples were made for each knitting condition. The design of experiment and details of the sample specifications are provided in Tables 1 and 2. A random inlay orientation means that the inlaid foam rods are randomly placed in all directions, while localized orientation means a continuous inlay as shown in Figure 2. The knitting notations of the fabric are provided in Table 3.

**Table 1.** Design of experiment.

| Factor | Level | | | |
|---|---|---|---|---|
| Inlay density (ratio of spacer and inlay structure) | 4:1 | 3:2 | 1:1 | 2:3 |
| Inlay Orientation | Random | Localized | | |
| Number of spacer yarn per spacer structure | 1 course | 2 courses | | |

**Table 2.** Sample specifications of inlaid spacer fabrics.

| pslp | Knitted Structure | Inlay Density (Course per cm) | Orientation of Inlay | Spacer Yarn Density (Course per cm) |
|---|---|---|---|---|
| AS1 | A | 1.20 | Random | 5 |
| AS2 | B | 1.20 | Random | 10 |
| BC1 | C | 2.00 | Localized | 5 |
| BC2 | D | 2.00 | Localized | 10 |
| BS1 | E | 2.00 | Random | 5 |
| BS2 | F | 2.00 | Random | 10 |
| CS1 | G | 2.30 | Random | 4 |
| CS2 | H | 2.30 | Random | 8 |
| DC2 | I | 2.50 | Localized | 10 |
| DS2 | J | 2.50 | Random | 10 |
| C1 | C1 | Nil | Nil | 13 |
| C2 | C2 | Nil | Nil | 26 |

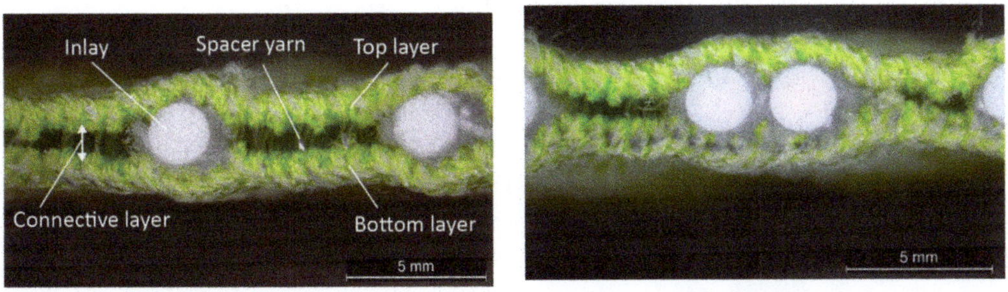


(a) (b)

**Figure 2.** Microscopic view of spacer fabric with (**a**) random inlays and (**b**) inlays with localized orientation.

**Table 3.** Knitting notations and images of fabrics with different knitted structures.

| Fabric | AS1 | AS2 | BC1 | BC2 | BS1 | BS2 | CS1 | CS2 | DC2 | DS2 |
|---|---|---|---|---|---|---|---|---|---|---|
| Structure | A | B | C | D | E | F | G | H | I | J |
| Knitting notation | 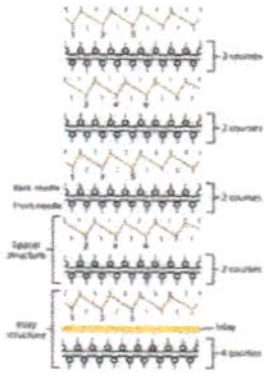 | 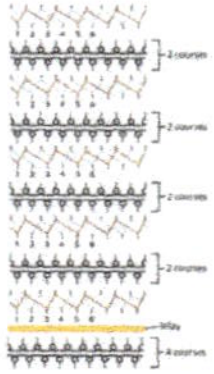 | 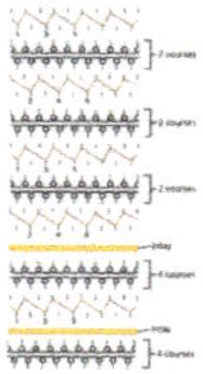 | 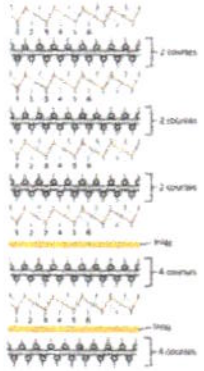 | 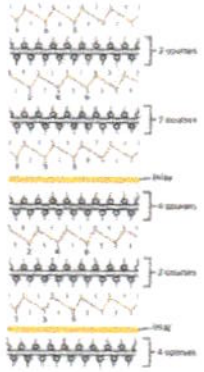 | 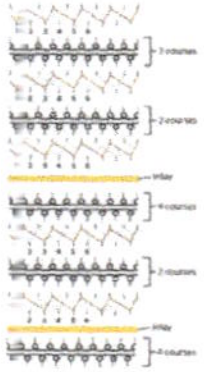 | 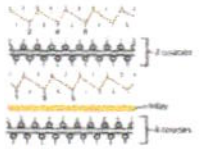 | 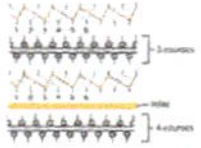 | 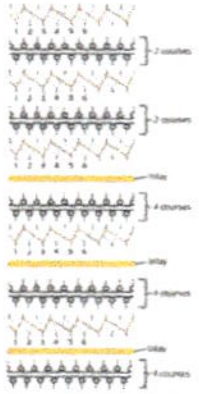 | 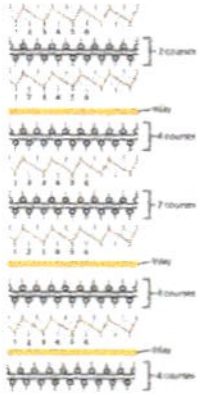 |
| Image |  |  |  |  |  |  |  |  |  |  |

### 2.3. Evaluation of Mechanical and Thermal Behaviours of Fabric

With reference to the key requirements and end-uses of insoles including cushioning purposes, resistance to body weight and wear comfort during walking, tests on the physical, compression and thermal comfort properties of the samples were conducted (Table 4). The fabric samples were conditioned for 24 h at a temperature of 20 ± 1 °C and relative humidity of 65 ± 5% before the testing took place. Depending on the sample size, the tests were repeated 4 to 6 times on different areas of the samples, and their mean value was calculated and used.

**Table 4.** Summary of test methods.

| Property | Device | Testing Standard |
|---|---|---|
| Thickness | Dial thickness gauge (Model H, Peacock OZAKI MFG. Co., Ltd, Tokyo, Japan) | ASTM D1777 Standard Test Method for Thickness of Textile Materials |
| Hardness | Durometer (GS-744G, Type: FO, TECLOCK Co., Ltd., Nagano, Japan) | ASTM D2240-05: 2010 Standard Test Method for Rubber Property—Durometer Hardness |
| Surface unevenness | 3D-optical microscope (VR-3000, KEYENCE, Osaka, Japan) | ISO4287:1997 Surface unevenness-Definitions |
| Air permeability | Air permeability tester (SDL M021S, SDL International Textile Testing Solutions, Rock Hill, SC, USA) | ASTM-D737 Standard Test Method for Air Permeability of Textile Fabrics |
| Compression | Compression tester (Instron 4411, Instron, Norwood, MA, USA) | ASTM D575 Standard Test Methods for Rubber Properties in Compression |
| Thermal and evaporative resistance | Sweating guarded hot plate (YG(B)606G, Wenzhou, China) | ASTM F1868-17 Standard Test Method for thermal and evaporative resistance of clothing materials |

#### 2.3.1. Physical Properties

The fabric thickness was measured under a pressure of 4 gf/cm$^2$ with an accuracy of 0.01 mm. A 3D-optical microscope was used to examine the variations in the surface unevenness of the samples. The air permeability test was carried out under a water pressure difference of 125 Pa.

#### 2.3.2. Compression and Impact Force Reduction

Each fabric sample was tested by using a compression tester with a flat circular indenter that is 150 mm in diameter. The compression test was conducted at a rate of 12 mm/min and the samples were compressed up to 80% of their initial thickness. As shown in Figure 3, the compression process of the spacer fabrics can be divided into three stages including the linear elasticity (Stage I), plateau (Stage II), and densification (Stage III) stages [12,25]. In Stage I, the presence of a smaller slope is observed as the compressed monofilaments start to buckle and shear constantly at a larger scale. When the fabrics are further compressed in Stage II, a nearly constant stress can be observed. This shows the collapse of the spacer fabric structure and inlaid material as the compression strain of the fabric is largely increased under a small increase in stress. In Stage III, the structure completely collapses in which the monofilaments and the inlays come into contact with each other and the outer layers. The sample becomes stiff and rigid so the slope of the curve becomes steep again. A higher maximum compression stress means that the fabric can withstand a higher compression force. The strain is calculated as the distance of the compression by the indenter divided by the original thickness of the specimen. A steeper curve in Stage I represents stiffer mechanical behaviour of the fabric with a higher Young's modulus.

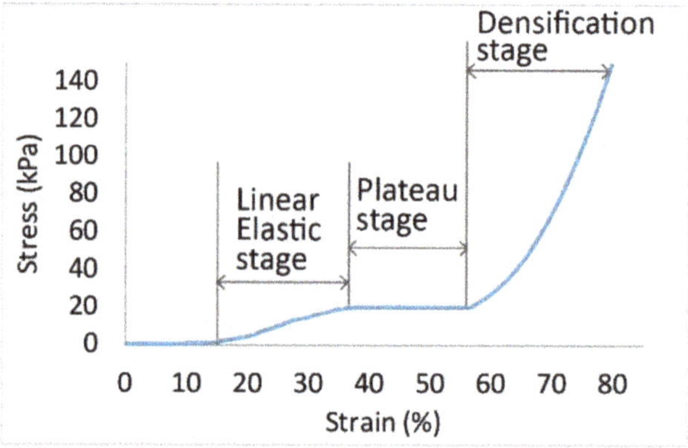


**Figure 3.** Plotted compression stress-strain of spacer fabric.

When measuring the impact force reduction, two specimens of each sample were stacked on top of each other with dimensions of 100 mm × 100 mm before the testing was carried out. A ball bearing was released and dropped onto the sample through a straight tube at a height of 400 mm. The highest impact force was measured by using a load cell placed at the bottom of the instrument. The impact force reduction capacity of the sample is defined as a percentage of the maximum impact force with the sample and the ground surface [26]:

$$FR_x = \left(1 - \left(\frac{F_x}{F_o}\right)\right) \times 100\%, \quad (1)$$

where $FR_x$ is the impact force reduction of the sample (%), $F_x$ is the peak force measured for the sample (N), and $F_o$ is the peak force measured for the ground surface.

2.3.3. Thermal and Evaporative Resistance

The thermal and evaporative resistance tests were conducted in an air temperature of 25 °C at a relative humidity of 65% with air flowing over the hot plate and air velocity of 1 m/s. The test began when the backside of the fabric was placed onto the hot plate to simulate contact with human skin. All wrinkles were eliminated from the fabric to prevent air bubbles from forming between the fabric sample and the hot plate. The average intrinsic thermal resistance of the fabric is calculated by using:

$$R_{cf} = ((T_{plate} - T_{air})\, A/H_c) - R_b, \quad (2)$$

where $R_{cf}$ is the intrinsic resistance to dry heat transfer provided by the fabric system ($m^2{\cdot}K/W$), $T_{plate}$ is the surface temperature of the hot plate (°C), $T_{air}$ is the air temperature (°C), $A$ is the area of the tested section on the hot plate ($m^2$), $H_c$ is the power input (W), and $R_b$ is the thermal resistance of the still air where no fabric was placed onto the plate ($m^2{\cdot}K/W$).

For the evaporative resistance test, the sweat pores of the hot plate emitted water to the surface of the plate and a liquid barrier was used to cover the plate to prevent wetting of the fabric specimens. The liquid barrier was adhered closely to the hot plate and guard section with no wrinkles or air bubbles present. Then, a fabric sample was placed on the hot plate. The average intrinsic evaporative resistance of the fabric sample is calculated as follows:

$$R_{ef} = ((P_{plate} - P_{air})\, A/H_e) - R_e, \quad (3)$$

where $R_{ef}$ is the intrinsic resistance to evaporative heat transfer provided by the fabric system ($m^2{\cdot}K/W$), $P_{plate}$ is the water vapor pressure at the plate surface (kPa), $P_{air}$ is the water vapor pressure in the air (kPa), $A$ is the area of the section in which the testing was carried out on the hot plate ($m^2$), $H_e$ is the power input (W), and $R_e$ is the evaporative resistance value measured for the air layer and liquid barrier with no fabric placed onto the hot plate ($m^2 \cdot K/W$).

2.3.4. Statistical Analysis

The experimental data were analysed by using SPSS 23 (IBM Corp., Armonk, NY, USA). A multivariate analysis of variance (MANOVA) was used to examine the mean differences among the three different independent variables: (1) inlay density, (2) orientation of inlays and (3) spacer yarn density on five dependent variables (maximum compressive stress, impact force reduction, air permeability, and thermal and evaporative resistance). Prior to the analysis, the values were evaluated to ensure that the assumptions for the multivariate tests are valid. Measures of skewness and kurtosis, histograms and normal Q-Q plots were used for the dependent and independent variables. Observation of these measures and plots showed a normal distribution at the different levels of these variables. The significance level of the statistical analysis was set at 0.05.

## 3. Results and Discussion

### *3.1. Air Permeability*

This study found that over 99 percent of the variance in air permeability is accounted for by the spacer yarn density ($\eta^2 = 0.996$) and inlay density ($\eta^2 = 0.991$) but no significant difference is found with the orientation of the inlays ($p > 0.05$). It is interesting that the fabrics with a higher spacer yarn density (AS2, BC2, BS2, CS2 and C2) have a relatively higher air permeability than those with less spacer yarn (AS1, BC1, BS1, CS1 and C1). Additional pulling forces are applied to the surface layer when the spacer yarn density of fabric is increased. The knitted loops of the surface layer are then extended so that the restriction of air flow through the fabric is reduced. This is confirmed by the lower stitch density and fabric weight of the spacer fabrics with a higher spacer yarn density (AS2, BS2, CS2 and C2) (Table 5). When comparing the various inlay densities (AS2, BS2, CS2 and DS2), the samples with higher inlay density exhibit relatively lower air permeability (Figure 4). The reason is that the inlaid foam rods block the circulation of trapped air between the two surface layers.

**Table 5.** Physical properties of the inlaid spacer fabrics.

| Sample Code | Weight (g/m$^2$) | | Thickness (mm) | | Stitch Density (loop/cm$^2$) | | Hardness (Shore A) | |
|---|---|---|---|---|---|---|---|---|
| | Mean | SD | Mean | SD | Mean | SD | Mean | SD |
| AS1 | 1483.32 | 6.12 | 4.30 | 0.03 | 93.33 | 4.04 | 83.17 | 1.33 |
| AS2 | 1417.25 | 0.46 | 4.31 | 0.03 | 91.00 | 0.00 | 86.17 | 1.72 |
| BC1 | 1649.50 | 18.62 | 4.34 | 0.03 | 93.33 | 4.04 | 86.17 | 0.93 |
| BC2 | 1683.28 | 15.94 | 4.39 | 0.05 | 91.00 | 0.00 | 86.50 | 1.22 |
| BS1 | 1669.15 | 6.38 | 4.40 | 0.04 | 91.00 | 0.00 | 87.33 | 0.82 |
| BS2 | 1620.13 | 21.14 | 4.39 | 0.05 | 86.50 | 3.91 | 87.67 | 0.82 |
| CS1 | 1759.97 | 17.31 | 4.35 | 0.04 | 88.83 | 3.75 | 88.67 | 1.21 |
| CS2 | 1689.88 | 17.47 | 4.39 | 0.03 | 86.33 | 4.04 | 88.83 | 0.41 |
| DC2 | 1866.20 | 7.60 | 4.34 | 0.05 | 82.33 | 3.75 | 88.92 | 0.92 |
| DS2 | 1861.70 | 22.59 | 4.33 | 0.04 | 82.33 | 3.75 | 89.00 | 0.84 |
| C1 | 1094.38 | 5.65 | 3.51 | 0.05 | 108.00 | 4.00 | 88.67 | 1.37 |
| C2 | 1068.20 | 16.54 | 3.75 | 0.14 | 106.67 | 4.62 | 89.00 | 0.63 |

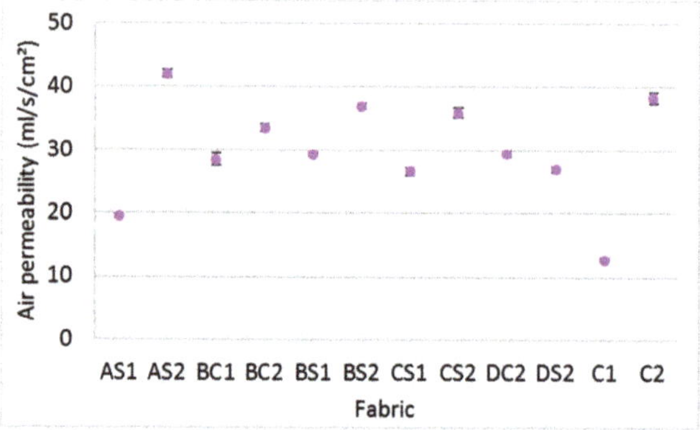


**Figure 4.** Air permeability of the inlaid spacer fabric and controls.

### 3.2. Compression Behaviour

The results of the factorial MANOVA show an overall significant difference between the maximum compressive stress, various inlay densities (F = 486.97, $p < 0.001$, and $\eta^2 = 0.993$), spacer yarn density (F = 335.18, $p < 0.001$, and $\eta^2 = 0.971$) and orientation of the inlays (F = 65.14, $p < 0.001$, and $\eta^2 = 0.867$). The inlay and spacer yarn densities account for over 97 percent of the variance in the maximum compressive stress while the orientation of the inlays has the least impact and accounts for only 87% of the variance. As shown in Figure 5, all of the spacer fabric samples with inlay have a higher maximum compressive stress versus the fabric samples without inlay (C1 and C2). This is agreement with the findings in Yu et al. [12], in that inlays largely strengthen spacer fabric to withstand the compressive stress.

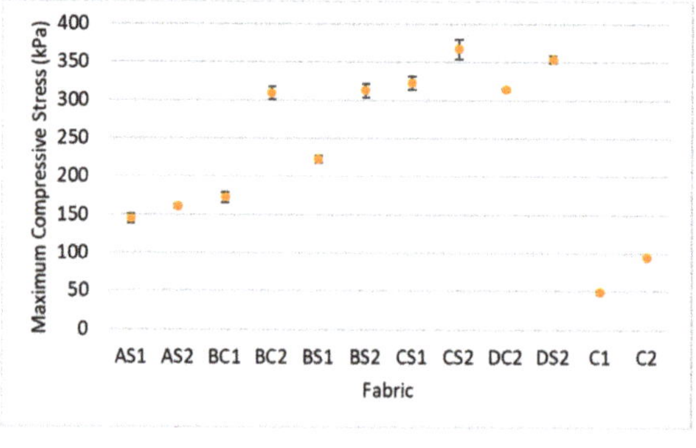


**Figure 5.** Maximum compressive stress of inlaid spacer fabrics and controls.

#### 3.2.1. Effect of Inlay Density

A higher maximum compressive stress can be observed when the inlay density increases from 1.2 course/cm (AS2) to 2.5 course/cm (DS2); see Figure 6a. Meanwhile, the Young's modulus of the fabric generally increases with inlay density (Figure 7). This means that the spacer fabric with more inlays has stiffer mechanical behaviour and higher compression resistance. The stress-strain curve in Figure 6 shows that the area under the curve is the energy absorbed by the fabric. In Stage III, the area under the curve of DS2 is more than twice that of C2, the fabric sample without inlay (Figure 6a). This shows that the inlaid foam rods absorb a greater amount of energy when they deform and increase the strength of the spacer fabric to resist compression stress.

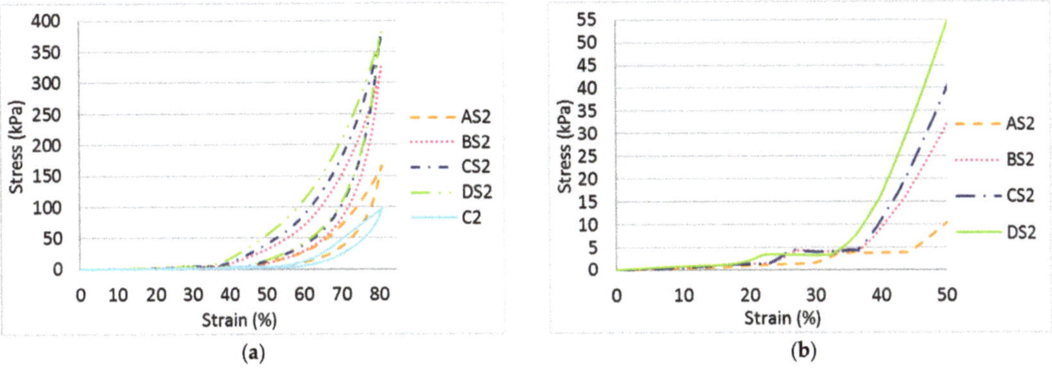


**Figure 6.** Compression stress-strain curves of the inlaid spacer fabrics with different inlay densities: (**a**) at 80% strain and (**b**) at 50% strain.

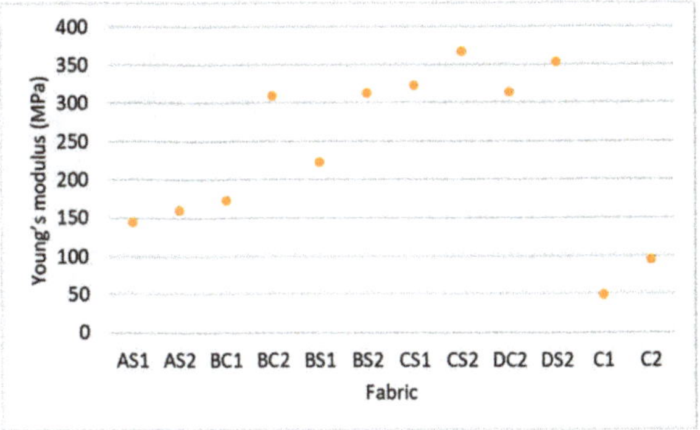


**Figure 7.** Young's modulus of the inlaid spacer fabric and controls.

It is interesting that Stage II appears at a similar compression stress of around 3.81 kPa and a similar range of compression strain (around 10.45%) is obtained when the inlay density is increased; see Figure 6b. However, Stage II appears at a lower level of strain for fabric with more inlays (1.36 mm in DS2, and 1.86 mm in AS2). In Stage II, the wider range of compression strain and higher stress mean that the material has a better cushioning effect by absorbing more energy before reaching Stage III. Stage II which appears at a lower level of strain means less distortion of the fabric occurs under the same compression stress. Although the fabric samples with various inlay densities absorb a similar amount of energy in Stage II under a low stress level, fabric with the highest inlay density (DS2) has less deformation under the same stress. This may be due to the surface unevenness of the fabric.

Rc, Rt and Ra have a small value which indicates that the surface layer is uneven. As shown in Figure 8, the fabric unevenness can be reduced by increasing the inlay density. The surface of DS2 is relatively flat with a lower Rc, Rt and Ra while AS2 is comparatively more wavy (Figure 9). The compressive forces can be evenly distributed throughout the entire surface of DS2 which increases the compression resistance but this is only true for the convex areas of AS2. Therefore, the linear relationship between surface unevenness and maximum compressive stress can be found in Figure 10. The reduced surface unevenness (Rc, Rt and Ra) of the fabric sample results in an increased maximum compressive stress. Therefore, the spacer fabric with more inlays can withstand higher compressive forces due to the even distribution of the force on the flat surface and the increased fabric stiffness. The materials used for insoles should be less susceptible to deformation under the same stress to enhance standing and walking stability to the same degree as a hard insole.

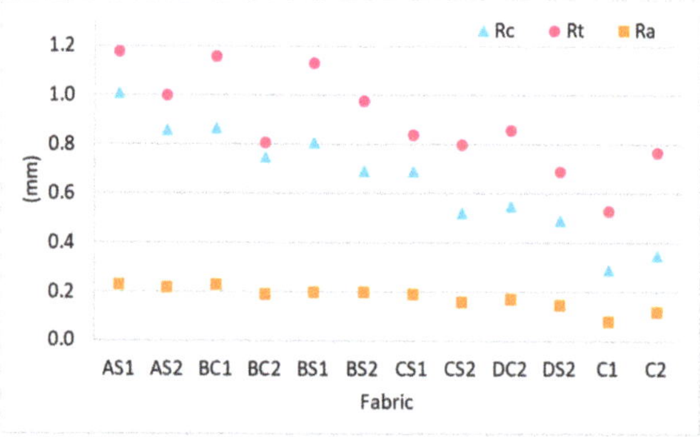


**Figure 8.** Surface unevenness Rc, Rt and Ra of the inlaid spacer fabrics and controls.

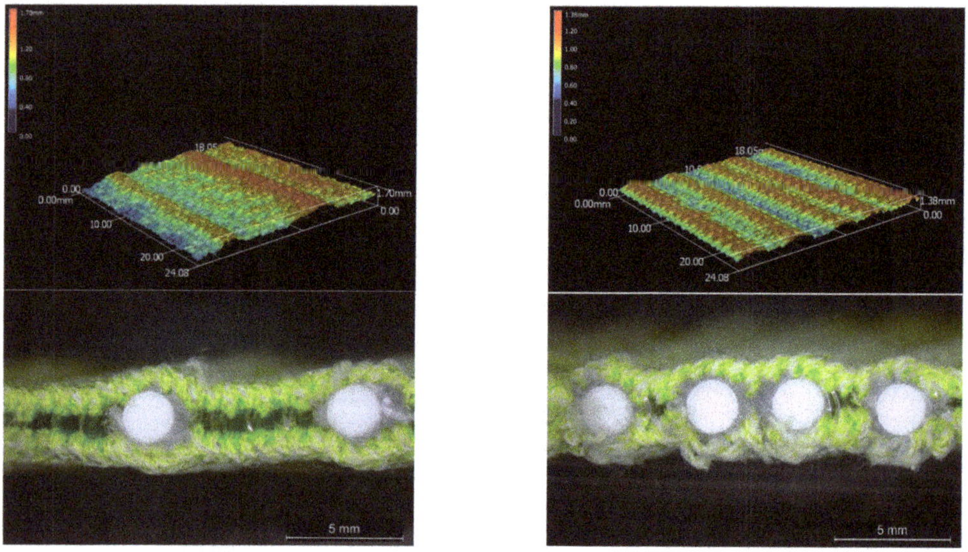


(a) (b)

**Figure 9.** Microscopic view and surface thickness variations of inlaid spacer fabrics (**a**) AS2 and (**b**) DS2 along the wale direction.

3.2.2. Effect of Spacer Yarn Density

As shown in Figures 7 and 11, the spacer fabric samples with a higher spacer yarn density (BS2, CS2 and C2) have a higher maximum compression stress and higher Young's modulus than the samples with a lower spacer yarn density (BS1, CS1 and C1) respectively. This shows that the additional spacer yarns in the connective layer provide extra support to the spacer fabric structure, as more stress is needed to compress the fabric and the extra spacer yarns also increase the stiffness of the fabric. However, the curves of BS2 and CS1 are similar (Figure 11a), which means that the fabric sample with 2 inlay courses and 10 spacer yarn courses per cm (BS2) has similar compression properties as the fabric sample with 2.3 inlay courses and 4 spacer yarn courses per cm (CS1).

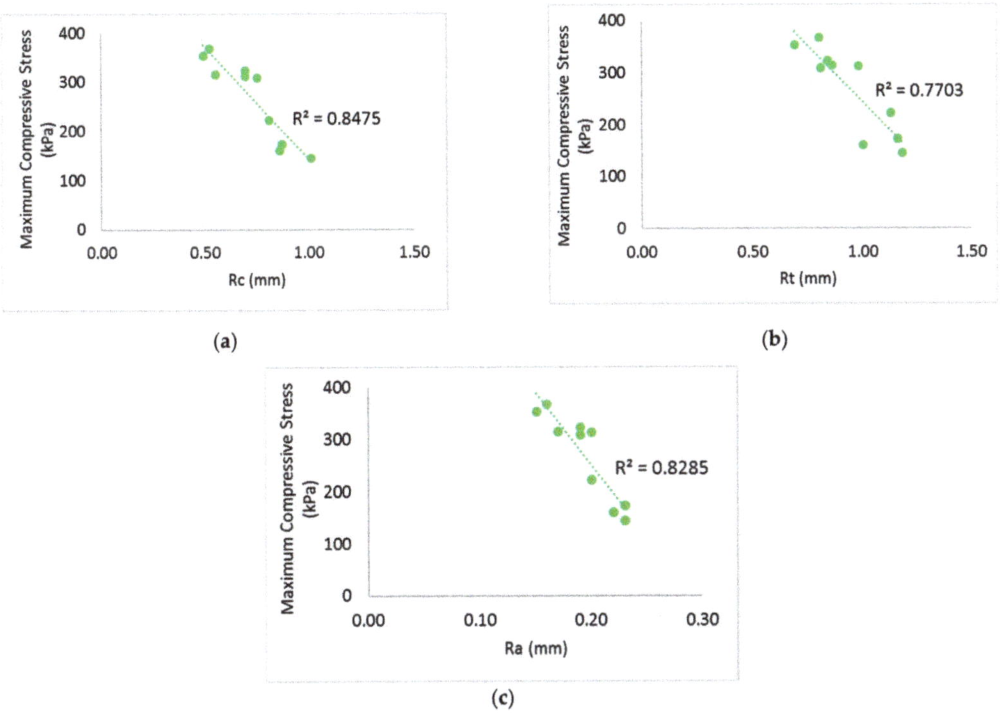


**Figure 10.** Linear relationship between maximum compressive stress and surface unevenness: (**a**) Rc, (**b**) Rt and (**c**) Ra.

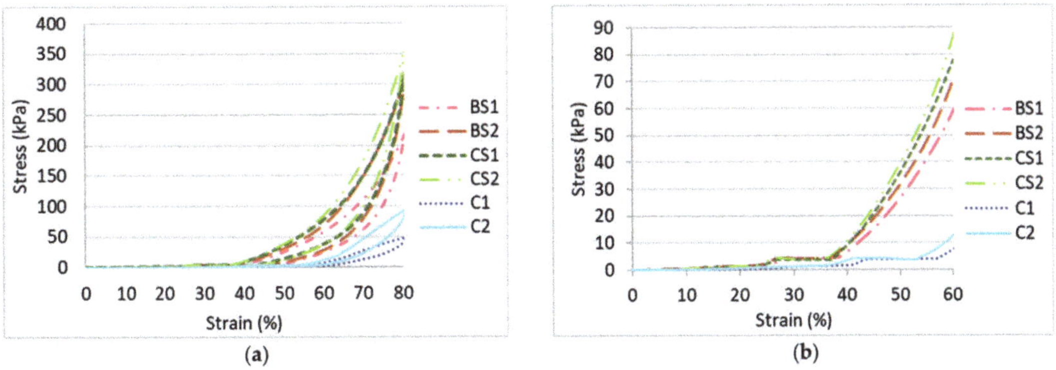


**Figure 11.** Compression stress-strain curves of inlaid spacer fabrics with different spacer yarn densities: (**a**) at 80% strain and (**b**) at 60% strain.

Similar to the effect of the inlay density, Stage II of the fabric samples with various spacer yarn and inlay densities (BS1, BS2, CS1, CS2, C1 and C2) appears at a similar compression stress (around 4 kPa) with similar range of strain (around 11.11%) (Figure 11b). However, Stage II of the inlaid spacer fabric appears at a compression strain of 27% which is smaller than the control fabrics (42.26%). Although the inlaid spacer fabric absorbs a similar amount of energy under a low stress condition, the deformation of the spacer fabric structure with inlays is 15.26% less than that of the fabric without inlays under a compression stress of 4 kPa.

3.2.3. Effect of Inlay Orientation

As shown in Figures 7 and 12, the curve of BS1 indicates a higher maximum compressive stress and Young's modulus as opposed to BC1. However, BC2 and BS2 have similar compression properties as they have a similar curve. This means that the compression properties are correlated with both inlay orientation and spacer yarn density which validates the results of the MANOVA. The fabric sample with inlays in a random arrangement with a lower spacer yarn density (BS1) is more rigid and has a higher compression resistance in comparison to the inlay in a localized arrangement (BC1). As for the fabric samples with a higher spacer yarn density (BC2 and BS2), the effect of the inlay orientation on the compression properties becomes less significant.

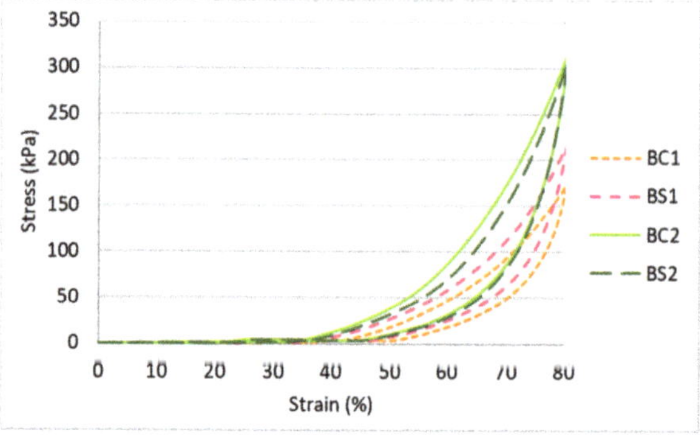


**Figure 12.** Compression stress-strain curves of inlaid spacer fabrics with different orientation of the inlays.

### 3.3. *Impact Force Reduction*

Fabric that can sustain a higher impact force is a more suitable material for insoles which would reduce the plantar pressure of the foot. As shown in Figure 13, all of the inlaid spacer fabrics have a higher force reduction than the spacer fabrics without inlays (C1 and C2). This shows that the inlaid rods enhance the ability of the fabric to reduce the impact forces. Among the inlaid spacer fabric samples, the sample with the higher inlay density can generally reduce higher impact forces. The fabric samples with a higher spacer yarn density (AS2, BS2, CS2 and C2) show a higher reduction of impact forces than those with less spacer yarn (AS1, BS1, CS1 and C1). The ability to reduce higher impact forces is a function of the higher inlay and spacer yarn densities. These results are confirmed by a MANOVA analysis, and significant differences are observed only between the inlay density (F = 10.89, $p < 0.005$, and $\eta^2 = 0.766$) and spacer yarn density (F = 5.32, $p < 0.05$, and $\eta^2 = 0.347$) on the impact force reduction. Around 77% of the variance in force reduction is accounted by the inlay density while the spacer yarn density only accounts for 35% of the variance. This implies that the inlay density is the main factor in reducing the impact forces.

### 3.4. *Thermal Comfort Properties*

Unlike the knitted flat fabrics [27], the results of the Pearson's correlations indicated that the surface unevenness—Rc, Rt, and Ra, has no significant correlation with thermal and evaporative resistance ($p > 0.05$).

3.4.1. Evaporative Resistance

Unlike air permeability, the evaporative resistance of the fabric samples depends on the inlay density (F = 32.32, $p < 0.001$, and $\eta^2 = 0.907$) and orientation of the inlays (F = 14.98, $p < 0.005$, and $\eta^2 = 0.600$) but not the spacer yarn density ($p > 0.05$). When comparing AS1, BS1 and CS1, the evaporative resistance of the inlaid spacer fabric samples increases with inlay density (Figure 14). The fabric samples with inlays in a random arrangement

(BS1, BS2, and DS2) have lower evaporative resistance than those with a continuous inlay arrangement of the foam rods (BC1, BC2, and DC2). This is likely because the inlaid material arranged in a localized area provides extra resistance to inhibit the moisture from infiltrating the top surface layer to the bottom surface layer. In insole applications, it is recommended that the foam rods are inlaid in a random arrangement to allow optimal moisture evaporation and reduce the likelihood of ulceration and discomfort from humidity inside the shoes.

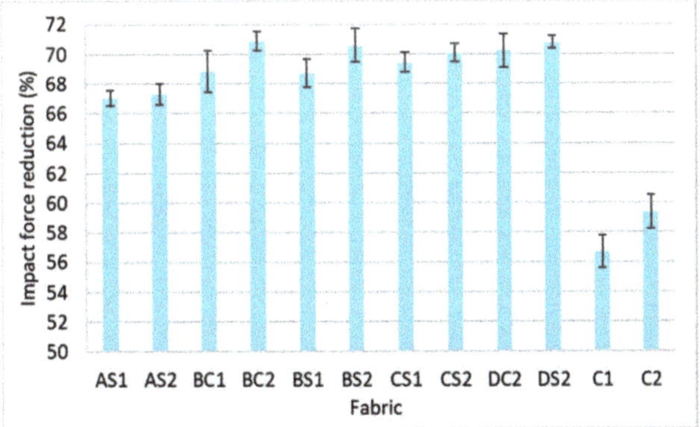


**Figure 13.** Percentage of impact force reduction of inlaid spacer fabrics and controls.

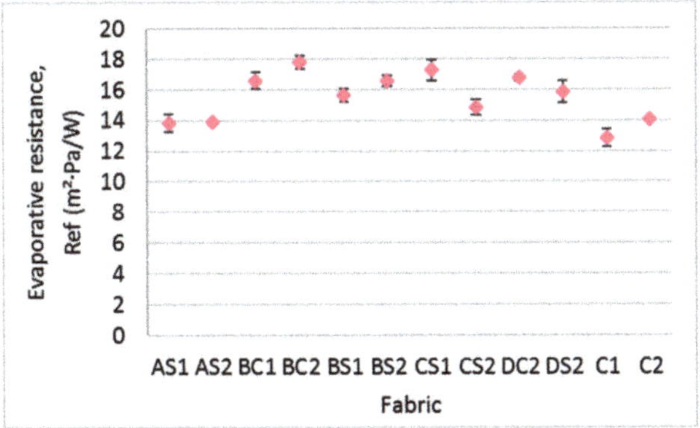


**Figure 14.** Evaporative resistance of the inlaid spacer fabrics and controls.

### 3.4.2. Thermal Resistance

The thermal properties of insole material should accommodate the environment of the end use of the insole. In cold environments, insole material with higher thermal resistance should be used to keep the wearer warm. As with air permeability, the thermal resistance of the spacer fabric samples is significantly affected by the inlay density ($F = 6.27$, $p < 0.05$, and $\eta^2 = 0.653$) and spacer yarn density ($F = 14.286$, $p < 0.005$, and $\eta^2 = 0.588$) only; however, they account for less than 66% of the variance. When comparing the fabric samples with a low spacer yarn density (AS1, BS1 and CS1) (Figure 15), the fabric sample with a higher inlay density exhibits a higher thermal resistance as the inlaid foam rods prevent air from passing through the fabric. However, for the fabric samples with a high spacer yarn density, the thermal resistance generally decreases with increased inlay density (AS2, BS2 and CS2). This is due to the reduced stitch density of the surface layers resultant of additional spacer courses. In agreement with the findings of Muthu Kumar et al. (2020), ease of passage of air is facilitated with lower stitch density which leads to lower thermal resistance [8]. Along the same line of reasoning, the fabric samples with a higher spacer yarn density (BC2, BS2 and CS2) have lower thermal resistance in comparison to BC1, BS1 and CS1.

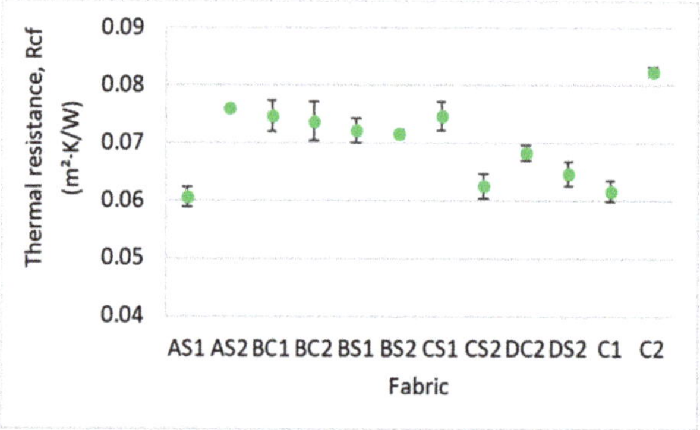


**Figure 15.** Thermal resistance of the inlaid spacer fabric samples and controls.

## 4. Conclusions

Traditional insoles are primarily designed and made of a plurality of materials with various stiffnesses, compression resistances and thermal comfort properties that correspond to the 3D shape and localised pressure at different regions of the plantar of the foot. The process of constructing such insoles is highly complex and time-consuming. A novel knitted spacer fabric structure with inlays is proposed in this study which can greatly simplify the insole fabrication process with a one-off knitting approach. A thorough understanding of the structural parameters that affect the functional performance and suitability of the fabric during practical use could have clinical significance for advancing the design process of insoles with better foot protection and wear comfort. In this study, CS2 is the optimal inlaid spacer fabric for insoles application. It can withstand high compressive stress, reduce high impact forces, air permeable and allow optimal heat and moisture evaporation. The effect of the inlay density, orientation of the inlays and spacer yarn density on the mechanical and thermal behaviours of the fabric has been systematically investigated. The following conclusions are drawn based on the experimental results.

- The high spacer yarn density of inlaid spacer fabric offers good air permeability. The size of the knitted loops on the surface layer are extended by the additional spacer yarn that allow air to flow through the material. Therefore, spacer fabric with a higher spacer yarn density has higher air permeability, and lower stitch density and fabric weight;
- For the mechanical properties, the compression resistance of the inlaid spacer fabric is largely increased by an increased number of inlays and spacer yarns while the absorbed compression energy in Stage II remains unchanged. The increased compression resistance is associated with the even distribution of forces exerted onto the flat surface of the fabric and the increased fabric stiffness. The effect of the inlay orientation on the compressive stress is less significant in this study. The fabric samples made with inlaid yarns embedded in a random orientation are stiffer than those with inlaid yarns embedded in a localized orientation;
- Meanwhile, spacer fabrics with higher inlay density and more spacer yarns can reduce higher impact forces. The inlay density is the key factor as the inlay foam rods can effectively reduce and absorb the impact forces for cushioning;
- The evaporative resistance performance of the inlay spacer fabrics is increased with inlay density and embedded in a localized orientation. On the other hand, the fabric thermal resistance is decreased with the higher inlay density for fabric with a high spacer yarn density. Samples with more spacer yarn have a lower stitch density in

the surface layers which facilitates air passage, thus leading to low thermal resistance. Therefore, samples with more spacer yarns while embedded in a random orientation allow optimal moisture evaporation and reduce thermal discomfort of humidity to the wearers.

**Author Contributions:** N.-W.L. conceived, designed, performed experiments, analysed the results, and wrote the manuscript. K.-L.Y. supervised, acquired funding, gave advice, reviewed and edited the paper writing. A.Y. gave advice, reviewed and edited the paper writing. S.N. performed experiments. All authors have read and agreed to the published version of the manuscript.

**Funding:** This research is funded by the Laboratory for Artificial Intelligence in Design (Project code: RP1-2) under the InnoHK Research Clusters, Hong Kong Special Administrative Region Government.

**Institutional Review Board Statement:** Not applicable.

**Informed Consent Statement:** Not applicable.

**Conflicts of Interest:** The authors declare no conflict of interest.

## References

1. Du, L.; Li, T.; Wu, S.; Zhu, H.F.; Zou, F.Y. Electrospun composite nanofibre fabrics containing green reduced Ag nanoparticles as an innovative type of antimicrobial insole. *RSC Adv.* **2019**, *9*, 2244–2251. [CrossRef]
2. Martinez-Santos, A.; Preece, S.; Nester, C.J. Evaluation of orthotic insoles for people with diabetes who are at-risk of first ulceration. *J. Foot Ankle Res.* **2019**, *12*, 35. [CrossRef] [PubMed]
3. Niu, J.; Liu, J.; Zheng, Y.; Ran, L.; Chang, Z. Are arch-conforming insoles a good fit for diabetic foot? Insole customized design by using finite element analysis. *Hum. Factors Ergon. Manuf. Serv. Ind.* **2020**, *30*, 303–310. [CrossRef]
4. Paton, J.; Glasser, S.; Collings, R.; Marsden, J. Getting the right balance: Insole design alters the static balance of people with diabetes and neuropathy. *J. Foot Ankle Res.* **2016**, *9*, 40. [CrossRef]
5. Cham, M.B.; Mohseni-Bandpei, M.A.; Bahramizadeh, M.; Forogh, B.; Kalbasi, S.; Biglarian, A. Effects of vibro-medical insoles with and without vibrations on balance control in diabetic patients with mild-to-moderate peripheral neuropathy. *J. Biomech.* **2020**, *103*, 109656. [CrossRef]
6. Messaoud, M.; Vaesken, A.; Aneja, A.; Schacher, L.; Adolphe, D.; Schaffhauser, J.-B.; Strehle, P. Physical and mechanical characterizations of recyclable insole product based on new 3D textile structure developed by the use of a patented vertical-lapping process. *J. Ind. Text.* **2015**, *44*, 497–512. [CrossRef]
7. Rajan, T.P.; Sundaresan, S. Thermal comfort properties of plasma-treated warp-knitted spacer fabric for the shoe insole. *J. Ind. Text.* **2020**, *49*, 1218–1232. [CrossRef]
8. Kumar, N.M.; Thilagavathi, G.; Periasamy, S. Development and characterization of warp knitted spacer fabrics for helmet comfort liner application. *J. Ind. Text.* **2020**, 152808372093921. [CrossRef]
9. Yang, Y.; Hu, H. Application of Superabsorbent Spacer Fabrics as Exuding Wound Dressing. *Polymers* **2018**, *10*, 210. [CrossRef]
10. Datta, M.K.; Behera, B.; Goyal, A. Characterization of Warp Knitted Spacer Fabric for Application in Sports Bra. *Fibers Polym.* **2019**, *20*, 1983–1991. [CrossRef]
11. Du, Z.; Wu, Y.; Wu, Y.; He, L. Determination of pressure indices to characterize the pressure-relief property of spacer fabric based on a pressure pad system. *Text. Res. J.* **2016**, *86*, 1443–1451. [CrossRef]
12. Yu, A.; Sukigara, S.; Yick, K.-L.; Li, P.-L. Novel weft-knitted spacer structure with silicone tube inlay for enhancing mechanical behavior. *Mech. Adv. Mater. Struct.* **2020**, 1–12. [CrossRef]
13. Du, Z.; Wu, Y.; Li, M.; He, L. Analysis of structure of warp-knitted spacer fabric on pressure indices. *Fibers Polym.* **2015**, *16*, 2491–2496. [CrossRef]
14. Turki, S.; Ben Abdallah, S.; Ben Abdessalem, S. Development of composite materials reinforced with flat-knitted spacer fabrics. *J. Text. Inst.* **2018**, *109*, 1315–1321. [CrossRef]
15. Zhao, T.; Long, H.; Yang, T.; Liu, Y. Cushioning properties of weft-knitted spacer fabrics. *Text. Res. J.* **2018**, *88*, 1628–1640. [CrossRef]
16. Hamedi, M.; Salimi, P.; Jamshidi, N. Improving cushioning properties of a 3D weft knitted spacer fabric in a novel design with NiTi monofilaments. *J. Ind. Text.* **2020**, *49*, 1389–1410. [CrossRef]
17. Yu, A.; Sukigara, S.; Shirakihara, M. Effect of Silicone Inlaid Materials on Reinforcing Compressive Strength of Weft-Knitted Spacer Fabric for Cushioning Applications. *Polymers* **2021**, *13*, 3645. [CrossRef] [PubMed]
18. Li, N.W.; Ho, C.P.; Yick, K.L.; Zhou, J.Y. Influence of inlaid material, yarn and knitted structure on the net buoyant force and mechanical properties of inlaid knitted fabric for buoyant swimwear. *Text. Res. J.* **2021**, *91*, 1452–1466. [CrossRef]
19. Azam, Z.; Jamshaid, H.; Nawab, Y.; Mishra, R.; Muller, M.; Choteborsky, R.; Kolar, V.; Tichy, M.; Petru, M. Influence of inlay yarn type and stacking sequence on mechanical performance of knitted uni-directional thermoplastic composite prepregs. *J. Ind. Text.* **2020**, 152808372094772. [CrossRef]

20. Siddique, H.F.; Mazari, A.A.; Havelka, A.; Mansoor, T.; Ali, A.; Azeem, M. Development of V-Shaped Compression Socks on Conventional Socks Knitting Machine. *Autex Res. J.* **2018**, *18*, 377–384. [CrossRef]
21. Shabaridharan; Das, A. Study on heat and moisture vapour transmission characteristics through multilayered fabric ensembles. *Fibers Polym.* **2012**, *13*, 522–528. [CrossRef]
22. Oswald, C.; Denhartog, E. Transient heat loss analysis of fabrics using a dynamic sweating guarded hot plate protocol. *Text. Res. J.* **2020**, *90*, 1130–1140. [CrossRef]
23. Bagherzadeh, R.; Latifi, M.; Najar, S.S.; Tehran, M.A.; Gorji, M.; Kong, L. Transport properties of multi-layer fabric based on electrospun nanofiber mats as a breathable barrier textile material. *Text. Res. J.* **2011**, *82*, 70–76. [CrossRef]
24. Li, N.W.; Yick, K.L.; Yu, A. Novel weft-knitted spacer structure with silicone tube and foam inlays for cushioning insoles. *J. Ind. Text.* **2022**. [CrossRef]
25. Zhang, X.; Ma, P. Compression fatigue resistance of three-dimensional warp-knitted spacer structure for car cushion. *Fibers Polym.* **2017**, *18*, 605–610. [CrossRef]
26. Lo, W.T.; Yick, K.L.; Ng, S.P.; Yip, J. New methods for evaluating physical and thermal comfort properties of orthotic materials used in insoles for patients with diabetes. *J. Rehabil. Res. Dev.* **2014**, *51*, 311–324. [CrossRef] [PubMed]
27. Erdumlu, N.; Saricam, C. Investigating the effect of some fabric parameters on the thermal comfort properties of flat knitted acrylic fabrics for winter wear. *Text. Res. J.* **2017**, *87*, 1349–1359. [CrossRef]

polymers

MDPI

Article

# Ultrathin Multilayer Textile Structure with Enhanced EMI Shielding and Air-Permeable Properties

**Shi Hu [1,*], Dan Wang [1], Aravin Prince Periyasamy [1,2], Dana Kremenakova [1], Jiri Militky [1] and Maros Tunak [3]**

1 Department of Material Engineering, Faculty of Textile Engineering, Technical University of Liberec, Studenska, 1402/2, 46117 Liberec, Czech Republic; dan.wang@tul.cz (D.W.); aravinprince@gmail.com (A.P.P.); dana.kremenakova@tul.cz (D.K.); jiri.militky@tul.cz (J.M.)

2 Department of Bioproducts and Biosystems, School of Chemical Engineering, Aalto University, 02150 Espoo, Finland

3 Department of Textile Evaluation, Faculty of Textile Engineering, Technical University of Liberec, Studenska, 1402/2, 46117 Liberec, Czech Republic; maros.tunak@tul.cz

* Correspondence: shi.hu@tul.cz

**Abstract:** A textile material's electromagnetic interference (EMI) shielding effectiveness mainly depends on the material's electrical conductivity and porosity. Enhancing the conductivity of the material surface can effectively improve the electromagnetic shielding effectiveness. However, the use of highly conductive materials increases production cost, and limits the enhancement of electromagnetic shielding effectiveness. This work aims to improve the EMI shielding effectiveness (EMSE) by using an ultrathin multilayer structure and the air-permeable textile MEFTEX. MEFTEX is a copper-coated non-woven ultrathin fabric. The single-layer MEFTEX SE test results show that the higher its mass per unit area (MEFTEX 30), the better its SE property between 56.14 dB and 62.53 dB in the frequency band 30 MHz–1.5 GHz. Through comparative testing of three groups samples, a higher electromagnetic shielding effect is obtained via multilayer structures due to the increase in thickness and decrease of volume electrical resistivity. Compared to a single layer, the EMI shielding effectiveness of five layers of MEFTEX increases by 44.27–83.8%. Due to its ultrathin and porous structure, and considering the balance from porosity and SE, MEFTEX 10 with three to four layers can still maintain air permeability from 2942 $L/m^2/s$–3658 $L/m^2/s$.

**Keywords:** electromagnetic shielding effectiveness; multilayer structure; porosity; air permeability

**Citation:** Hu, S.; Wang, D.; Periyasamy, A.P.; Kremenakova, D.; Militky, J.; Tunak, M. Ultrathin Multilayer Textile Structure with Enhanced EMI Shielding and Air-Permeable Properties. *Polymers* **2021**, *13*, 4176. https://doi.org/10.3390/polym13234176

Academic Editor: Alexey V. Lyulin

Received: 22 October 2021
Accepted: 24 November 2021
Published: 29 November 2021



## 1. Introduction

The influence of electromagnetic waves on human health is contentious, but to date, scientific consensus has been that excessive electromagnetic radiation impacts human health [1,2]. In addition, radiation due to electromagnetic effects causes unnecessary electromagnetic interference [3]. The electric field in radiation can interact with the electrons in the metal conductors, and this interference can cause some sensitive electronics to malfunction [4]. To prevent the extra electromagnetic wave, due to the electrostatic equipotentiality of metals, electromagnetic interference (EMI) shielding material can effectively shield external electric fields from electromagnetic interference. The setup of EMI shielding protective clothing should comply with the minimum health and safety requirements regarding the exposure of workers to the risks arising from physical agents (18th individual directive within the meaning of Article 16(1) of Directive 89/391/EEC), which require the employee to use suitable protective equipment to work in high electromagnetic radiation environments [5]. The research on metal-coated textiles has proven the effectiveness of EMI shielding in many applications. Compared to other EMI shielding materials like metal, carbon-polymer composites, and nanofibrils, conductive textiles not only provide effective EMI shielding due to the textile-based structure of metal-coated textile material, but also have good wearable properties, such as air permeability and thermal properties [6].

Three phenomena determine how electromagnetic field strength is lost as it interacts the shielding objects. These phenomena are absorption, attenuation due to reflection, and attenuation due to internal reflection [7]. The primary mechanisms of EMI shielding material are reflection and absorption [8]. The loss due to the multiple reflections usually can be ignored due to the distance between the reflecting surfaces, or the interface is large compared to the skin depth [9]. When a textile material has a low thickness, the primary mechanism for EMI shielding metal-coated textiles is reflection. In this case, to enhance the EMI shielding effectiveness (SE) of textile shielding clothing, the most common method is coating the low electrical resistance metals to enhance the material conductivity, for example, by using silver or gold as coating metal. This concept is not widely implemented, which may be because these metals are too expensive for large-scale industrial production and the consumer market, especially for technical clothing use [10,11]. The MXene/metal oxides nanostructured coating for textile EMI shielding is being quickly developed [12]. Regarding nano-sized metal or heavy metal coated textile, the human body's health risks are still of concern [13].

As well as metal conductivity, the porosity and thickness of EMI shielding textiles significantly influence electromagnetic shielding effectiveness. The test results show that, by using aluminum foils with pores after increasing the pore size from 1 mm to 3 mm, the SE reduced by 20–37.5%. Similar results were observed from different pore sizes of EMI shielding textiles [14]. Considering the influence of thickness after increasing the fabrics' thicknesses by using the same material, and the SE improved by 5–10 dB for each increased fabric on different frequencies [3]. Researcher S. Palanisamy studied the influence factor of textile SE. From the findings of the design of experiment (DoE) screening design, the main influence factors are the thickness of the materials, the apertures, and the strips laid angle, which have a statistically significant effect on electromagnetic shielding effectiveness [15].

Generally, compared to other EMI shielding materials such as nanofibrous membranes, porous light fabric with high SE shows that there is promise for technical textiles to achieve the balance of high air permeability and high SE with low thickness. For this reason, a particular type of composite nonwoven polyester fabric MILIFE® with a surface covering by a particulate-based copper layer produced by Bochemie Ltd., Bohumin, Czech Republic, named MEFTEX® used in this research.

The purpose of this study is to investigate if the EMI shielding fabrics by using multilayer combination and mutually aligned with each other, fulfills the expectation of increased SE in the frequency band from 30 MHz to 1.5 GHz, which is specified in ASTM D 4935-10 [16], simultaneously ensure a certain degree of air permeability.

## 2. Materials and Methods

### 2.1. Material Geometric and Properties

The copper-coated nonwoven polyester fabric, named MEFTEX, is purchased from Bochemie Ltd., Bohumin, Czech Republic. The greige fabrics are nonwoven polyester-based fabrics (commercial name MILIFE) from JX Nippon ANCI, Tokyo, Japan. MILIFE is composed of a dense net of monofilaments bonded by solid spots from locally melted filaments, making it a potentially useful material for surface metallization. Three different mass per unit area (GSM; w) [$g/m^2$] of greige MILIFE fabric were used for copper coating, 10 $g/m^2$, 20 $g/m^2$, and 30 $g/m^2$. The fabric mass per unit area [$g/m^2$] was measured using the standard ASTM D 377633 [17] and the sample size was 600 $cm^2$. Fabric thickness (t) was measured using a thickness gauge [mm], as per standard ASTM D 5729 [18] (nonwoven samples), and pressure of 1 kPa was used. A total of 10 times measurements were performed; the mean values and 95% confidence intervals for the means are summarized in Table 1.

### 2.2. Metal Coating

MEFTEX is produced via a unique roll-to-roll technological process (patent pending) based on subsequent chemical and electroplating processes. The surface activation process was carried out using an activation solution (CATAPOSIT44, supplied by Rohm and Hass company, Hoek, The Netherlands) at 45 °C for 5 min, then immersed in 10% hydrochloric acid for 1 min. Pre-activated MILIFE fabric is passing through a bath containing salt $CuSO_4$ and copper nanoparticles on fiber surface obtained by the action of reducing bath (based on borohydride). By this process, a fragile metal particles dense layer (copper) was created on MILIFE fabrics' surface (see Figure 1).

**Table 1.** Characteristics of the samples.

| Sample Number | Sample Details | Thickness (1 Layer) (mm) | Mass Per Unit Area (1 Layer) (g/m$^2$) | Deposit of Cu Per Unit Area (g/m$^2$) |
|---|---|---|---|---|
| MEFTEX 10 | 100% PET nonwoven | 0.042 ± 0.11 | 11.84 | 1.84 |
| MEFTEX 20 | 100% PET nonwoven | 0.074 ± 0.008 | 24.01 | 4.01 |
| MEFTEX 30 | 100% PET nonwoven | 0.112 ± 0.01 | 41.67 | 11.67 |

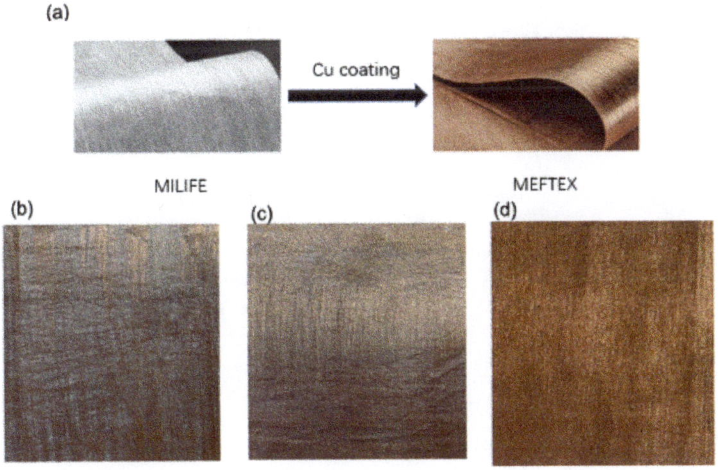


**Figure 1.** Sample appearance image (**a**) Copper coated MEFTEX (**b**) MEFTEX 10 (20 cm × 20 cm) (**c**) MEFTEX 20 (20 cm × 20 cm) (**d**) MEFTEX 30 (20 cm × 20 cm).

### 2.3. *Characterization*

#### 2.3.1. SEM and EDX

The surface morphology of the MEFTEX fabric was observed under the scanning electron microscopy (SEM) (VEGA TESCAN Inc. Brno, Czech Republic) at 20 kV. The cross-section and the surface of the Cu-coated fabric were observed. Elemental analysis was performed on the metal composition on the surface of MEFTEX (energy-dispersive X-ray spectroscopy-EDX) using an Oxford Instruments analyzer and AZTEC software Version 3.2.

#### 2.3.2. EMSE Measurements

The effectiveness of the electromagnetic shielding of textiles was evaluated by the insertion loss method according to ASTM 4935-10, which is designed for the evaluation of flat materials. This standard assumes a plane wave impacts on a shielding barrier in the near zone of the electromagnetic field at a frequency of 30 MHz to 1.5 GHz. The measuring fixture consisted of a coaxial specimen holder (manufactured by Electro-Metrics, Inc., model

EM-2107A), whose input and output were connected to a perimeter analyzer. A Rhode & Schwarz ZNC3 circuit analyzer was used to generate and receive an electromagnetic signal. The test principle of this method is illustrated in Figure 2. The *SE* can be interpreted by the forward transmission coefficient *S21*, which is the ratio of power without and with shielding material; the calculation method is as Equation (1):

$$SE(S21) = -10log\frac{P_1}{P_2} = 10log\frac{P_2}{P_1} \quad (1)$$

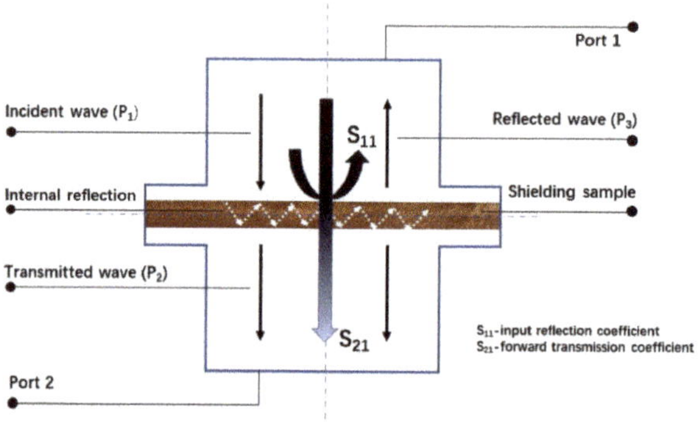


**Figure 2.** Measurement of SE for MEFTEX.

P1 is the received power without the fabric present, and P2 is the received power with the fabric present. To interpret Equation (1), by the sign "-", the higher the value of SE, the smaller the number should be that is preceded by "-." To perform a straightforward inference, in the following, the result of the SE value is presented without the sign "-" [19]. The electromagnetic wave reflection coefficient interpreting the electromagnetic wave signal from the transmitting antenna is reflected by the sample and received by Port 1. The ratio of the receiving reflected power ($P_3$) and the received power without the fabric present ($P_1$) calculates the input reflection coefficient by Equation (2):

$$S11 = 10log\frac{P_3}{P_1} \quad (2)$$

The measurements were performed under the following laboratory conditions: T = 23.9 °C ± 2 °C, RH = 48% ± 5%. Each sample was measured five times at different locations.

#### 2.3.3. Volume Resistivity Test

The volume resistivity $\rho_v$ [Ω mm] of three groups of the sample was calculated according to standard ASTM D257-14 [20] at a temperature 23.2 ± 2 °C and a relative humidity RH of 50.7% ± 5%. The electrodes were connected with a 100 V direct current (DC) power supply. The sample was placed in an air-conditioned room for 24 h before testing. Volume resistivity $\rho_v$ [Ω mm] was then calculated from Equation (3):

$$\rho_v = R_v(\frac{S}{t}) \quad (3)$$

$R_v$ is the reading of volume resistance measurement, *t* [m] is the material thickness and *S* [m$^2$] is the surface area of measurement electrodes.

#### 2.3.4. Porosity (Optical) Analysis and Air Permeability

A fabric's porosity has a strong positive correlation with its air permeability, which is one of the essential wearing comforts for evaluating fabrics. One characteristic of the distribution of a nonwoven fabric's porosity is unevenness. In image analysis, a nonwoven fabric's porosity can be expressed by the optical porosity, which is a percentage value calculated from the aperture area of the observation part divided by area of the observation part. For the porosity investigation, images of MEFTEX were obtained using a Nikon Eclipse E200 microscope in transmitted light. Images were captured as RGB image matrices of size 1200 × 1600 px. All images were captured using the same microscope. Following image analysis, for better visual representation, the original picture was cropped to 1200 × 1200 px after adjusting the intensity to size 100 × 100 px (0.64 × 0.64 mm), converting this image into a binary image. Finally, by using the MATLAB program processing this image, each grid porosity of 12 × 12 mesh structured image was generated (see Figure 3). The numbers in each grid represent the porosity, and porosity distribution can be expressed via box plot and observed via the whole generated image.

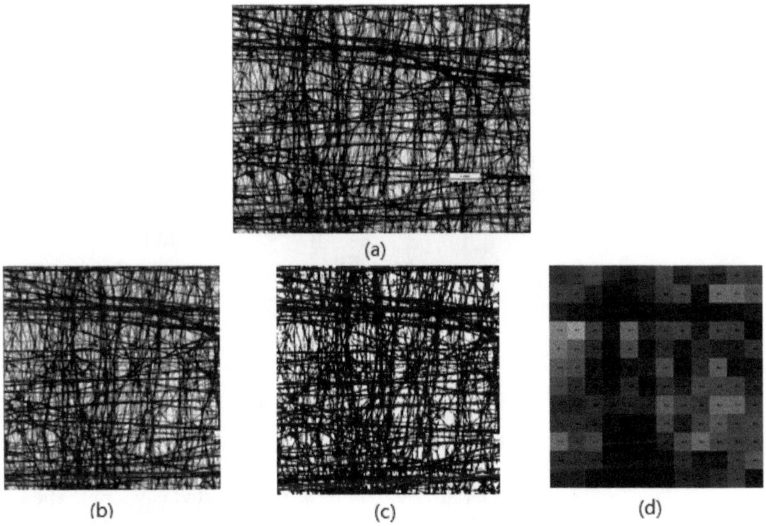


**Figure 3.** (**a**) Original image of sample MEFTEX 10 (1200 × 1600 px; 7.69 mm × 10.26 mm), (**b**) image crop and sub windows of size 100 × 100 px (0.64 mm × 0.64 mm), (**c**) binary image (**d**) map of porosity.

An essential factor in the comfort of fabric is air permeability (AP), which is influenced by its porosity. An AP tester (FX 3300, TEXTEST Instruments, Hertogenbosch, The Netherlands) was used at 200 Pa to conduct the AP test according to the ISO 9237 standard [21]. The AP was measured under the laboratory conditions T = 21.3 °C ± 2 °C, RH = 50% ± 5%. To test air permeability, each of three different samples was measured ten times.

## 3. Results and Discussion

### 3.1. Morphology Analysis and Elemental Analysis

The structure of single-layer MEFTEX 10, MEFTEX 20, and MEFTEX 30 was observed from SEM images (Figure 4). From the image, it is clear the porous structure and critical point of the nonwoven fabric MILIFE. MILIFE is composed of a dense net of monofilaments from which critical points produce locally melted filaments, making it a promising material for surface metallization. To highlight the unique processing technology of MILIFE, the porous structure is presented in the image. The porous structure decreases with the increasing mass per unit area of MILIFE nonwoven fabrics. MEFTEX 20 and MEFTEX 30's arrangement of fibers are dense in the same area, while MEFTEX 10 has more overlapping fibers. This is the reason MEFTEX 20 and MEFTEX 30 have higher thickness and areal density. Figure 4d,e shows the surface morphologies of MEFTEX. It is apparent from the micrographs that a thin copper layer can be clearly distinguished on the MEFTEX surface.

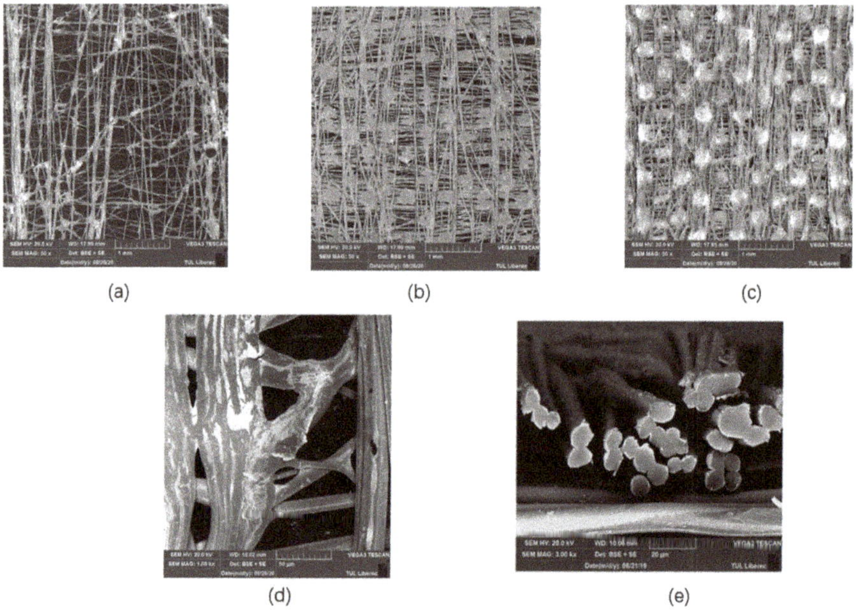

**Figure 4.** Scanning electron micrographs images of (**a**) MEFTEX 10 (50×) (**b**) MEFTEX 20 (50×) and (**c**) MEFTEX 30 (50×) (**d**) Meftex surface coated with copper micro view (**e**) A cross section view of MEFTEX.

Figure 5 displays the EDX analysis results of Cu deposition elements in the area of 450 μm × 450 μm. Five times measurement were applied; Table 2 displays the mean value and standard deviation for the detected elements. The high proportion of copper with high surface density is visible on the surface of each sample. In addition to Cu, oxygen (O), carbon (C), calcium (Ca), and titanium (Ti) are visible, and their concentration is presented in Figure 5. The presence of titanium is due to the matting of polyester by $TiO_2$ before creating a MILIFE structure [22]. It is observed that the copper coating is not continuous, but distances between coated parts are over the percolation threshold. The electromagnetic SE, in particular, is extraordinarily high.

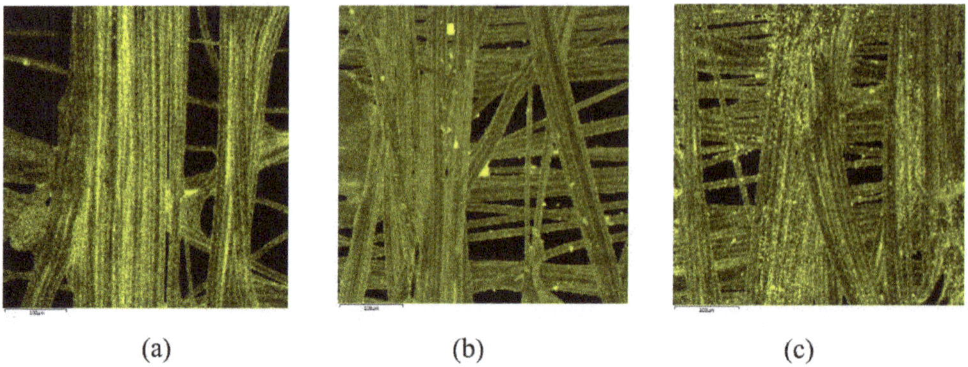

**Figure 5.** X-ray spectroscopy–EDX analysis of (**a**) MEFTEX 10 (**b**) MEFTEX 20 (**c**) MEFTEX 30 element Cu distribution.

**Table 2.** EDX analysis of MEFTEX surface.

| Sample Number | Element | Wt [%] | Standard Deviation | Element | Wt [%] | Standard Deviation |
|---|---|---|---|---|---|---|
| Meftex 10 | Cu | 75.87 | 2.12 | C | 16.32 | 1.03 |
| | O | 7.28 | 0.87 | Ti | 0.36 | 0.08 |
| | Ca | 0.16 | 0.32 | | | |
| Meftex 20 | Cu | 45.2 | 1.21 | C | 33.49 | 1.09 |
| | O | 20.18 | 0.38 | Ti | 0.47 | 0.04 |
| | Ca | 0.67 | 0.15 | | | |
| Meftex 30 | Cu | 63.52 | 2.01 | C | 23.94 | 0.9 |
| | O | 11.94 | 1.08 | Ti | 0.35 | 0.02 |
| | Ca | 0.24 | 0.14 | | | |

### 3.2. Single Layer EMSE of MEFTEX

The results of single-layer MEFTEX's EMI shielding effectiveness are displayed in Figure 6. Figure 6 shows that the MEFTEX 10 reported the lowest SE from 42.5 dB to 48.2 dB compared to other samples in the frequency band 30 MHz–1.5 GHz. MEFTEX 30 provides higher SE between 56.14 dB and 62.53 dB in the frequency band 30 MHz–1.5 GHz. According to the results, the SE of MEFTEX with a higher mass per unit area of 30 $g/m^2$ (sample MEFTEX 30) performs better EMI shielding due to a larger amount of metalized fibers than MEFTEX with a basis weight of 10 $g/m^2$ or 20 $g/m^2$ (samples MEFTEX 10, MEFTEX 20). All samples of single-layer MEFTEX, according to the classification, evaluated in the "Excellent" category for Class II general use. For Class I Professional use, MEFTEX 10 fulfills the SE requirements of grade "AAA" (Good), while MEFTEX 20 and MEFTEX 30 fulfill the SE requirements for between a "AAAA" (Very good) and "AAAAA" (Excellent) grade [23] (see Table A1).

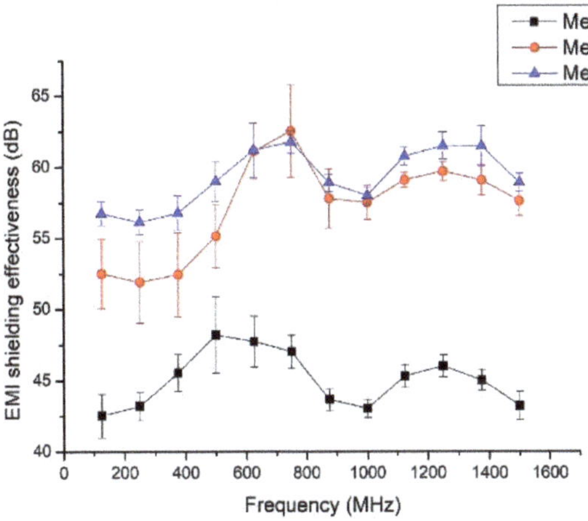


**Figure 6.** MEFTEX EMI shielding effectiveness in the frequency band from 30 MHz–1.5 GHz.

For conductor plate shielding materials, $SE$ can be calculated by the Schelkunoff equation (Equation (4)) based on the transmission theory, with $SE_A$ [dB] being the absorbing loss of the shielding materials, $SE_R$ [dB] the single reflection loss on the surface of the shielding materials, and $SE_M$ [dB] the multi-reflection loss inside the shielding materials [24].

$$SE = SE_A + SE_R + SE_M \quad (4)$$

The electromagnetic shielding efficiency of an element is characterized by its electric conductivity, permittivity, permeability, parameters of source, and properties of the ambient surroundings [9]. For EMI shielding materials with an $SE_A$ of more than 6 dB, the multi-reflection loss inside the shielding material can be ignored. When $SE_A$ is less than 10 dB, the correction term $SE_M$ should be considered. The multiple reflection correction term can be calculated via the following equation [25,26]:

$$SE_M = 20log_{10}(1 - \frac{(C-1)^2}{(C+1)^2}10^{-SE_A/10}) \tag{5}$$

where $C = Z_S/Z_H$, $Z_s$ is the shield impedance and $Z_H$ is the impedance of the incident magnetic field.

According to White's model, the *SE* can be explained by Equation (6) [26,27]:

$$SE = 168 - 10log(\frac{K_c f}{K}) + 1.315t\sqrt{\frac{fK}{K_c}} \tag{6}$$

where *Kc* [S/cm] is the copper conductivity (5.82 × 105 S/cm), *f* [MHz] is the frequency, *t* [cm] is the thickness, and *K* [S/cm] is the volume conductivity of the conductive material, which can be calculated via volume electrical resistance via Equation (7):

$$K = \frac{1}{\rho_v} \tag{7}$$

Equation (6) can be re-written as:

$$SE = 168 - 10log(K_c f \rho_v) + 1.315t\sqrt{\frac{f}{\rho_v K_c}} \tag{8}$$

The volume electrical resistivity and *SE* at 1.5 Hz frequency for all samples are listed in Table 3. To evaluate the results of different samples, the SE obtained at 1.5 GHz frequency was used. This frequency was found to be important because it is close to the frequency used by many working devices (e.g., cell phones, GPS, and Wi-Fi routers) [26]. The results are very close to the experiment results at 1.5 GHz, especially for MEFTEX 20 and MEFTEX 30. It was clear that the experimental SE value of the single-layer MEFTEX 10 differs greatly from the theoretical value. This is the reason for providing the theoretical value even though it is not designed and derived for the ultrathin fabrics used in this work. The thickness of MEFTEX 10 is significantly less (0.042 ± 0.11 mm; cf. Table 1) than the other fabrics used in this work, which contributes to the great difference in the theoretical EM SE at 1.5 GHz (cf. Table 3).

**Table 3.** Volume electrical resistivity and EMSE, $SE_R$, $SE_A$, $SE_M$ at 1.5 GHz.

| Sample Number | Thickness [mm] | Volume Electrical Resistivity [Ω·mm] | The Experiment Result of EM SE on 1.5 GHz [dB] | $SE_R$ [dB] | $SE_A$ [dB] | $SE_M$ [dB] | The Theoretical Calculated Result of EM SE on 1.5 GHz [dB] |
|---|---|---|---|---|---|---|---|
| MEFTEX 10 | 0.042 | 5022.78 | 45.2 | 36.79 | 8.41 | −1.47 | 51.58 |
| MEFTEX 20 | 0.074 | 1676.40 | 57.64 | 44.71 | 12.93 | - | 56.34 |
| MEFTEX 30 | 0.112 | 991.50 | 58.91 | 43.76 | 15.14 | - | 58.62 |

### 3.3. Multilayer MEFTEX SE and SEG

Adding a layer of MEFTEX caused SE to significantly increase. These results can be observed in Figure 7. Compared to single-layer MEFTEX, the multilayer MEFTEX provides significantly increased EMI shielding properties. After calculating the average SE in the frequency band from 30 MHz to 1.5 GHz, the SE of five layers of MEFTEX 10 (10-5) increased 83.3% compared to one-layer MEFTEX 10 (10-1). Five-layer MEFTEX 20 (20-5) and five-layer MEFTEX 30 (30-5) improved the electromagnetic shielding effect by 49.13% and 44.27% respectively, in comparison to one-layer MEFTEX 10. To explain this result, Equation (8) clarifies that, by increasing thickness and decreasing volume electrical resistivity, the SE can be increased. The test results reinforce this. By increasing layers, the thickness is increased and the volume electrical resistance decreases, as shown in Table 4.

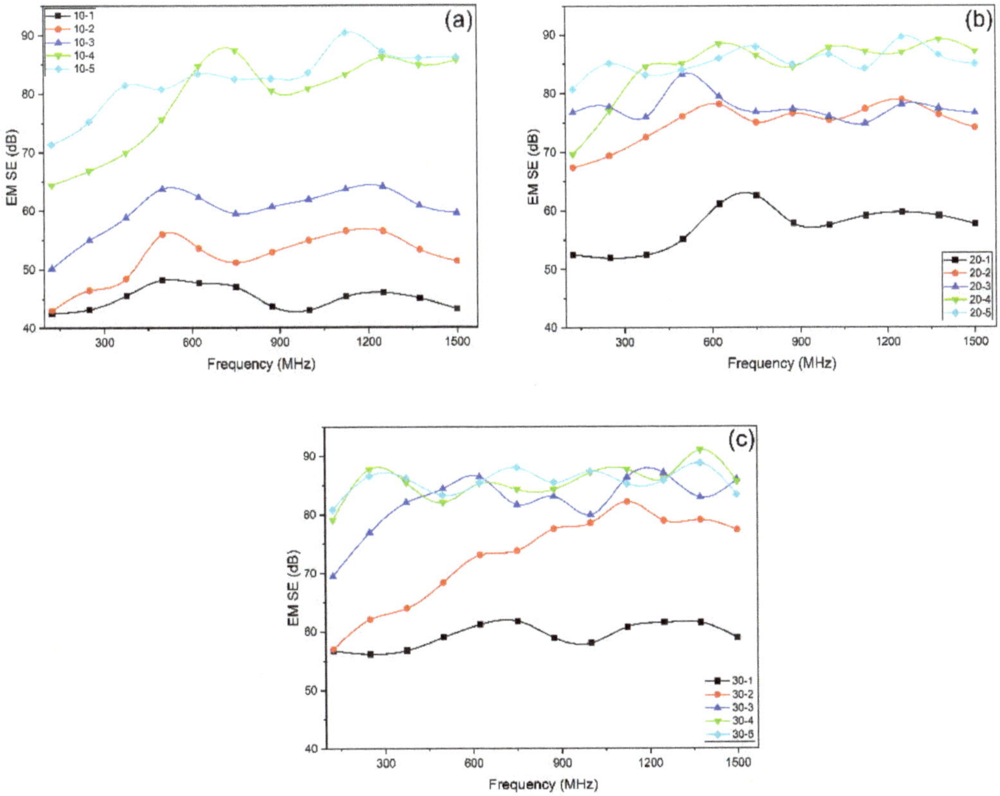


**Figure 7.** EMI shielding effectiveness (EMSE) of increasing layers for MEFTEX. (**a**) MEFTEX 10 (**b**) MEFTEX 20 (**c**) MEFTEX 30.

**Table 4.** Volume electrical resistivity change and thickness change with increasing of layers.

| Sample Code * | Volume Electrical Resistivity [Ω·mm] | Sample Code * | Volume Electrical Resistivity [Ω·mm] | Sample Code * | Volume Electrical Resistivity [Ω·mm] |
|---|---|---|---|---|---|
| 10-1 | 5022.78 | 20-1 | 1676.40 | 30-1 | 991.50 |
| 10-2 | 1640.93 | 20-2 | 663.40 | 30-2 | 430.43 |
| 10-3 | 963.80 | 20-3 | 433.67 | 30-3 | 297.05 |
| 10-4 | 126.70 | 20-4 | 325.73 | 30-4 | 248.19 |
| 10-5 | 60.04 | 20-5 | 249.12 | 30-5 | 204.11 |

* 10-1: 1 layer of MEFTEX 10, 10-2: 2 layers of MEFTEX 10. Same rule for other samples.

For multilayer MEFTEX, the *SE* at 1.5 GHz was always above 40 dB, which is classifies it in the "Excellent" category for professional use. The maximum SE at 1.5 GHz can be reached at 87.14 dB by five layers of MEFTEX 20. This performance can be classified as an "Excellent" grade. From one layer to two layers, the SE increase rate compared to the previous layer was from 18.91% to 31.26%, two layers to 3 layers are between 3.37% and 15.98%, and from 3 layers to 4 layers is 2.67–43.64%. However, the increasing rate from the fourth layer to the fifth layer is only 0.37–2.47%.

To explain this result, the previously described Schelkunoff equation (Equation (4)) based on the transmission theory was used.

$$SE_R = 106 + 10lg(\delta_r / fu_r) \tag{9}$$

$$SE_A = 131.43t\sqrt{fu_r\delta_r} \tag{10}$$

where $f$ is frequency [Hz], $u_r$ is the permeability of material relative to copper, $\delta_r$ is the electrical conductivity of the material relative to copper, and t is the thickness [m] [27].

For multilayer structured MEFTEX, the reflection loss $SE_R$ is relatively constant compared to a single layer (Equation (9)). When the number of layers is increased, the absorbing loss $SE_A$ increases due to the change in thickness (Equation (10)). However, the transmitted EM wave will decrease due to the decreasing porosity. When the thickness increases from the fourth layer to the fifth layer, the number of pores that electromagnetic waves can penetrate slightly reduces. Therefore, when the number of layers increases, the increase rate of the shielding effectiveness compared to the previous layer will decrease.

For different applications of EMI shielding materials, the weight of the electromagnetic shield material is significant as well. In these cases, the specific electromagnetic EMSE, *SEG* [dB·m$^2$/g] can be calculated according to the following equation:

$$SEG = \frac{SE}{w} \tag{11}$$

where $w$ (g/m$^2$) is the planar mass.

The material with a higher *SEG* is desirable. It is often possible to make the shield thicker for a higher shielding ability. For the multilayer MEFTEX, the SEG decreases with the increasing of layers (see Figure 8). The SEG value for aluminum foil is 1.42 (*SE* = 78 dB, w = 55 g/m$^2$), which was the best material tested in reference [28,29]. A single layer of MEFTEX with all specifications provided higher *SEG* than aluminum foil, and multilayer MEFTEX 10 with up to four layers provided higher *SEG* than 1.42. This is because of the ultra-thin characteristics and high porosity of MEFTEX 10. For the statistical approach, the *SE* and planner mass can be fitted using the exponential function, and the final fitting model had R2 > 0.9 for all three samples, which confidently proves an exponential relationship between mass per unit area and *SE*. In contrast, the increase of planner mass via layer increasing is linear. The difference of increasing trend between mass per unit area and *SE* causes *SEG* to decrease as layers are added.

### 3.4. Air Permeability Is Influenced by the Multilayer Structure

As the number of layers increases, the porosity of the nonwoven structured sample decreased. Using the MATLAB® image analysis program, the optical porosity influenced by the layer layout is presented via the distribution of porosity in boxplot Figure 9, which also proves the aforementioned conclusion. The total volume of void space within a specified area of the MEFTEX 30 fabric is smaller than the other fabrics. For this reason, MEFTEX 30 has the lowest porosity (cf. Figure 4a–c).

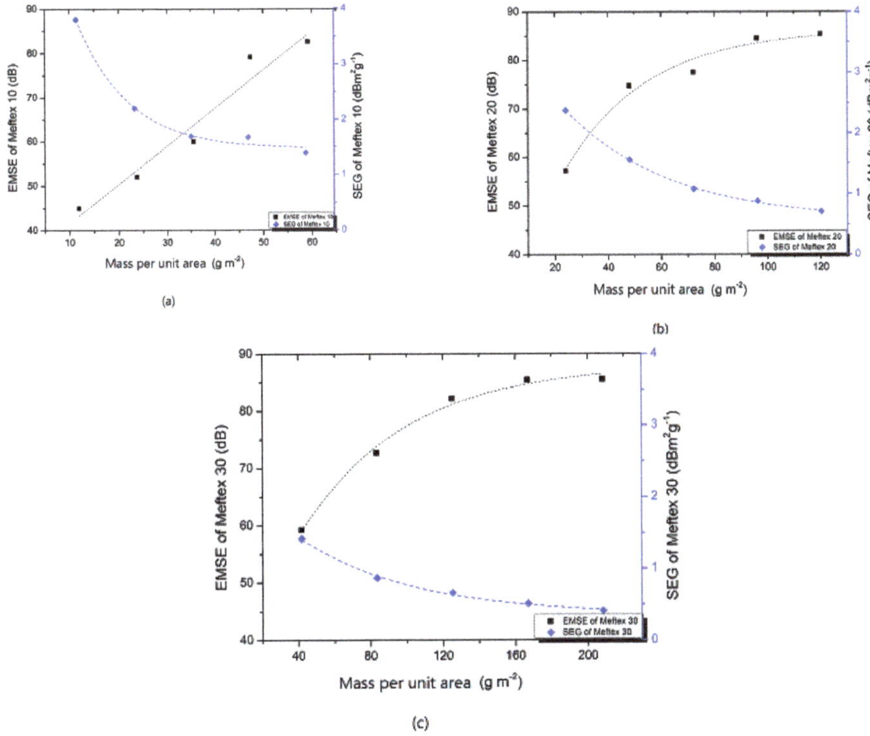


**Figure 8.** SEG value and relationship between SE and planner mass (**a**) MEFTEX 10 (**b**) MEFTEX 20 (**c**) MEFTEX 30.

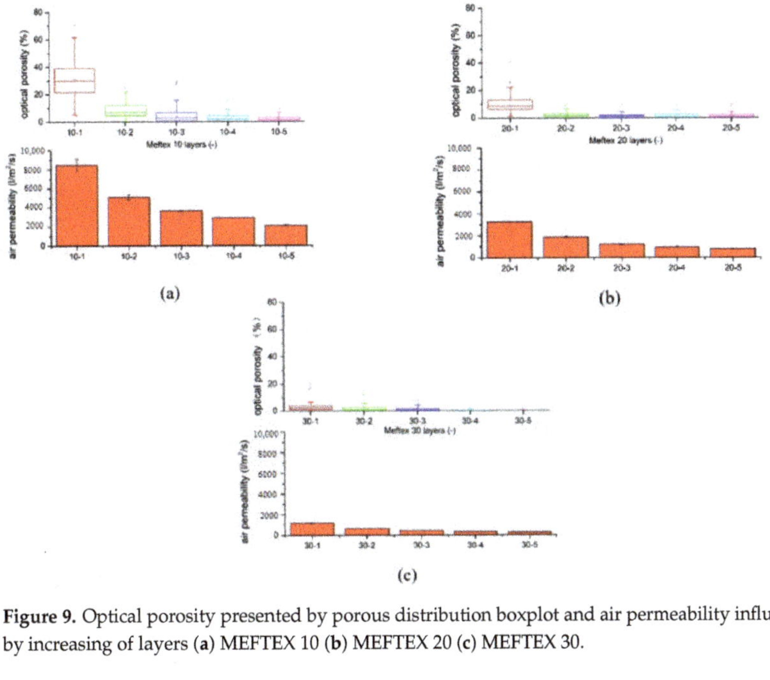


**Figure 9.** Optical porosity presented by porous distribution boxplot and air permeability influenced by increasing of layers (**a**) MEFTEX 10 (**b**) MEFTEX 20 (**c**) MEFTEX 30.

As we know, porosity has a significant influence on the air permeability of textile material. The diameter of pores influences the air permeability of EMI shielding textiles and affects their SE. In the case of metal foil, the SE can be calculated by the following empirical equation (Equation (12)) [28]:

$$SE = 20lg(\frac{1}{T_t + T_h}) = 20lg(\frac{1}{T_t + 4n(q/F)^{\frac{3}{2}}}) \quad (12)$$

$T_t$ is the transmission coefficient of the total shielding metal, $T_h$ is the transmission coefficient of the pores on the shielding metal foil, $n$ is the number of pores, $q$ is the area of each pore, and $F$ is the total area of the shielding metal foil.

According to this equation, if the metal foil pores are getting larger, the SE becomes lower. This conclusion may indicate a contradiction for air permeability and SE. To illustrate this problem, the relationship of the air permeability of multilayered MEFTEX and its SE (average SE from frequency 30 MHz to 1.5 GHz) is presented in Figure 10. All three groups of MEFTEX indicate that with increasing air permeability, the SE decreases, creating an exponential relationship (R2 > 0.9). The increase in air permeability can be seen as a decrease in sample thickness. By combining the conclusions from Equations (8) and (10), it can be inferred that when the number of layers is increased, the sample's porosity will decrease. These two factors can increase SE.

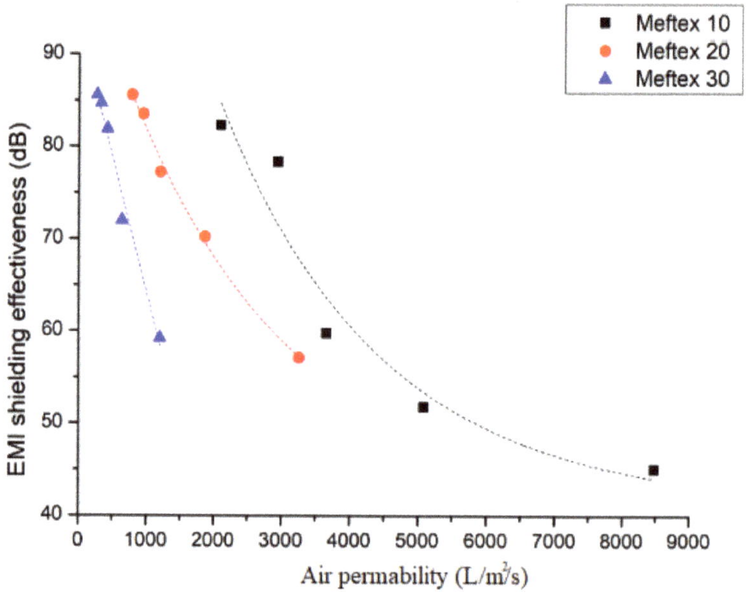


**Figure 10.** The relationship between air permeability and EMSE of MEFTEX.

There is one point that should be noticed. Three layers of MEFTEX 10 have a similar thickness, mass per unit area (cf. Table 1), and volume resistivity (cf. Table 3) compared to one-layer MEFTEX 30. The same principle applies when comparing two layers of MEFTEX to one layer of MEFTEX 20, indicating a similar EMI shielding property. The reason is that, for the same material, increasing the number of layers at the same time causes the volume resistivity will decrease. When the thickness and resistivity are similar, the SE performance should not be significantly different, which has been proven from the SE test result. (cf. Figure 7) However, the air permeability of three-layer MEFTEX 10 is much better than one-layer MEFTEX 30. The same applies when comparing two layers of MEFTEX 10 to one-layer MEFTEX 20 and four layers of MEFTEX 10 to two layers of MEFTEX 20. The difference in air permeability is caused by the difference in porosity, which can be observed via optical porosity. The optical porosity of 3 layers of MEFTEX 10 is higher than one-layer MEFTEX 30. The mass per unit area and thickness of one layer of MEFTEX 30 and three layers of MEFTEX 10 are similar, but the fiber arrangement of MEFTEX 10 and MEFTEX 30 can vary. It cannot simply be interpreted that the porosity of three layers of MEFTEX 10 is similar to one layer of MEFTEX 30, but it can be inferred from this research.

One advantage of copper-coated material is combining the suitable EMI shielding property and textile wearable properties, including air permeability. In this study, a higher mass per unit area fabric (MEFTEX 30) provided better SE. However, on the same SE level, the lower mass per unit area fabric (MEFTEX 10) shows better air permeability. Considering the balance of SE and air permeability for MEFTEX (Figure 10), three to four layers of MEFTEX 10 will perform average SE of 59.79 dB–78.38 dB in the frequency band 30 MHz to 1.5 GHz, and average air permeability of 2942 $L/m^2/s$–3658$L/m^2/s$ [30].

## 4. Conclusions

In this research, three different mass per unit area EMI shielding materials, MEFTEX 10, MEFTEX 20, MEFTEX 30, were studied to determine whether multilayered structure influences SE and air permeability. Observed by SEM and EDX, a uniform copper coating is plated on the surface of the polyester fiber, which gives the material excellent electromagnetic shielding performance and maintains the porosity of the base material itself. MEFTEX 30 has the best SE effect among single-layer materials, which is around 58.92 dB, because of its higher thickness and lower volume electrical resistivity. According to the shielding principle, a thicker material increases the absorption attenuation capacity of the shield under the same shielding material. In this research, comparing the SE with two materials of similar thicknesses, two layers of MEFTEX 10 and one layer of MEFTEX 20, as well as three layers of MEFTEX 10 and one layer of MEFTEX 30, there is no significant difference in SE performance. As the number of material layers increases, the shielding effect of five layers of MEFTEX is significantly higher (44.27–83.8%) than one-layer MEFTEX. After the number of layers is increased, the porosity is significantly reduced, and the air permeability is also reduced. Nevertheless, within a considerable EMI shielding range around 59.79 d–78.38 dB, the air permeability of three to four layers of MEFTEX 10 material was maintained from 2942 $L/m^2/s$–3658 $L/m^2/s$.

**Author Contributions:** Software, M.T.; Supervision, D.K. and J.M.; Writing—original draft, S.H.; Writing—review & editing, D.W. and A.P.P. All authors have read and agreed to the published version of the manuscript.

**Funding:** This research was funded by Frames of Operational Program Research, Development and Education; and project Hybrid Materials for Hierarchical Structures (HyHi, Reg. No. CZ.02.1.01/0.0/0.0/16_019/0000843) at Technical University of Liberec.

**Institutional Review Board Statement:** Not applicable.

**Informed Consent Statement:** Not applicable.

**Data Availability Statement:** Not applicable.

**Acknowledgments:** This work was supported by the Ministry of Education, Youth and Sports of the Czech Republic and the European Union; European Structural and Investment Funds in the Frames of Operational Program Research, Development and Education; and project Hybrid Materials for Hierarchical Structures (HyHi, Reg. No. CZ.02.1.01/0.0/0.0/16_019/0000843) at Technical University of Liberec.

**Conflicts of Interest:** The authors declare no conflict of interest.

## Appendix A

**Table A1.** Classification of EM SE values on textiles for general use [29].

| Type | Grade | Shielding Effectiveness (dB) | Classification | Percentage of Electromagnetic Shielding (%) |
|---|---|---|---|---|
| Class I Professional use | AAAAA | SE > 60 dB | Excellent | ES > 99.9999% |
| | AAAA | 60 dB ≥ SE > 50 dB | Very good | 99.9999% ≥ ES > 99.999% |
| | AAA | 50 dB ≥ SE > 40 dB | Good | 99.999% ≥ ES > 99.99% |
| | AA | 40 dB ≥ SE > 30 dB | Moderate | 99.99% ≥ ES > 99.9% |
| | A | 30 dB ≥ SE > 20 dB | Fair | 99.9% ≥ ES > 99.0% |
| Class II General use | AAAAA | SE > 30 dB | Excellent | ES > 99.9% |

## References


1. Garvanova, M.; Garvanov, I.; Borissova, D. The influence of electromagnetic fields on human brain. In Proceedings of the 2020 21st International Symposium on Electrical Apparatus & Technologies (SIELA), Bourgas, Bulgaria, 6 June 2020. [CrossRef]
2. Rahimpour, S.; Kiyani, M.; Hodges, S.E.; Turner, D.A. Deep brain stimulation and electromagnetic interference. *Clin. Neurol. Neurosurg.* **2021**, *203*, 106577. [CrossRef]
3. Chen, H.C.; Lee, K.C.; Lin, J.H.; Koch, M. Fabrication of conductive woven fabric and analysis of electromagnetic shielding via measurement and empirical equation. *J. Mater. Process. Technol.* **2007**, *184*, 124–130. [CrossRef]
4. Chung, D.D.L. Materials for electromagnetic interference shielding. *Mater. Chem. Phys.* **2020**, *255*, 123587. [CrossRef]
5. Rathebe, P.; Weyers, C.; Raphela, F. A health and safety model for occupational exposure to radiofrequency fields and static magnetic fields from 1.5 and 3 T MRI scanners. *Health Technol.* **2019**, *10*, 39–50. [CrossRef]
6. Cao, W.; Ma, C.; Tan, S.; Ma, M.; Wan, P.; Chen, F. Ultrathin and Flexible CNTs/MXene/Cellulose Nanofibrils Composite Paper for Electromagnetic Interference Shielding. *Nano-Micro. Lett.* **2019**, *11*, 1. [CrossRef] [PubMed]
7. Roh, J.-S.; Chi, Y.-S.; Kang, T.J.; Nam, S.-W. Electromagnetic Shielding Effectiveness of Multifunctional Metal Composite Fabrics. *Text. Res. J.* **2008**, *78*, 825–835. [CrossRef]
8. Geetha, S.; Satheesh Kumar, K.K.; Rao, C.R.K.; Vijayan, M.; Trivedi, D.C. EMI shielding: Methods and materials—A review. *J. Appl. Polym. Sci.* **2009**, *112*, 2073–2086. [CrossRef]
9. Militký, J.; Šafářová, V. Numerical and experimental study of the shielding effectiveness of hybrid fabrics. *Vlak. A Text.* **2012**, *19*, 21–27.
10. Zeng, W.; Tao, X.M.; Chen, S.; Shang, S.; Chan, H.L.W.; Choy, S.H. Highly durable all-fiber nanogenerator for mechanical energy harvesting. *Energy Environ. Sci.* **2013**, *6*, 2631–2638. [CrossRef]
11. Moazzenchi, B.; Montazer, M. Click electroless plating of nickel nanoparticles on polyester fabric: Electrical conductivity, magnetic and EMI shielding properties. *Colloids Surfaces A Physicochem. Eng. Asp.* **2019**, *571*, 110–124. [CrossRef]
12. Rajavel, K.; Hu, Y.; Zhu, P.; Sun, R.; Wong, C. MXene/metal oxides-Ag ternary nanostructures for electromagnetic interference shielding. *Chem. Eng. J.* **2020**, *399*, 125791. [CrossRef]
13. Mantecca, P.; Kasemets, K.; Deokar, A.; Perelshtein, I.; Gedanken, A.; Bahk, Y.K.; Kianfar, B.; Wang, J. Airborne Nanoparticle Release and Toxicological Risk from Metal-Oxide-Coated Textiles: Toward a Multiscale Safe-by-Design Approach. *Environ. Sci. Technol.* **2017**, *51*, 9305–9317. [CrossRef]
14. Xiao, H.; Shi, M.W.; Wang, Q.; Liu, Q. The electromagnetic shielding and reflective properties of electromagnetic textiles with pores, planar periodic units and space structures. *Text. Res. J.* **2014**, *84*, 1679–1691. [CrossRef]
15. Palanisamy, S.; Tunakova, V.; Hu, S.; Yang, T.; Kremenakova, D.; Venkataraman, M.; Petru, M.; Militky, J. Electromagnetic Interference Shielding of Metal Coated Ultrathin Nonwoven Fabrics and Their Factorial Design. *Polymers* **2021**, *13*, 484. [CrossRef] [PubMed]
16. ASTM International ASTM D 4935. *Standard Test Method for Measuring the Electromagnetic Shielding Effectiveness of Planar Materials*; ASTM International: West Conshohocken, PA, USA, 1999; Volume 10, pp. 1–10. [CrossRef]
17. ASTM D3776/D3776M-20. Standard Test Methods for Mass Per Unit Area (Weight) of Fabric. Available online: https://www.astm.org/Standards/D3776.htm (accessed on 11 March 2021).
18. ASTM D5729-97(2004)e1. Standard Test Method for Thickness of Nonwoven Fabrics (Withdrawn 2008). Available online: https://www.astm.org/Standards/D5729.htm (accessed on 11 March 2021).
19. Marciniak, K.; Grabowska, K.E.; Stempień, Z.; Stempień, S.; Luiza, I.; Bel, C.-W. Shielding of electromagnetic radiation by multilayer textile sets. *Text. Res. J.* **2019**, *89*, 948–958. [CrossRef]
20. ASTM D257-14. Standard Test Methods for DC Resistance or Conductance of Insulating Materials. *Standard* **2012**, *1*, 1–18. [CrossRef]

21. ISO. *ISO 9237: 1995 Textiles—Determination of the Permeability of Fabrics to Air*; ISO: Geneva, Switzerland, 1995.
22. Militký, J.; Křemenáková, D.; Venkataraman, M.; Večerník, J. Exceptional Electromagnetic Shielding Properties of Lightweight and Porous Multifunctional Layers. *ACS Appl. Electron. Mater.* **2020**, *2*, 1138–1144. [CrossRef]
23. Tezel, S.; Kavuşturan, Y.; Vandenbosch, G.A.; Volski, V. Comparison of electromagnetic shielding effectiveness of conductive single jersey fabrics with coaxial transmission line and free space measurement techniques. *Text. Res. J.* **2014**, *84*, 461–476. [CrossRef]
24. Schelkunoff, S.A. Transmission theory of plane electromagnetic waves. *Proc. Inst. Radio Eng.* **1937**, *25*, 1457–1492. [CrossRef]
25. Perumalraj, R.; Dasaradan, B.S.; Anbarasu, R.; Arokiaraj, P.; Harish, S.L. Electromagnetic shielding effectiveness of copper core-woven fabrics. *J. Text. Inst.* **2009**, *100*, 512–524. [CrossRef]
26. Tong, X.C. *Advanced Materials and Design for Electromagnetic Interference Shielding*; CRC Press: Boca Raton, FL, USA, 2009; p. 324.
27. Shinagawa, S.; Kumagai, Y.; Urabe, K. Conductive papers containing metallized polyester fibers for electromagnetic interference shielding. *J. Porous Mater.* **1999**, *6*, 185–190. [CrossRef]
28. White, D.R.J. *A Handbook Series on Electromagnetic Interference and Compatibility: Electrical Noise and EMI Specifications*; Interference Control Technologies, Inc.: Germantown, MD, USA, 1971.
29. Palanisamy, S.; Tunakova, V.; Militky, J. Fiber-based structures for electromagnetic shielding—Comparison of different materials and textile structures. *Text. Res. J.* **2018**, *88*, 1992–2012. [CrossRef]
30. Liu, J.L. *Electromagnetic Wave Shielding and Absorbing Materials*, 1st ed.; Xing, T., Ed.; Beijing Chemical Industry Press: Beijing, China, 2006; ISBN 978-7-5025-9341-4.

*polymers*

*Article*

# Shaping in the Third Direction; Synthesis of Patterned Colloidal Crystals by Polyester Fabric-Guided Self-Assembly

**Ion Sandu, Claudiu Teodor Fleaca, Florian Dumitrache, Bogdan Alexandru Sava *, Iuliana Urzica, Iulia Antohe, Simona Brajnicov and Marius Dumitru ***

Lasers Department of National Institute for Lasers, Plasma and Radiation Physics, 409 Atomistilor Street, 077125 Bucharest, Romania; ion.sandu@inflpr.ro (I.S.); claudiu.fleaca@inflpr.ro (C.T.F.); florian.dumitrache@inflpr.ro (F.D.); iuliana.iordache@inflpr.ro (I.U.); iulia.antohe@inflpr.ro (I.A.); simona.brajnicov@inflpr.ro (S.B.)

* Correspondence: bogdan.sava@inflpr.ro (B.A.S.); marius.dumitru@inflpr.ro (M.D.); Tel.: +40-7280-621-60 (B.A.S.)

**Abstract:** A polyester fabric with rectangular openings was used as a sacrificial template for the guiding of a sub-micron sphere (polystyrene (PS) and silica) aqueous colloid self-assembly process during evaporation as a patterned colloidal crystal (PCC). This simple process is also a robust one, being less sensitive to external parameters (ambient pressure, temperature, humidity, vibrations). The most interesting feature of the concave-shape-pattern unit cell (350 μm × 400 μm × 3 μm) of this crystal is the presence of triangular prisms at its border, each prism having a one-dimensional sphere array at its top edge. The high-quality ordered single layer found inside of each unit cell presents the super-prism effect and left-handed behavior. Wider yet elongated deposits with ordered walls and disordered top surfaces were formed under the fabric knots. Rectangular patterning was obtained even for 20 μm PS spheres. Polyester fabrics with other opening geometries and sizes (~300–1000 μm) or with higher fiber elasticity also allowed the formation of similar PCCs, some having curved prismatic walls. A higher colloid concentration (10–20%) induces the formation of thicker walls with fiber-negative replica morphology. Additionally, thick-wall PCCs (~100 μm) with semi-cylindrical morphology were obtained using $SiO_2$ sub-microspheres and a wavy fabric. The colloidal pattern was used as a lithographic mask for natural lithography and as a template for the synthesis of triangular-prism-shaped inverted opals.

**Keywords:** colloidal crystals; polyester fabric; polystyrene; silica; sub-micron spheres; self-assembly; negative diffraction; super-prism effect

**Citation:** Sandu, I.; Fleaca, C.T.; Dumitrache, F.; Sava, B.A.; Urzica, I.; Antohe, I.; Brajnicov, S.; Dumitru, M. Shaping in the Third Direction; Synthesis of Patterned Colloidal Crystals by Polyester Fabric-Guided Self-Assembly. *Polymers* **2021**, *13*, 4081. https://doi.org/10.3390/polym13234081

Academic Editors: Muhammad Tayyab Noman and Michal Petrů

Received: 27 October 2021
Accepted: 22 November 2021
Published: 24 November 2021



## 1. Introduction

Colloidal solutions containing spheres of sub-micron dimensions can form through the phenomenon of self-assembly following the loss of their liquid by evaporation, an ordered porous solid known as colloidal crystal (CC) [1–4]. As one-dimensional (1D) or bi-dimensional (2D) materials, they are challenging, by virtue of their low cost, to sophisticated but expensive technologies, such as photolithography [5], direct laser writing [6], two-photon polymerization [7], and holographic methods [8], when competing for the production of photonic crystals (PCs) [9–11]. There are two winning strategies in this competition. In the first, CCs are used as templates by filling their interstitial spaces with a fluid precursor capable of solidification. Then, the templates are removed to obtain a porous inverse replica [12]. In the second strategy, CCs are used as masks in colloidal lithography [13], where a mono or bilayer of colloidal crystals is exposed to reactive ion etching to produce surfaces with heterogeneous chemistry or to metal evaporation to produce arrays with nanometric features over a large area. In both cases, the resulting structures can be defined as PCs, namely periodic dielectric structures on the light wavelength scale, which can manipulate light in the same manner as a crystal lattice manipulates electrons [9].

As in the case of PCs, their most important property is Bragg diffraction of light [14]. Thus, they present optical band gaps (range of frequencies and directions through which light cannot pass) [15]. As is the case with PCs, CCs can be used as opto-chemical and opto-biological sensors [16,17] to detect chemical and living events, as band-stop and band-pass filters [18,19] for integrated optics, as micro-antennas [20] for satellite and mobile communications, or as laser components [21].

As three-dimensional (3D) materials, CCs compete with three-dimensionally-ordered macro-porous (3DOM) materials obtained by means other than self-assembly methods [22]. In this case, voids in the volume of CCs and the interconnectivity of their voids are useful in phenomena involving the movement of fluids through pores. Such movement of fluids is not facilitated by CCs themselves but by their porous inverse replica, which results when a 3D CC is filled with a fluid precursor capable of solidification, followed by removal of the template. As 3DOM materials, they can serve as catalysts [23,24], porous electrodes [25], or membranes for smart filtration [26], among other applications. Even if they look similar to microporous and mesoporous ordered materials (such as zeolites [27] and metal organic frameworks (MOFs) [28]), the main difference consists in their pore size. Whether CCs are produced as PCs or 3DOM materials, they must present a high level of ordering both for Bragg diffraction to exists and for a decrease in the tortuosity of materials (diffusion and fluid-flow easiness in porous media) [29].

Usually, simple techniques, such as sedimentation, LB technique, floating packing, spin coating, drop drying, vertical deposition, and others [30], produce colloidal crystals with flat extended surfaces flat parallel with the substrate, having sub-micron spheres arranged only in a hexagonal or cubic packing, which repeats in all three directions. Expensive and sophisticated technologies [5–8] can synthesize almost any highly ordered, reproducible array or pattern that can be designed and fabricated as thin film. However, both approaches fail in the third-direction periodical structuring of PCs. Simple techniques can present only hexagonal or cubic structuring, and expensive techniques can repeat the structuring of the extending surfaces only along a few layers. Optical transmission properties of PCs and CCs as PCs depend on the number of structured layers (between one and dozens) in the third direction [31]. In the great majority of cases, self-assembled colloidal crystals exhibit only (111) dense-packed planes at the surface [32] or in some cases, on the fcc (100) or bcc (100) [33], (001), or (110) planes [34]. The simultaneous controlled presence of multiple types of surface planes in these photonic crystals would greatly enrich their photonic bands and, consequently, their potential applications. A curved surface exposes different packed planes, and this can be seen in round colloidal crystals, such as spheres, hemispheres, cylinders and rings [35–38]. Compared with the flat surfaces of CCs, which present different colors when observed from different angles, the curved surfaces show the same color, which is independent of the viewing angle [39,40]. Unfortunately, other interestingly shaped colloidal crystals, such as flat surfaces that are not parallel with the substrate surfaces, like pyramids or triangular prisms, or curved but concave (as opposed to convex) surfaces, as in the case of spheres, hemispheres, cylinders, or rings, were rarely obtained by self-assembly. However, there are a few approaches through which we can increase the physical or chemical properties of an ordered, porous thin film. The first is to fabricate thin crystals onto a curved substrate [41]. The second is, as mentioned before, to shape the film as a three-dimensional object. The third approach is to produce hierarchical order by patterning a substrate with small crystal units. These patterns have minimized volumes and integrated functions, compared to traditional film devices, and provide better performance than their structural units [32]. Usually, the patterns are designed by using the same techniques employed for obtaining CCs units; thus, the absence of complex structuring of CC units in the third dimension persists.

There are two approaches that are usually used for the fabrication of patterned colloidal crystals (PCCs). In the first, sub-micron spheres self-assemble as a film, which are patterned using plasma etching [42], ultra-sonication [43], lift-up soft lithography [44], and other methods. In the second, the substrate is patterned first, and on this patterned substrate, the sub-micron spheres self-assemble in the proper zones. These zones are restricted from the rest of the substrate by chemical wettability [45,46], physical confinement [47,48], the most used method of which is ink-jet printing [49], and surface relief [50,51]. However, in almost all cases, the resulting patterns are rather poor, limited to simple bi-dimensional arrays of flat lines and dots. Moreover, by using expensive technologies for substrate pre-patterning, the overall cost greatly increases. In the first approach, it is impossible to sculpture sub-micron sphere film other than normally to the surface. In the second approach, even when sub-micron sphere film can be precisely cut or etched in the lateral extension, it is limited to the spherical [52] or flat liquid/air interface shape of the self-assembly zones. This induces a spherical or flat shape of the final colloidal crystals, leading to the poor structuration of colloidal crystals in all three dimensions. A pyramid, for example, is much more difficult to achieve, especially trough a pure self-assembly phenomenon. However, self-assembly of sub-micron spheres on a surface-modified relief seems to offer some possibilities. If "positive-relief" rectangle-shaped colloidal crystals could be produced [50], pyramid-shaped crystals could only be obtained only by using a "negative" relief [53,54]. In both cases, the most important problem is that after self-assembly, the colloidal crystals remain trapped in their relief. Their most important surfaces— those which are not parallel with the substrate—remain closed to the external medium.

However, assuming that we can fabricate a single-micrometer prism by self-assembly—meaning to succeed in shaping the capillary forces involved in the process—its multiplying and ordering in a pattern requires repetition on a large scale of the local conditions necessary for the self-assembly of a single prism. This means that the capillary forces that act on the micrometer scale must be able to repeat in a pattern on the macroscopic scale. We found that a commonplace material, more precisely a polyester microfiber woven fabric [55], can fulfill these conditions. The fabric can be seen as a lithographic mask, but comparing with it a fabric presents some remarkable properties regarding its use in the patterning of substrates. It has knots [56]. These knots allow the fibers of the fabrics to settle towards the substrate at a certain distance. Thus, the colloidal liquid can wet the substrate below each fiber. A fabric in close contact with a smooth solid substrate will form a kind of three-dimensional lithographic mask. A colloidal solution deposited between substrate and fabric will allow the movement of the liquid-dispersed nanometer or micron solid particles over a much longer distance than a single unit, as in the case of a lithographic mask. Moreover, between the hydrophobic polyester fibers and the hydrophilic substrate, a special air/liquid interface will form in the zone of each opening, which will act not only to induce their own shape to the final solid pattern, but these interfaces will strongly confine the sub-micron spheres during solvent losses, providing the necessary force for the self-assembly of a quality colloidal crystal. However, the most important thing is that after sub-micron spheres self-assemble in the dry-patterned colloidal crystal, the fabric can be completely peeled off. It is worth mentioning that a specific research domain in materials physics, namely "capillary-bridge-mediated-assembly" or "liquid-bridge-induced assembly (LBIA)" [57–59], may offer some suggestions on the synthesis of patterned and shaped colloidal crystals. Even though their main purpose is to obtain 1D nanostructures to be used as waveguides and they usually use nanometric particles and not sub-micron spheres, the fact that they are based on capillary forces that confine nanoparticles through liquid bridges built into a different architecture than the methods cited in reference [30] could help us. Unfortunately, this approach also uses expensive technologies.

In this work, we present, for the first time, a cheap and fast method by which a new kind of patterned colloidal crystal can be fabricated by using a hydrophilic substrate, a water-based sub-micron-sphere colloidal solution, and a hydrophobic polyester fabric. A sub-micron sphere can thus self-assemble with remarkable three-dimensional ordered concavities, much similar to those possessed by some living creatures [60].

## 2. Materials and Methods

### 2.1. Materials

Polystyrene (PS) sub-micron-sphere aqueous colloidal solutions with 0.150 μm, 0.300 μm, 0.488 μm, and 20.0 μm and $SiO_2$ with 0.245 μm mean diameter, 5% *w*/*v*, were purchased from microParticles GmbH, Berlin, Germany, and used as they were or diluted with deionized water when needed. Commercially available microscope glass slides, optical polished silicon, and steel samples were used as substrates after a few minutes cleaning by ultrasonication in acetone, distilled water, and ethanol, followed by natural drying. Commercially available polyester fabric sheets (polyester veil) were cut in 1 cm × 1 cm pieces and used as they were. However, most appropriate materials can be found on the Internet by using the keywords: polyester, plain, precision, fabric.

### 2.2. Synthesis of Patterned Colloidal Crystals

A few droplets of sub-micron-sphere colloidal solution were deposited onto the surface of a microscope glass slide (Figure 1a).

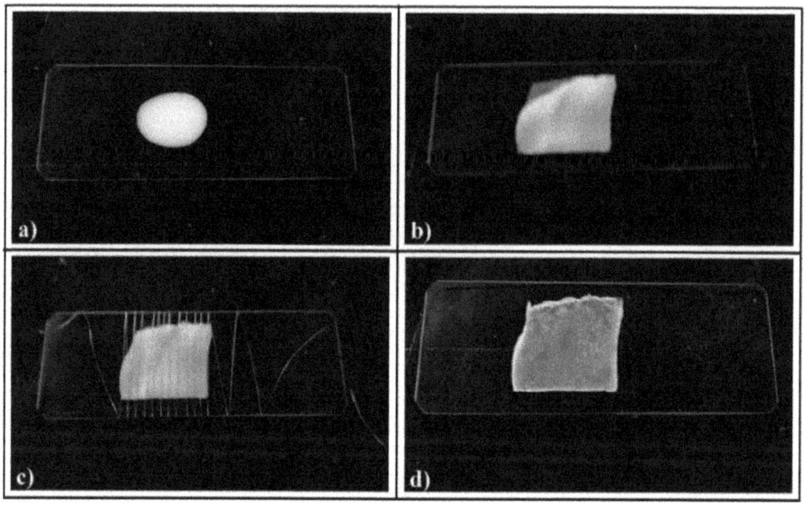


**Figure 1.** Experimental set-up and the necessary steps for synthesis of colloidal crystal self-assembly guided by polyester fabric: (**a**) colloidal solution drop on the microscope glass slide substrate; (**b**) polyester fabric pressing the colloidal drop; (**c**) metallic wire rolled around the fabric and the colloidal film; (**d**) the final dried PCCs after the wire and the fabric were removed.

A piece of fabric sheet of ~1 $cm^2$ was gently deposited onto the colloidal drop and lightly pressed to stick it to the glass substrate (Figure 1b). A thin metallic wire (100 μm diameter) was rolled tight around the fabric and colloid (the distance between metallic spires was 2–4 mm) (Figure 1c). The role of the metal wire is to keep the fabric in tight contact with the substrate throughout the evaporation of the solvent and, at the same time, to introduce as little disturbance as possible to the natural evaporation process. Several types of wires were tried—metallic, polymeric (from polyester or cellulose)—but the copper wire was proven to best meet these conditions. Close contact between the fabric and the substrate throughout the evaporation of water is necessary because the fabric, if left free, undergoes some square-millimeter detachments from the substrate in the final phase of evaporation (inhomogeneous evaporation). After a few minutes (~30 min in the normal conditions of temperature and humidity of the laboratory, T = 25 °C, relative humidity (RH) = 40–60%), once the liquid completely evaporated, the wire and fabric were removed and the colloidal crystal can be further used in experiments (Figure 1d). We mention that the same result was obtained if we first placed the fabric on the empty substrate, rolled the metal wire and then deposited a few drops of colloidal solution on top, slightly tapping it.

Thin-film colloidal crystals of 488 nm PS spheres were also fabricated by the spin-coating technique using a WS-400BX-6NPP/LITE model (Laurell Technologies, New Wales, PA, USA). A few drops of colloidal solution (PS 488 nm, 5% *w*/*v*) were poured onto the microscope glass slides used as substrates, which were spun at 1500 rpm for 30 s.

### 2.3. Investigation

Macro-scale observations of the as-synthesized patterned colloidal crystals were performed by using optical microscopy and a digital photo camera. A scanning electron microscope (SEM) (Apreo S Thermo Fisher Scientific, Auburn, AL, USA) and an atomic force microscope (AFM) XE-100 (Park Systems Inc., Santa Clara, CA, USA) were used to observe the structures and morphologies of the self-assembled patterned colloidal crystals at sub-micron and nanometric scales. A thin layer of gold was sputtered onto the samples prior to imaging. UV—vis transmittance spectra were acquired by using an optical-fiber-connected AvaLight-DHc light source (spot size ~200 µm) and an AvaSpec-ULS2048CL-EVO—high-resolution spectrometer, all from Avantes, Apeldoorn, The Netherlands.

## 3. Results and Discussion

By looking through a magnifying lens, we were able to observe a regular grid (Figure 1d), which seems to reproduce the fabric design (Figure 2a). SEM imaging (Figure 2b) shows regular square cells of around 0.35 mm in size with a small, round opening (a few dozen micrometers) in the center of each cell. Sparkling, solid walls and a pale yellow or brown surface between walls and the centered opening was observed by reflection optical microscopy (Figure 2c). Upon closer examination (SEM), magnificent triangular prisms comprising well-ordered sub-micron spheres in a hexagonal array stay straight for hundreds of micrometers in length (Figure 2d–f), each prism finishing on its top with one dimensional array of sub-micron spheres (Figure 2f). AFM investigations show that the prism surfaces are slightly concave, and the width and height of prisms are around 10 µm and 3 µm, respectively (Figure 2g).

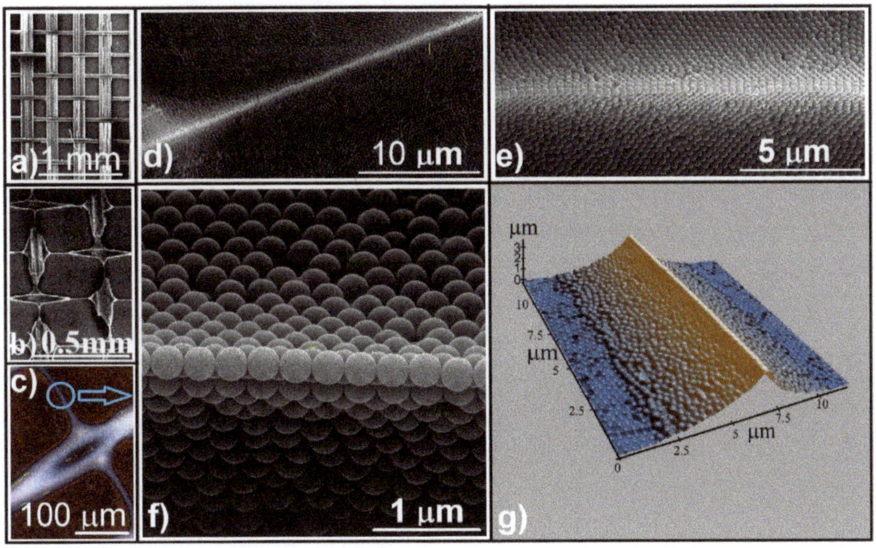


**Figure 2.** Patterned and shaped colloidal crystal: (**a**) SEM image of the fabric structure; (**b**) SEM image of the patterned crystal; (**c**) optical microscopy image of a cells-unit intersection; (**d–f**) SEM images of triangular prisms of colloidal crystals; (**g**) AFM image of a triangular prism.

The ordering and the reproducibility of these structures are remarkable, obtained every time the experiments were performed under the same conditions. However, during a large number of experiments, we noticed the following general trends:

(a) Colloidal concentrations greater than 1% are needed for prisms to appear, and concentrations greater than 3% are needed for a line of spheres to settle on their upper edge. These values should not be taken as absolute but rather as starting values. They can vary within certain limits with all the parameters upon which self-assembly of the spheres depends.
(b) The ordering quality increases with the increase in the hydrophilicity of the substrate. Although prisms can be obtained even on polymeric substrates, the best results are obtained on glass substrates.
(c) The overall quality of the grids slightly increases as the size of the spheres decreases.
(d) The shape of the unit cell and the CC pattern are imposed by the shape of the fabric cell and its pattern, at least within the limits of our experiments (polyester fiber diameter of tens of micrometers, fabric openings between 300 and 1000 μm).
(e) The quality of the obtained grids depends on the hydrophilicity of the fabric. Less hydrophilic fabrics obtain better results.
(f) The self-assembly phenomenon using fabric is less sensitive to external parameters, such as ambient pressure, temperature, humidity, or minor vibrations or shocks.
(g) The structure of the grids and the ordering quality of the sub-micron spheres do not change with the variation in a wide band of liquid volatility. Experiments performed in which the liquid medium volatility (evaporation rate) was extremely low (RH ~99%) or very fast (evaporation inside the oven, T = 30–80 °C, RH < 10%) did not show changes in the ordering quality of the sub-micron spheres, although the evaporation time of the same volume of colloidal solution varied between several days and a few minutes.

The unit cells could be described by means of three basic components:

(a) Triangular prisms (Figure 3a), which can be straight or curved in 2D, depending on the shape of the fabric opening.
(b) Knots (Figure 3c), whose shape also depends on the structure of the fabric and the number and thickness of fibers that intersect at each node of the fabric. Although spheres are tightly packed in the volume of these structures, the surface layers are often disordered. The cause could be the "late evaporation" [61,62] of the solvent, but as long as ordered surfaces are sometimes obtained, there is the possibility that this phenomenon can be controlled.
(c) The interior of the unit cells has an almost flat bottom (Figure 3d) and consists mainly of a monolayer of tightly packed spheres with an empty space of a few tens of micrometers, in the center (Figure 3e). The degree of ordering is good enough (Figure 3f) that the spheres form a polycrystalline structure, each mono-crystalline domain having several tens of micrometers in size, large and packed enough to produce the Bragg light diffraction phenomenon. However, bilayers zones, amorphous regions, and packing defects are often found. Remarkably, no crack was observed at any level.
(d) The quantification of the order quality was done through the parameter "range order, RO", defined by us and published in a previous paper as "the result of multiplication between the mean area of a perfect domain and the number of nanospheres that it contains" [63]. By analyzing the SEM images of our patterned colloidal crystals, we measured some values of the range-order parameter (Table 1).

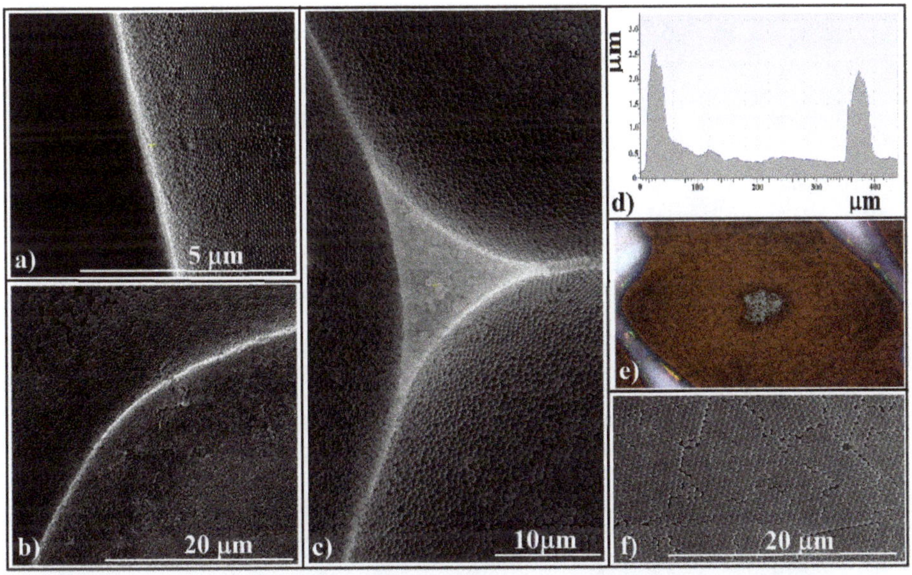


**Figure 3.** SEM images of: (**a**) Straight triangular prism; (**b**) curved triangular prism; (**c**) knot; (**d**) profile of a unit cell; (**e**) reflection optical microscopy of a unit cell; (**f**) SEM image of PS single layer from the inside of unit cell.

**Table 1.** Quantification of the order quality of self-assembly of sub-micron spheres.

| | RO [$mm^2$] | | | |
|---|---|---|---|---|
| Structure | PS (0.15 μm) | PS (0.30 μm) | PS (0.488 μm) | $SiO_2$ (0.245 μm) |
| Prisms | 13.0 | 3.0 | 2.0 | 6.0 |
| Knots | ~0 | ~0 | ~0 | ~0 |
| The interior of the unit cell | 6.0 | 2.0 | 1.0 | 3.0 |

These were the three structural components we encountered each time, no matter what fabric, spheres, solvent, or substrate we used. Thus, we have colloidal crystals shaped as triangular prisms, somehow curved truncated pyramids (knots), and cavities, which present large areas of slightly concave surface. However, an increase in shaping complexity could be achieved by performing some specific experiments. The simplest one is to use colloidal solutions with high concentration. If by using colloidal solutions with concentrations of 3–5%, the structures mentioned above are obtained (shown for comparison in Figure 4a), then by using concentrations higher than 10–20%, the solid structures self-organized using fabric and change their shape. Although they retain the organization imposed by the fabric, the solid walls are no longer prismatic but take on the shape of the fibers that make up the fabric (Figure 4b), a kind of negative shape of the fibers. The fabric film becomes a mold. The monolayers inside each cell become multilayered, and the central area, empty of spheres, is completely covered.

Unfortunately, the degree of ordering decreases sharply, and future experiments would be needed to find the conditions under which we can keep both the ordering of the spheres and the shapes induced by the fibers.

A second approach by which we can change the shape of self-organized structures is to unfold the fabric itself into a new dimension. The normal fabric (50 μm in thickness) that is usually stuck to the substrate (schematic in Figure 4c) has been wavy by rolling a parallel metallic wire of 100 μm in diameter onto the substrate, followed by a second wire, which was rolled onto the sample after the colloid and fabric were deposited (Figure 4d).

The distance between two consecutive parallel metallic wires was of 2 mm, whether they are above or below the fabric. Thus, a variable distance between fabric and the substrate is obtained. In this case, the self-assembled structure keeps the grid pattern (Figure 4e), but the prisms no longer form. Instead, semi-cylindrical walls border each unit. The wall diameter reaches 100 μm (Figure 4f), and their surfaces present a high quality of nanosphere ordering.

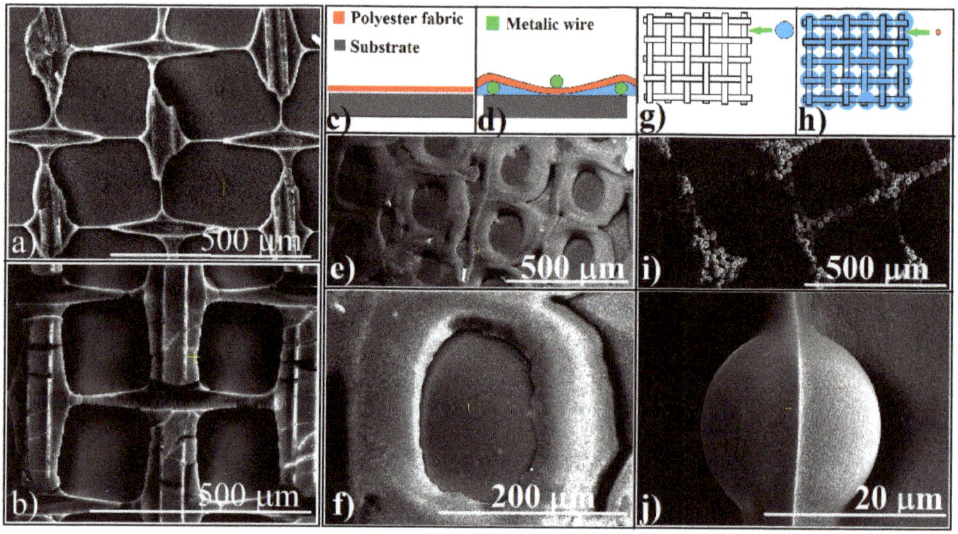


**Figure 4.** (**a**) SEM image of a colloidal crystal grid where 5% PS sphere concentration was used; (**b**) SEM image of a colloidal crystal grid where 20% PS sphere concentration was used; (**c**) schematic of fabric onto the substrate in a normal configuration; (**d**) schematic of wavy fabric onto the substrate; (**e**,**f**) SEM images of the grid and unit cell resulted by using silica spheres and wavy fabric; (**g**,**h**) schematics of the stepped process for colloidal-crystal self-assembly guided by polyester fabric; (**i**) 20 μm PS spheres forming a colloidal crystal grid; (**j**) 0.150 μm PS spheres covering layer onto 20 μm PS spheres.

A third viable strategy by which we can change the shape of self-organized structures is to take advantage of a method characteristic, namely that multiple deposition can be performed in a sequential manner as long as the previous structure is not dislocated or dissolved. Thus, multi-layered structures can be achieved. As an example, if we first deposit a 20 μm microsphere suspension, although it is at the maximum limit of the size of the spheres that can still be organized (Figure 4g), after evaporation of the solvent, a quantity of colloidal solution containing much smaller spheres can be applied—in our case, 0.15 μm PS spheres (Figure 4h)—which begin to conformally deposit over the prisms composed of large spheres (Figure 4i), forming a coating with a ridge on their surface with 1D spheres on its upper edge (Figure 4j).

We present three possible applications of patterned colloidal crystals formed by self-assembly and guided by a fabric, although the actual number of applications is probably much higher.

As we mentioned before, Bragg diffraction of light is a common phenomenon in colloidal crystals. Usually, the quality of such colloidal crystals is given by the height and narrowness of the reflection bands or by the depth and narrowness of the transmission bands. The number of features and their position are imposed, on the one hand, on the sphere's nature and size, and on the other hand, on the sphere's packing. The most important part of the self-assembly method forms a hexagonal close-packed (hcp) monolayer, and thicker film can be seen as stacks of this single-layer. There are three ways of stacking the layers: hexagonal, cubic-centered face (cfc), and double hexagonal [64]. Each method presents distinct features in reflected or transmitted light. UV-vis. spectrometry measurements performed on our samples show the existence of three dips in transmitted light (Figure 5a).

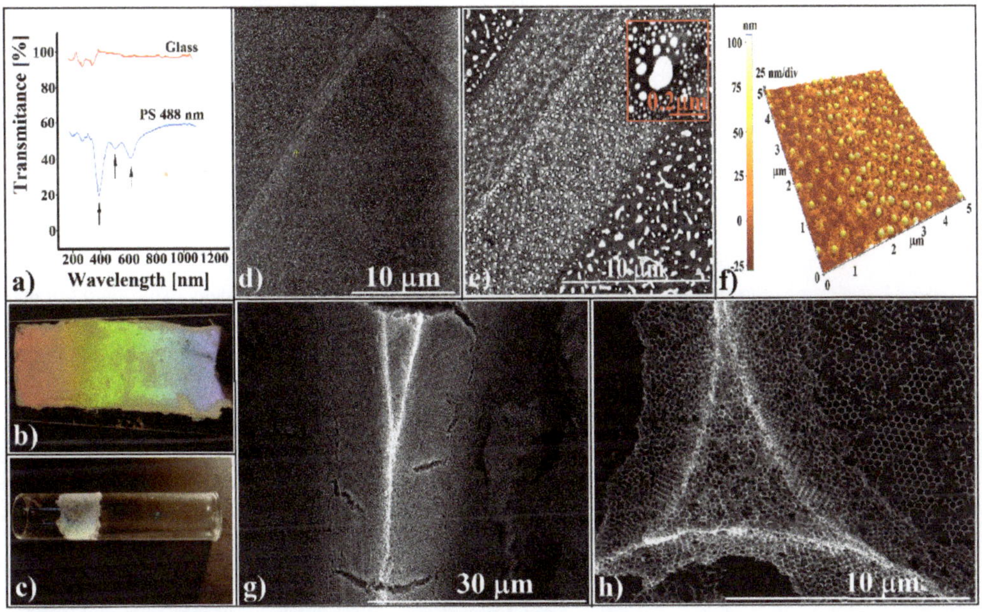


**Figure 5.** (**a**) UV-Vis spectra of white light transmission trough a patterned, shaped colloidal crystal; (**b**) image of super-prism effect of a patterned and shaped colloidal crystal formed onto a flat substrate; (**c**) image of super-prism effect of a patterned and shaped colloidal crystal formed onto a curved substrate; (**d**,**e**) SEM images of the Au nanometric array resulted by using a triangular prism colloidal crystal as lithographic mask; (**f**) AFM image of the Au nanometric array; (**g**,**h**) SEM images of a triangular prism and a knot, inverted opals.

The dips are placed at: $\lambda_{EXP}$ = 620 nm; 510 nm; and 390 nm. First, we must remark on the absence of the most important dip corresponding to 3D CCs with a cfc packing. The Bragg light diffraction on its (111) planes [65] it should have generated a minimum placed at 1172 nm, and in the case of polystyrene spheres, 488 nm. Second, the transmission bands from Figure 5a fit very well with the band gap's positions found by other authors [66,67] when they studied Bragg light diffraction on a polystyrene hexagonal close-packed sphere as a monolayer. They showed that dips in the transmission spectra arise at the spectral positions where parameter $Z = \frac{\sqrt{3}d}{2\lambda}$ satisfies condition Z = 0.71; 0.85; 1.00; 1.34, or 1.55 [66], where $\lambda$ is the dip position and d is the sphere diameter. Verifying our wavelength positions for a 488 nm polystyrene sphere, we found, for Z, the next values: Z = 0.68; 0.85; 1.08. All of these, as well as the profilometry image from Figure 5d, suggest that inside of each unit cell of the patterned CCs, a single layer of sub-micron spheres is formed. The difference between the theoretical values and experimental values can be attributed to the stacking faults and to the mixture of single layers with double layers in some proportion. Because an identical spectrum was obtained on the samples produced by spin coating where no prism was present (not seen at SEM images), we can conclude that the role of triangular prisms is absent or minor in Bragg diffraction of light on our patterned CCs.

An important property of CCs is their negative refractive index, which induces left-handed behavior of the diffracted light [68]. Because accurate measurements of this phenomenon can be performed only through a special optical experimental setup, in its absence, we resorted to simple photos, which can be seen in Figure 5b (the white light dispersion when, it passes through a patterned CC, forms 488 nm PS spheres onto a flat substrate on microscope glass slide). We also used rough geometrical measurements (angles at which a spot from the sample changes its color from dark blue to light red). The incident angle was higher than 20°. Thus, we measured an angular dispersion of around 6 nm/degree. However, the samples synthesized by spin coating showed the same phenomenon and the same angular dispersion value; thus, with the super-prism effect, we cannot assign an influence of the triangular prisms. Even so, accurate measurements and/or an intelligent designed experimental setup might evidence their possible properties of light manipulation, such as waveguides, for example.

The super-prism effect could also be seen in our patterned colloidal crystals formed onto a curved substrate (glass cylinder) (Figure 5c). The self-assembly of sub-micron spheres onto curved substrates is a great challenge [69]. However, we found that this is possible if the fabric is highly elastic (a sheet of women's stocking, for example). We found that even by using a normal polyester fabric sheet rolled onto a curved substrate, grids and prisms formed; the sphere single layers became completely disordered but kept their ordering if the fabric sheet makes a tight contact with the curved substrate.

A second application of the sub-micron sphere grid that generates some interesting results is its use as a lithographic mask. As mentioned in the introduction, arrays of nanometric structures can be obtained by using the interstice between the solid spheres by chemical etching or by depositing various materials from the vapor phase. A close-packed colloidal assembly usually forms an array of triangular-shaped structures below the single layer or a hexagonal array of a quasi-hexagonal nanodots in the double-layer case [13]. We present, for the first time, what results if a 3D prismatic opal is used as template for colloidal lithography. In fact, the sample used for close viewing under SEM (Figure 5d,e), the surface on which a thin gold film was deposited by sputtering, was calcinated for PS sub-micron-sphere removal and reanalyzed by SEM. Somewhat surprisingly, after calcination, a shaded material structure could be seen in the SEM images (Figure 5d), which seems to reproduce the triangular prisms from Figure 3. We assume that this shadow is induced by the difference in dot surface density generated by the angle at which the gold source irradiates the sample (60° relative to the substrate) [70]. If the irradiation angle is different from 90°, the Au atoms will first hit the nearest colloidal prism surface, depositing a higher Au density than in the zone of the opposite surface. Upon closer inspection, we can see a hexagonal array of slightly ellipsoidal gold dots (non-circular shape could be also an argument of angle deposition effect), each one of them surrounded by a few smaller satellites, and so on, in a fractal-like organization (Figure 5e inset). By using PS sub-micron spheres of 488 nm in diameter (14 nm standard deviation) as a mask, the mean sizes of Au dots measured from SEM images (100 measurements) were of 126 nm for the first-order dots (26 nm standard deviation) and 50 nm (10 nm standard deviation) for the second-order dots (satellites). AFM investigations (Figure 5f) reveal a height of around 50 nm of the first-order gold dots. The intriguing question (for which we have no answer) is how the gold vapors penetrated so many sub-micron-sphere layers (more than six) to fix and form Au dots at the bottom of and in the central part of the prisms.

The third application of our colloidal crystal is using it as a template for the synthesis of inverted opal structures. By infiltrating a 25 wt.% $Na_2S_iO_3$ water-based solution in the patterned colloidal crystal obtained by using a fabric and drying and calcinating it at 400 °C, we obtained an inverted replica of the initial patterned colloidal crystal (Figure 5g,h). However, the most important finding was to obtain a skin-free structure during solution infiltration because helping techniques such as microtome cutting or etching processes do not work to remove the overlayer deposited onto the colloid prism's surface. The quality of our inverted opals must be further improved (we are still working at this), and interesting applications of such shaped crystals must be imagined. The most important quality of a triangular-prism-shaped material is its capacity to disperse an incident-coherent beam in many more secondary, angular-dependent, refracted beams when the beam interacts with the non-parallel surfaces of the prism. For non-optical phenomena, such as atoms, molecules, or nano-objects flowing through the prism, the different distances that they cross between the two non-parallel walls can create inhomogeneities, anisotropies, or useful gradients after their exit from the prism. The triangular shape of colloidal prisms can be used for improving some parameters of applications that are already used for the "microprism array" key concept. Thus, they might be used for directional transmission of light [71] or for improving the efficiency of solar cells [72]. Interesting fields of application might be opto-fluidics [73] and micro-fluidics [74] if the triangular prisms are fabricated as inverted opals, as wave guiding 1D structures [75], if they could be fabricated in the nanometric domain or as cell and tissue scaffolds [76], or if they could be fabricated in the millimetric domain.

However, this work is a proof of concept. We prefer to suggest a number of interesting future applications rather than to systematically investigate a single one of them. We consider it a huge challenge to correlate the fabric architecture with experimental setup parameters and final properties of the patterned colloidal crystals and to predict the colloidal-crystal unit shape and the packing structure of the sub-micron spheres in each unit. The third dimension of colloidal crystals must be conquest!

At the end of this paper, we will try to briefly show how these triangular prisms may form. The mathematics and physics behind this apparently simple phenomenon are of a high complexity, many of the specific equations requiring numerical simulations. Therefore, first of all, we consider that fabric-guided self-assembly takes place on modified-relief patterned surfaces. On such surfaces, the capillary forces, which act for sub-micron spheres confinement, are the result of interplay between the hydrophobicity of substrate and fabric fibers. The adhesion force between water and fibers or substrate is much higher than the cohesion force between water molecules. During solvent evaporation, water from the center of each unit cell will be moved closer to each fiber, thus leaving an empty space in the center of each cell. When only a small part of the initial colloid solution remains in the system, the liquid bridges that form between the substrate and fiber surfaces impose the shape of the final solid deposits. A water droplet that rests on a flat substrate and one that hangs on a fiber can take (in conditions close to ours) the shape shown in Figure 6a. A droplet that simultaneously wets two close, flat, parallel, and chemically similar substrates takes the shape shown in Figure 6b. If the substrates present dissimilar wetting properties, the liquid-bridge shape changes, as in Figure 6c. Thus, the shape of the colloid liquid bridge between the substrate and the polymer fibers of the fabric could look like Figure 6d. In this case, the dynamics of the liquid bridge shape during evaporation could be close to those in Figure 6e. The capillary forces that appear at the interface between the liquid and the two solid bodies, polyester fiber and glass substrate, which have different wettability properties, have different orientations and sizes so that their result generates a volume of liquid in the shape of a laterally elongated clepsydra. As the solvent evaporates, the liquid retains its shape and, decreasing in volume, confines more and more the sub-micron spheres into a colloidal crystal in the shape of a double triangular prism (Figure 6e). Second, the extremely thin neck of the clepsydra can have the thickness of a single sphere so that when the fabric is removed, even if the formed crystal adheres to the fabric, this neck becomes the ideal cleavage line. Thus, the remaining part on the substrate keeps its integrity (Figure 6e). This hypothesis is suggested by SEM images in which we can see a very similar prismatic colloidal crystal (Figure 6g) on some fiber surfaces very similar to that formed on the substrate (Figure 6f). These observations could give us some ideas about the complexity of the self-assembly phenomenon, the complexity of which we could use to make shaped colloidal crystals.

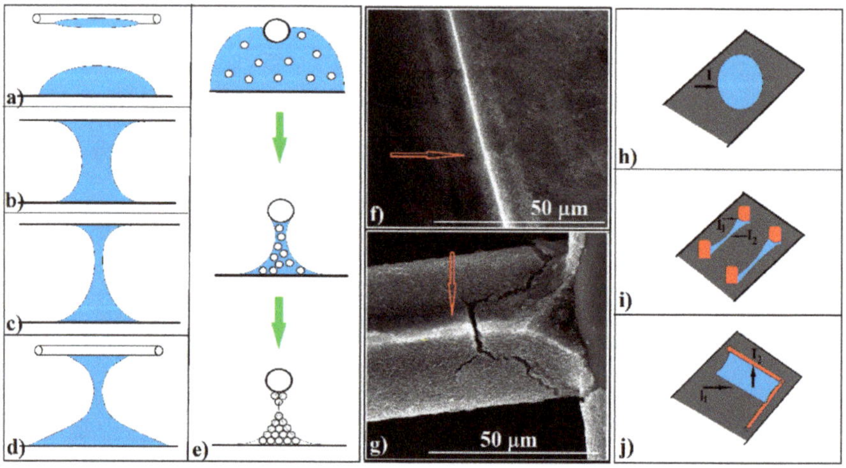


**Figure 6.** (**a**) Schematic of water droplet resting on a flat substrate or hanging on a fiber; (**b**) schematic of a liquid bridge formed between two parallel, flat, similar substrates; (**c**) schematic of a liquid bridge formed between two parallel, flat, dissimilar substrates; (**d**) schematic of the possible shape of a liquid bridge formed between a flat substrate and a fiber; (**e**) liquid bridge shape changing during evaporation; (**f**) SEM image of the colloidal crystal shaped as a triangular prism onto the flat substrate; (**g**) SEM image of the colloidal crystal shaped as a triangular prism onto the curved fiber surface; (**h**) schematic of solid/liquid/gas interface, which appears in the usual colloidal-crystal self-assembly methods; (**i**) schematic of solid/liquid/gas interfaces, which appear in the "capillary-bridge-mediated-assembly" technique; (**j**) schematic of solid/liquid/gas interfaces, which appear in the "fabric-guided self-assembly" technique.

The current methods of colloidal-crystal synthesis [30,35–38] work only with a single solid/liquid/gas interface (Figure 6h) and offer only flat or convex crystals. Triangular prisms are often produced when the "capillary-bridge-mediated-assembly" phenomenon/technique is used [57–59]. It is worth noting that in this architecture, two solid/liquid/gas interfaces are involved (Figure 6i). They induce the formation of a horizontal liquid bridge parallel with the substrate. By using a fabric sheet, we also have two solid/liquid/gas interfaces (Figure 6j). However, this time, the liquid bridge is in a vertical position, normal to the substrate. We can see that the liquid bridges between two (and why not more than two?) solids can offer complex shapes where capillary forces can lead to the proper self-assembly of colloids in complex three-dimensional shaped colloidal crystals.

## 4. Conclusions

Patterned colloidal crystals (PCCs) on smooth, flat, and few $cm^2$ wide hydrophilic substrates can be obtained by using a polyester fabric for guiding the self-assembly of colloidal sub-micron polystyrene (or silica) and even much larger (20 μm) PS spheres. Curved-substrate (such as cylindrical ones) deposition of PCCs was also achieved, especially when using an elastic fabric. This robust method for PCC synthesis is inexpensive and less sensitive to external parameters.

The unit cells of the patterned colloidal crystal have a concave shape, consisting of prisms (or elongated structures as negative-replica of the fabric fibers at higher sphere concentration and even flatted, donut-shaped structures when folded fabrics were employed) at their borders and deposits corresponding to the fabric knots at their intersections, whereas mainly monolayers of sub-micron spheres close-packed in a hexagonal array occupied most of their interior zones.

The mechanisms responsible for the formation of the border structures of the PCC unit cells imply the concentration of spheres in three-dimensional vertically liquid bridges formed between polyester fibers and hydrophilic substrate and their subsequent compacting during liquid evaporation, followed by tissue detachment from the upper part of the deposit after drying.

Patterned colloidal crystals present interesting optical properties that can be visually observed, such as negative refraction and the super-prism effect with an angular dispersion of ~5 nm/degree, an effect also observed for similar PCCs deposited onto cylindrical surfaces. We used the PCCs prisms as lithographic masks and obtained hierarchically assemblies of gold nanodots after sputtering and calcination. Furthermore, they were employed as templates for shaped inverted opals (IO) synthesized from PS sphere PCCs infiltrated with sodium silicate solutions, dried, and calcined, which preserved the main components if the initial opals (prisms, knots, and inner zones).

**Author Contributions:** Conceptualization and formal analysis, I.S.; methodology, I.S.; validation, I.S.; investigation, I.S., I.U., I.A., S.B., and M.D.; resources, F.D. and B.A.S.; writing—original draft preparation, I.S.; writing—review and editing, I.S. and C.T.F.; supervision, I.S.; project administration, F.D. and B.A.S.; funding acquisition, C.T.F. and B.A.S. All authors have read and agreed to the published version of the manuscript.

**Funding:** This research was funded by EU in the frame of Horizon 2020, project H2020-FETOPEN-2018-2020, no. 863227, "Photo-Piezo-ActUators based on Light SEnsitive COMposite", acronym PULSE-COM and by grants of the Romanian Ministry of Research, Innovation and Digitization, Agency of Romanian Ministry of Research and Innovation, CNCS/CCCDI—UEFISCDI, financial support through the project PN-III-P2-2.1-PED-2019-4951/Contract 304PED/2020, project 459 PED/2020, acronym AWISEM, within PNCDI III and Core Program PN 16N/2019 LAPLAS VI, grant no. PN 19 15 01 01.

**Institutional Review Board Statement:** Not applicable.

**Informed Consent Statement:** Not applicable.

**Conflicts of Interest:** The authors declare no conflict of interest. The funders had no role in the design of the study; in the collection, analyses, or interpretation of data; in the writing of the manuscript, or in the decision to publish the results.

## References

1. Mesegue, F. Colloidal crystals as photonic crystals. *Colloids Surf. A Physicochem. Eng. Asp.* **2005**, *270–271*, 1–7. [CrossRef]
2. Chiappini, A.; Ferrari, M.; Singh, M. Opal-type photonic crystals: Fabrication and application. *Adv. Sci. Technol.* **2010**, *71*, 50–57.
3. Cai, Z.; Smith, N.L.; Zhang, J.-T.; Asher, S.A. Two-Dimensional Photonic Crystal Chemical and Biomolecular Sensors. *Anal. Chem.* **2015**, *87*, 5013–5025. [CrossRef]
4. Xu, Y.; Schneider, G.; Wetzel, E.D.; Prather, D.W. Centrifugation and spin-coating method for fabrication of three-dimensional opal and inverse-opal structures as photonic crystal devices. *J. MicroNanolithogr. MEMS MOEMS* **2004**, *3*, 168–173. [CrossRef]
5. Yao, P.; Schneider, G.J.; Miao, B.; Prather, D.W. Fabrication of three-dimensional photonic crystals with multi-layer photolithography. In Proceedings of the Micromachining Technology for Micro-Optics and Nano-Optics III Conference MOEMS-MEMS Micro and Nanofabrication, San Jose, CA, USA, 22–27 January 2005; Volume 5720. [CrossRef]
6. Deubel, M.; von Freymann, G.; Wegener, M.; Pereira, S.; Busch, K.; Soukoulis, C.M. Direct laser writing of three-dimensional photonic-crystal templates for telecommunications. *Nat. Mater.* **2004**, *3*, 444–447. [CrossRef]
7. Sun, H.-B.; Matsuo, S.; Misawa, H. Three-dimensional photonic crystal structures achieved with two-photon-absorption photopolymerization of resin. *Appl. Phys. Lett.* **1999**, *74*, 786–788. [CrossRef]
8. Hsieh, M.-L.; John, S.; Lin, S.-Y. A holographic approach for a low cost and large-scale synthesis of 3D photonic crystal with SP2 lattice symmetry. Available online: https://www.spiedigitallibrary.org/conference-proceedings-of-spie/11498/1149808/A-holographic-approach-for-a-low-cost-and-large-scale/10.1117/12.2567840.full?SSO=1 (accessed on 15 August 2021).
9. Pawlak, D.A. Metamaterials and photonic crystals—potential applications for self-organized eutectic micro- and nanostructures. *Sci. Plena* **2008**, *4*, 014801.
10. Prather, D.W.; Murakowski, S.; Shi, J.; Schneider, G.J.; Sharkawy, A. Photonic Crystal Structures and Applications: Perspective, Overview, and Development. *IEEE J. Sel. Top. Quantum Electron* **2006**, *12*, 1416–1437. [CrossRef]
11. Nair, R.V.; Vijaya, R. Photonic crystal sensors: An overview. *Prog. Quantum Electron.* **2010**, *34*, 89–134. [CrossRef]
12. Stein, A.; Schroden, R.C. Colloidal crystal templating of three-dimensionally ordered macroporous solids: Materials for photonics and beyond. *Curr. Opin. Solid State Mater. Sci.* **2001**, *5*, 553–564. [CrossRef]
13. Wood, M.A. Colloidal lithography and current fabrication techniques producing in-plane nanotopography for biological applications. *J. R. Soc. Interface* **2007**, *4*, 1–17. [CrossRef]

14. Baryshev, A.V.; Kaplyanskii, A.A.; Kosobukin, V.A.; Limonov, M.F.; Samusev, K.B.; Usvyat, D.E. Bragg Diffraction of Light in Synthetic Opals. *Phys. Solid State* **2003**, *45*, 459–471. [CrossRef]
15. Meade, R.D.; Rappe, A.M.; Brommer, K.D.; Joannopoulos, J.D. Nature of the photonic band gap: Some insights from a field analysis. *J. Opt. Soc. Am.* **1993**, *10*, 328–332. [CrossRef]
16. Riyadh, B.A.; Hossain, M.; Mondal, H.S.; Rahaman, E.; Mondal, P.K.; Mahasin, M.H. Photonic Crystal Fibers for Sensing Applications. *J. Biosens. Bioelectron.* **2018**, *9*, 1000251. [CrossRef]
17. Dorfner, D.; Zabel, T.; Hürlimann, T.; Hauke, N.; Frandsen, L.; Rant, U.; Abstreiter, G.; Finley, J. Photonic crystal nanostructures for optical biosensing applications. *Biosens. Bioelectron.* **2009**, *24*, 3688–3692. [CrossRef] [PubMed]
18. Maigyte, L.; Staliunas, K. Spatial filtering with Photonic Crystals. *Appl. Phys. Rev.* **2015**, *2*, 011102. [CrossRef]
19. Sathyadevaki, R.; Sivanantha Raja, A.; Shanmugasundar, D. Photonic crystal-based optical filter: A brief investigation. *Photon. Netw. Commun.* **2017**, *33*, 77–84. [CrossRef]
20. Britto, E.C.; Danasegaran, S.K.; Johnson, W. Design of slotted patch antenna based on photonic crystal for wireless communication. *Int. J. Commun. Syst.* **2021**, *34*, e4662. [CrossRef]
21. Zhang, Y.; Khan, M.; Huang, Y.; Ryou, J.; Deotare, P.; Dupuis, R.; Lončar, M. Photonic crystal nanobeam lasers. *Appl. Phys. Lett.* **2010**, *97*, 051104. [CrossRef]
22. Pullar, R.C.; Novais, R.M. Ecoceramics. Cork-based biomimetic ceramic 3-DOM foams. *Mater. Today* **2017**, *20*, 45–46. [CrossRef]
23. Liu, Y.; Guo, R.; Duan, C.; Wu, G.; Miao, Y.; Gu, J.; Pan, W. Removal of gaseous pollutants by using 3DOM-based catalysts: A review. *Chemosphere* **2021**, *262*, 127886. [CrossRef]
24. Zhai, G.; Wang, J.; Chen, Z.; Yang, S.; Men, Y. Highly enhanced soot oxidation activity over 3DOM $Co_3O_4$-$CeO_2$ catalysts by synergistic promoting effect. *J. Hazard. Mater.* **2019**, *363*, 214–226. [CrossRef] [PubMed]
25. Reculusa, S.; Agricole, B.; Derré, A.; Couzi, M.; Sellier, E.; Delhaès, P.; Ravaine, S. Colloidal Crystals as Templates for Macroporous Carbon Electrodes of Controlled Thickness. *Electroanalysis* **2007**, *19*, 379–384. [CrossRef]
26. Yu, B.; Song, Q.; Cong, H.; Xu, X.; Han, D.; Geng, Z.; Zhang, X.; Usman, M. A smart thermo- and pH-responsive microfiltration membrane based on three dimensional inverse colloidal crystals. *Sci. Rep.* **2017**, *7*, 12112. [CrossRef] [PubMed]
27. Pan, T.; Wu, Z.; Yip, A.C.K. Advances in the Green Synthesis of Microporous and Hierarchical Zeolites: A Short Review. *Catalysts* **2019**, *9*, 274. [CrossRef]
28. Song, L.; Zhang, J.; Sun, L.; Xu, F.; Li, F.; Zhang, H.; Si, X.; Jiao, C.; Li, Z.; Liu, S.; et al. Mesoporous metal–organic frameworks: Design and applications. *Energy Environ. Sci.* **2012**, *5*, 7508–7520. [CrossRef]
29. Pham, Q.N.; Barako, M.T.; Tice, J.; Won, Y. Microscale Liquid Transport in Polycrystalline Inverse Opals across Grain Boundaries. *Sci. Rep.* **2017**, *7*, 10465. [CrossRef]
30. Cong, H.; Yu, B.; Tang, J.; Li, Z.; Liua, X. Current status and future developments in preparation and application of colloidal crystals. *Chem. Soc. Rev.* **2013**, *42*, 7774–7800. [CrossRef]
31. Bertone, J.F.; Jiang, P.; Hwang, K.S.; Mittleman, D.M.; Colvin, V.L. Thickness Dependence of the Optical Properties of Ordered Silica-Air and Air-Polymer Photonic Crystals. *Phys. Rev. Lett.* **1999**, *83*, 300–303. [CrossRef]
32. Dziomkina, N.V.; Vancso, G.J. Colloidal crystal assembly on topologically patterned templates. *Soft Matter.* **2005**, *1*, 265–279. [CrossRef]
33. Dziomkina, N.V.; Hempenius, M.A.; Vancso, J.G. Symmetry Control of Polymer Colloidal Monolayers and Crystals by Electrophoretic Deposition on Patterned Surfaces. *Adv. Mater.* **2005**, *17*, 237–240. [CrossRef]
34. Braun, P.V.; Zehner, R.W.; White, C.A.; Weldon, M.K.; Kloc, C.; Patel, S.S.; Wiltzius, P. Optical spectroscopy of high dielectric contrast 3D photonic crystals. *Europhys. Lett.* **2001**, *56*, 207–213. [CrossRef]
35. Wang, J.; Mbah, C.F.; Przybilla, T.; Zubiri, B.A.; Spiecker, E.; Engel, M.; Vogel, N. Magic number colloidal clusters as minimum free energy structures. *Nat. Commun.* **2018**, *9*, 5259. [CrossRef]
36. Choi, H.K.; Yang, Y.J.; Park, O.O. Hemispherical Arrays of Colloidal Crystals Fabricated by Transfer Printing. *Langmuir* **2014**, *30*, 103–109. [CrossRef]
37. Haibin, N.; Ming, W.; Wei, C. Sol-gel co-assembly of hollow cylindrical inverse opals and inverse opal columns. *Opt. Express* **2011**, *19*, 25900–25910. [CrossRef] [PubMed]
38. Sandu, I.; Fleacă, C.T.; Dumitrache, F.; Sava, B.; Urzică, I.; Dumitru, M. From thin "coffee rings" to thick colloidal crystals, through drop spreading inhibition by the substrate edge. *Appl. Phys. A* **2021**, *127*, 325. [CrossRef]
39. Zhao, Y.; Shang, L.; Cheng, Y.; Gu, Z. Spherical Colloidal Photonic Crystals. *Acc. Chem. Res.* **2014**, *47*, 3632–3642. [CrossRef] [PubMed]
40. Wang, J.; Zhu, J. Recent advances in spherical photonic crystals: Generation and applications in optics. *Eur. Polym. J.* **2013**, *49*, 3420–3433. [CrossRef]
41. Kohoutek, T.; Parchine, M.; Bardosova, M.; Pemble, M.E. Controlled self-assembly of Langmuir-Blodgett colloidal crystal films of monodispersed silica particles on non-planar substrates. *Colloids Surf. A Physicochem. Eng. Asp.* **2020**, *593*, 124625. [CrossRef]
42. Choi, D.-G.; Yu, H.K.; Jang, S.G.; Yang, S.-M. Colloidal Lithographic Nanopatterning via Reactive Ion Etching. *J. Am. Chem. Soc.* **2004**, *126*, 7019–7025. [CrossRef]
43. Ding, T.; Luo, L.; Wang, H.; Chen, L.; Liang, K.; Clays, K.; Song, K.; Yang, G.; Tung, C.-H. Patterning and pixelation of colloidal photonic crystals for addressable integrated photonics. *J. Mater. Chem.* **2011**, *21*, 11330–11334. [CrossRef]

44. Yao, J.; Yan, X.; Lu, G.; Zhang, K.; Chen, X.; Jiang, L.; Yang, B. Patterning colloidal crystals by lift-up soft lithography. *Adv. Mater.* **2004**, *16*, 81–84. [CrossRef]
45. Fustin, C.-A.; Glasser, G.; Spiess, H.W.; Jonas, U. Parameters influencing the templated growth of colloidal crystals on chemically patterned surfaces. *Langmuir* **2004**, *20*, 9114–9123. [CrossRef]
46. Choi, W.; Park, O. Micropatterns of colloidal assembly on chemically patterned surface. *Colloids Surf. A Physicochem. Eng. Asp.* **2006**, *277*, 131–135. [CrossRef]
47. Ng, E.C.H.; Chin, K.M.; Wong, C.C. Controlling Inplane Orientation of a Monolayer Colloidal Crystal by Meniscus Pinning. *Langmuir* **2011**, *27*, 2244–2249. [CrossRef]
48. Coll, A.; Bermejo, S.; Hernández, D.; Castaner, L. Colloidal crystals by electrospraying polystyrene nanofluids. *Nanoscale Res. Lett.* **2013**, *8*, 26. [CrossRef] [PubMed]
49. Nam, H.; Song, K.; Ha, D.; Kim, T. Inkjet Printing Based Mono-layered Photonic Crystal Patterning for Anti-counterfeiting Structural Colors. *Sci. Rep.* **2016**, *6*, 30885. [CrossRef] [PubMed]
50. Yi, D.K.; Kim, M.J.; Kim, D.-Y. Surface Relief Grating Induced Colloidal Crystal Structures. *Langmuir* **2002**, *18*, 2019–2023. [CrossRef]
51. Míguez, H.; Yang, S.M.; Ozin, G.A. Optical Properties of Colloidal Photonic Crystals Confined in Rectangular Microchannels. *Langmuir* **2003**, *19*, 3479–3485. [CrossRef]
52. Velev, O.D.; Lenhoff, A.M.; Kaler, E.W. A class of microstructured particles through colloidal crystallization. *Science* **2000**, *287*, 2240–2243. [CrossRef]
53. Fenollosa, R.; Rubio, S.; Meseguer, F.; Sanchez-Dehesa, J. Photonic crystal microprisms obtained by carving artificial opals. *J. Appl. Phys.* **2003**, *93*, 671–674. [CrossRef]
54. Matsuo, S.; Fujine, T.; Fukuda, K.; Juodkazis, S.; Misawa, H. Formation of free-standing micropyramidal colloidal crystals grown on silicon substrate. *Appl. Phys. Lett.* **2003**, *82*, 4283. [CrossRef]
55. Kaynak, H.K.; Babaarslan, O. *Polyester Microfilament Woven Fabrics, Woven Fabrics*; Jeon, H.-Y., Ed.; IntechOpen: London, UK, 2012. [CrossRef]
56. Grishanov, S.; Meshkov, V.; Omelchenko, A. A Topological Study of Textile Structures. Part I: An Introduction to Topological Methods. *Text. Res. J.* **2009**, *79*, 702–713. [CrossRef]
57. Yan, C.; Su, B.; Shi, Y.; Jiang, L. Liquid bridge induced assembly (LBIA) strategy: Controllable one-dimensional patterning from small molecules to macromolecules and nanomaterials. *Nano Today* **2019**, *25*, 13–26. [CrossRef]
58. Jiang, X.; Feng, J.; Huang, L.; Wu, Y.; Su, B.; Yang, W.; Mai, L.; Jiang, L. Bioinspired 1D Superparamagnetic Magnetite Arrays with Magnetic Field Perception. *Adv. Mater.* **2016**, *28*, 6952–6958. [CrossRef] [PubMed]
59. Su, B.; Zhang, C.; Chen, S.; Zhang, X.; Chen, L.; Wu, Y.; Nie, Y.; Kan, X.; Song, Y.; Jiang, L. A General Strategy for Assembling Nanoparticles in One Dimension. *Adv. Mater.* **2014**, *26*, 2501–2507. [CrossRef]
60. Sun, J.; Bhushan, B.; Tong, J. Structural coloration in nature. *RSC Adv.* **2013**, *3*, 14862–14889. [CrossRef]
61. Deegan, R.D.; Bakajin, O.; Dupont, T.; Huber, G.; Nagel, S.R.; Witten, T.A. Capillary flow as the cause of ring stains from dried liquid drops. *Nature* **1997**, *389*, 827–829. [CrossRef]
62. Li, Y.; Yang, Q.; Li, M.; Song, Y. Rate-dependent interface capture beyond the coffee-ring effect. *Sci. Rep.* **2016**, *6*, 24628. [CrossRef]
63. Sandu, I.; Dumitru, M.; Fleaca, C.T.; Dumitrache, F. Hanging colloidal drop: A new photonic crystal synthesis route. *Photonics Nanostruct.* **2018**, *29*, 42–48. [CrossRef]
64. Checoury, X.; Enoch, S.; López, C.; Blanco, A. Stacking patterns in self-assembly opal photonic crystals. *Appl. Phys. Lett.* **2007**, *90*, 161131. [CrossRef]
65. Rybin, M.V.; Samusev, K.B.; Limonov, M.F. High Miller-index photonic bands in synthetic opals. *Photonics Nanostruct.* **2007**, *5*, 119–124. [CrossRef]
66. Nguyen, T.V.; Pham, L.T.; Bui, K.X.; Nghiem, L.H.T.; Nguyen, N.T.; Vu, D.; Do, H.Q.; Vu, L.D.; Nguyen, H.M. Size Determination of Polystyrene Sub-Microspheres Using Transmission Spectroscopy. *Appl. Sci.* **2020**, *10*, 5232. [CrossRef]
67. Kurokawa, Y.; Miyazaki, H.; Jimba, Y. Light scattering from a monolayer of periodically arrayed dielectric spheres on dielectric substrates. *Phys. Rev. B* **2002**, *65*, 201102. [CrossRef]
68. Prasad, T.; Colvin, V.; Mittleman, D. Superprism phenomenon in three-dimensional macroporous polymer photonic crystals. *Phys. Rev. B* **2003**, *67*, 165103. [CrossRef]
69. Meng, G.; Paulose, J.; Nelson, D.R.; Manoharan, V.N. Elastic instability of a crystal growing on a curved surface. *Science* **2014**, *343*, 634–637. [CrossRef]
70. Ai, B.; Zhao, Y. Glancing angle deposition meets colloidal lithography: A new evolution in the design of nanostructures. *Nanophotonics* **2019**, *8*, 1–26. [CrossRef]
71. Lee, H.; Heo, S.G.; Bae, Y.; Lee, H.; Kim, J.; Yoon, H. Multiple guidance of light using asymmetric micro prism arrays for privacy protection of device displays. *Opt. Express* **2021**, *29*, 2884–2892. [CrossRef]
72. Kim, Y.-C.; Heo, Y.-J.; Lee, S.K.; Jung, Y.-S.; Jeong, H.-J.; Kim, S.-T.; Yoo, J.-S.; Kim, D.-Y.; Jang, J.-H. Improved light absorption in perovskite solar module employing nanostructured micro-prism array. *Sol. Energy Mater. Sol. Cells* **2021**, *226*, 111077. [CrossRef]
73. Park, D.S.; Young, B.M.; You, B.; Singh, V.; Soper, S.A.; Murphy, M.C.; An, A. An Integrated, Optofluidic System With Aligned Optical Waveguides, Microlenses, and Coupling Prisms for Fluorescence Sensing. *J. Microelectromec. Syst.* **2020**, *29*, 600–609. [CrossRef]

74. Lu, C.; Xu, J.; Han, J.; Li, X.; Xue, N.; Li, J.; Wu, W.; Sun, X.; Wang, Y.; Ouyang, Q.; et al. A novel microfluidic device integrating focus-separation speed reduction design and trap arrays for high-throughput capture of circulating tumor cells. *Lab Chip* **2020**, *20*, 4094–4105. [CrossRef] [PubMed]
75. Chen, L.; Su, B.; Jiang, L. Recent advances in one-dimensional assembly of nanoparticles. *Chem. Soc. Rev.* **2019**, *48*, 8–21. [CrossRef] [PubMed]
76. Jiang, Z.; Zhang, K.; Du, L.; Cheng, Z.; Zhang, T.; Ding, J.; Li, W.; Xu, B.; Zhu, M. Construction of chitosan scaffolds with controllable microchannel for tissue engineering and regenerative medicine. *Mater. Sci. Eng. C* **2021**, *126*, 11217. [CrossRef] [PubMed]

polymers

MDPI

*Article*

# Effect of Silicone Inlaid Materials on Reinforcing Compressive Strength of Weft-Knitted Spacer Fabric for Cushioning Applications

**Annie Yu *, Sachiko Sukigara and Miwa Shirakihara**

Faculty of Fiber Science and Engineering, Kyoto Institute of Technology, Matsugasaki, Sakyo-ku, Kyoto 606-8585, Japan; sukigara@kit.ac.jp (S.S.); m0651009@edu.kit.ac.jp (M.S.)
* Correspondence: annieyu@kit.ac.jp; Tel.: +81-75-724-7230

**Abstract:** Spacer fabrics are commonly used as cushioning materials. They can be reinforced by using a knitting method to inlay materials into the connective layer which reinforces the structure of the fabric. The compression properties of three samples that were fabricated by inlaying three different types of silicone-based elastic tubes and one sample without inlaid material have been investigated. The mechanical properties of the elastic tubes were evaluated and their relationship to the compression properties of the inlaid spacer fabrics was analysed. The compression behaviour of the spacer fabrics at an initial compressive strain of 10% is not affected by the presence of the inlaid tubes. The Young's modulus of the inlaid tubes shows a correlation with fabric compression. Amongst the inlaid fabric samples, the spacer fabric inlaid with highly elastic silicone foam tubes can absorb more compression energy, while that inlaid with silicone tubes of higher tensile strength has higher compressive stiffness.

**Keywords:** weft-knitted spacer fabric; cushioning; compression; tensile strength; silicone inlay; sandwich textile structure

**Citation:** Yu, A.; Sukigara, S.; Shirakihara, M. Effect of Silicone Inlaid Materials on Reinforcing Compressive Strength of Weft-Knitted Spacer Fabric for Cushioning Applications. *Polymers* **2021**, *13*, 3645. https://doi.org/10.3390/polym13213645

Academic Editor: Muhammad Tayyab Noman

Received: 5 October 2021
Accepted: 21 October 2021
Published: 22 October 2021



## 1. Introduction

Cushioning materials can be found in many different types of apparel and wearable items that provide shock absorption and wearer protection. Traditionally, polymeric foams are used in insoles, bra cups, and protective apparel to deliver the cushioning function [1–3]. Recently, spacer fabrics which are weft or warp knitted have been used as an alternative to foam materials. They are now a viable option because they have a unique three-dimensional (3D) knitted structure and provide the products with higher air and water vapour permeabilities and breathability [4–8]. Spacer fabrics can also be readily found in daily life items, such as in shoes, chairs, car seats, carpets, mattresses, backpacks, etc. [9].

Compression behaviour is an important criterion for determining whether or not a material can offer suitable cushioning functions and hand feel for different end-uses. In terms of polymeric foam, the compression properties can be controlled by varying the composition and the density to accommodate various applications [10,11]. Spacer fabric consists of two surface layers that are connected by spacer yarns which form a connective layer. Variations in any component of spacer fabric can affect its mechanical properties and wear comfort. The compression properties of spacer fabrics have often been studied. The elasticity of yarns used in the surface layer is one of the factors that contributes to the compression properties [12–14]. The compression properties of weft-knitted spacer fabric have been found to be affected by the number of tuck stitches in the surface layers [15]. However, it has also been shown that the compression properties of spacer fabric are related to the connective layer [16–18]. Monofilament and multifilament yarns are commonly used as spacer yarns and can impart entirely different properties to the fabric [19–21]. The thickness, composition, connecting distance, inclination angle, and pattern of the spacer yarns can be used to control the thickness and compressive stiffness of the spacer fabric [22–24]. The previous studies related to compression behaviour of spacer fabrics mainly focus on the structural properties. However, it can be challenging to produce a thin fabric with high compression strength and high energy absorption when using a conventional spacer fabric structure. A spacer fabric that is thin enough for use as insoles or protective garments (less than 1 cm in thickness) could easily collapse from stress produced by the human body [25]. However, if a more compact connective layer is used, the spacer fabric becomes heavy and stiff, thus reducing its cushioning ability. Hamedi et al., proposed the use of nickel–titanium alloy wires as the spacer yarn [26]. The spacer fabric shows an improvement on the compression energy absorption; however, the cost of the fabric is also largely increased. Moreover, investigation in applying materials other than polyester or polyamide filament yarns in the connective layer of the spacer fabric is still limited.

In this study, a composite structure that consists of additional silicone-based materials was investigated so as to enhance the cushioning properties of spacer fabric. A novel sandwich structure with inlaid silicone tubes in the connective layer of spacer fabric has previously been developed [27]. Silicone is a synthetic polymer with a silicon–oxygen main chain [28]. Silicone is flexible, flame retardant, and relatively inert. Silicone inlaid tubes offer extra support to reinforce the spacer structure so that the fabric can withstand pre-stress from the body during application without the flexibility and energy absorption properties being sacrificed. In order to further investigate the effect of inlaid materials on the properties of spacer fabric, samples made with three different kinds of silicone-based tubular materials were fabricated and evaluated. The main purpose was to understand the relationship between the mechanical properties of the inlaid tubular materials and the compression properties of spacer fabric. The findings can contribute to furthering the development of sandwich structured textile materials and enhance wearable cushioning products. The inlaid materials can become a new parameter in adjusting the compression and cushioning properties of knitted spacer fabrics which allows the fabric to provide the desired energy absorption ability for various end-uses.

## 2. Materials and Methodology

### *2.1. Materials*

The yarns for knitting the surface layers of the spacer fabric samples included 450D 3-ply 100% polyester drawn textured yarn (LS 1/20, Amossa, Osaka, Japan) and 140D spandex yarn (Heng Jing Limit, Jiangsu, China). The spacer yarn was 100% polyester monofilaments with a diameter of 0.12 mm (Nantong Ntec Monofilament Technology Co., Ltd., Nantong, China). Three types of silicone-based tubular materials, including silicone foam rods, silicone rods, and silicone hollow tubes (Yuema, Shanghai, China) were inlaid in the connective layer of the spacer fabrics, as listed in Table 1. The three types of inlay materials have a similar thickness but a different linear density. T1, which incorporates silicone foam as the tubular material, is relatively light in weight. The good elasticity and low density of silicone foams make it suitable to be used in challenging application such as shock absorbers, wound dressings, and joint sealants [29,30]. T2 and T3 both incorporate silicone rubber. Silicone rubbers can be made into tubes, hose, gaskets, and seals [31]. T2 incorporate silicone in a solid rod form, while T3 is a hollow tube form.

**Table 1.** Details of the three tubular samples.

| | Type | Diameter (mm) | Weight (g/m) | Longitudinal View | Cross-Sectional View |
|---|---|---|---|---|---|
| T1 | Silicone foam rod | 1.23 | 0.8 | 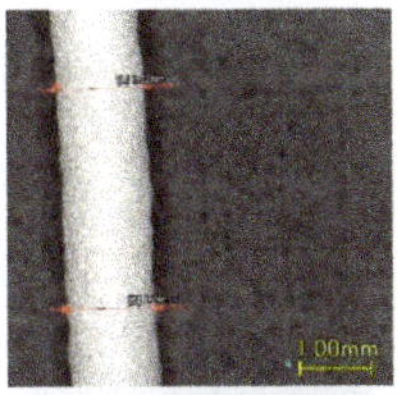 | 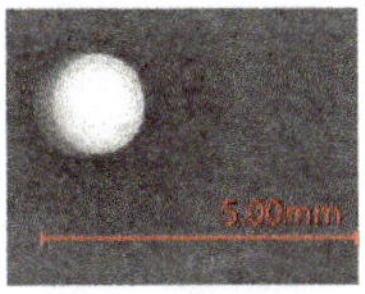 |
| T2 | Silicone rod | 1.22 | 1.4 | 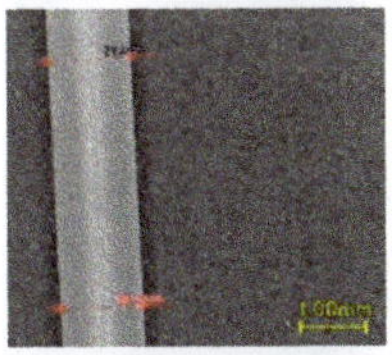 | 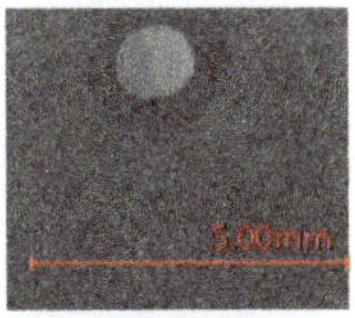 |
| T3 | Silicone hollow tube | 1.12 | 1.0 | 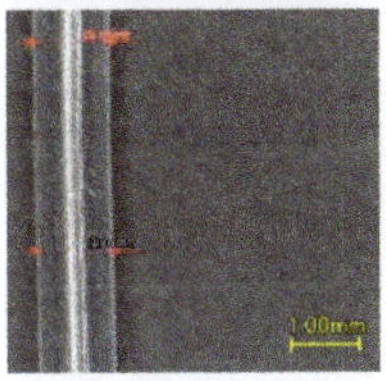 | 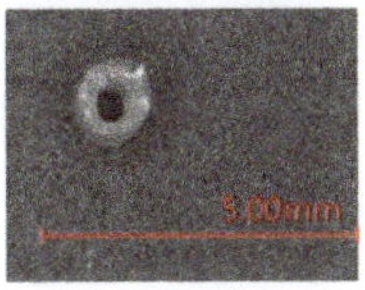 |

### *2.2. Preparation of the Inlaid Spacer Fabric*

Three samples inlaid with each type of tubular material and one sample without any inlaid material were produced by using a 10-gauge v-bed flat knitting machine (SWG091N210G, Shima Seiki, Wakayama, Japan). The two surface layers were knitted with a single jersey structure and the connective layer was a spacer structure with a linking distance of 6 needles for all the samples. One course of the spacer fabric consisted of 2 courses of knit stitches on the surface and 6 tuck stitch courses of spacer yarn. Following a previous study, the tubular materials were inlaid into the connective layer with miss stitches [27]. Therefore, the tubular materials did not come into contact with the knitting needles and floated between the front and back needle beds. The inlaid course was carried out between the tuck courses of spacer yarn and hence the spacer yarns acted as a net to hold the silicone tubes in place. One course of inlay was inserted in every 4th knitting course of the spacer fabric (Figure 1). The weight, thickness, and cross-sectional views of the four fabric samples are shown in Table 2.

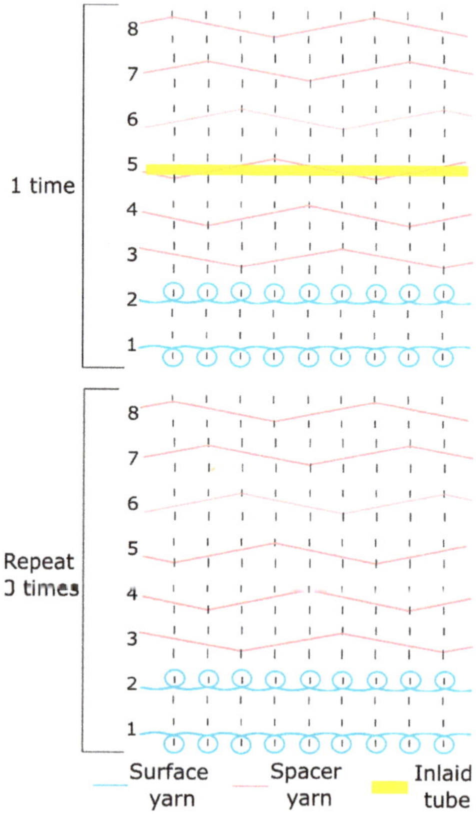


**Figure 1.** Yarn path diagram of the spacer fabric samples with inlaid tubes.

**Table 2.** Details of spacer fabric samples.

| Fabric Sample | Inlay | Weight (kg/m$^2$) | | Thickness (mm) | | Cross-Sectional View | |
|---|---|---|---|---|---|---|---|
| | | | | | | Course-Wise | Wale-Wise |
| F0 | No inlay | 468 | ±4.3 | 4.41 | ±0.13 |  | 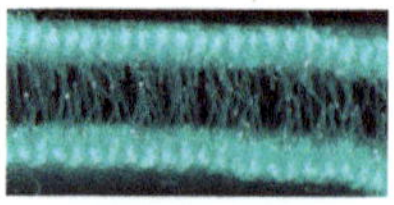 |
| FT1 | Inlaid with T1 | 681 | ±19.8 | 5.26 | ±0.03 | 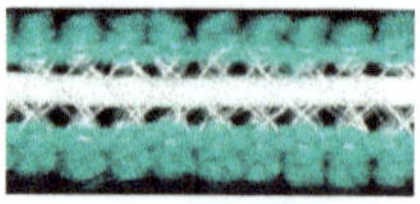 | 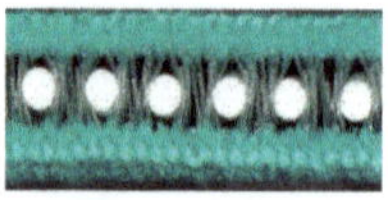 |
| FT2 | Inlaid with T2 | 865 | ±23.2 | 5.28 | ±0.13 | 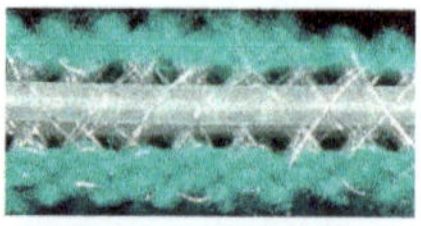 | 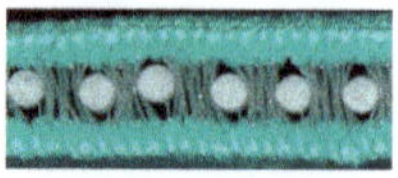 |
| FT3 | Inlaid with T3 | 784 | ±14.8 | 5.43 | ±0.06 | 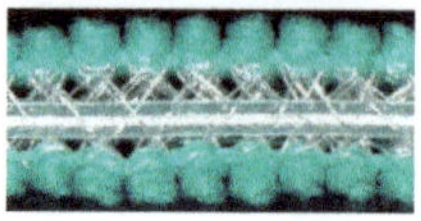 |  |

*2.3. Evaluation of Mechanical Properties of Inlaid Tubular Materials*

Tensile and compression tests were conducted on the three tubular samples by using a universal testing machine (EZ-S, Shimadzu, Kyoto, Japan). The tensile test was conducted in accordance with ASTM D2731-15, the standard test method for elastic properties of elastomeric yarns. The tubular sample was mounted between a pair of jaws (Figure 2a). The gauge length was set at 50 mm with a pre-tension of 2.55 g. The sample was subjected to 5 loading and unloading cycles. The sample was extended at a rate of 500 mm/min, held at the maximum extension limit for 30 s, and returned at a rate of 100 mm/min. The maximum extension was set at 300% of the gauge length. As T1 and T2 failed to extend to the 300% strain, 75% of the elongation at first break was used as the maximum extension instead. Therefore, T1 and T2 were extended up to 202% and 224% of the gauge length, respectively. The compression test was carried out by using a pair of compression plates. The sample was mounted onto the centre of the plate at a length of 118 mm. Double-sided tape was used to hold the sample in place during testing (Figure 2b). The compression speed was 20 mm/min with a maximum compression displacement of 0.6 mm. The samples were conditioned under a standard environment (20 ± 2 °C, 65 ± 2% relative humidity) for 24 h before they were tested.

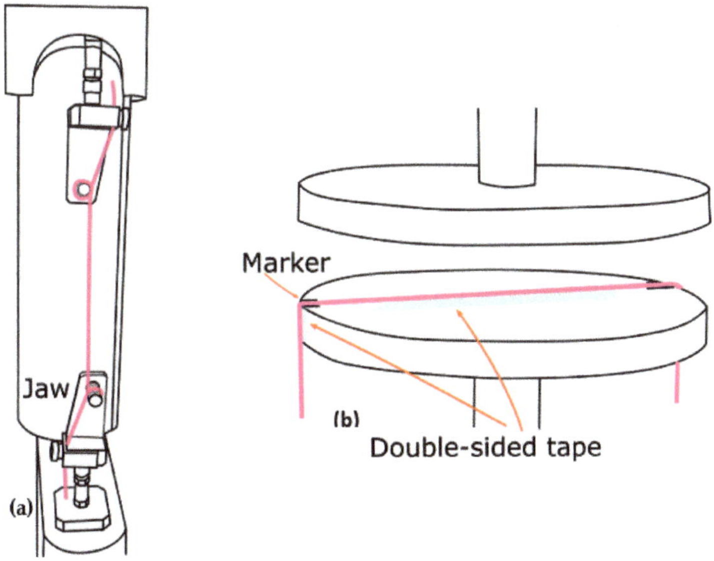


**Figure 2.** Setting of (**a**) tensile test and (**b**) compression test of the tubular samples.

*2.4. Evaluation of Compression Properties of the Spacer Fabrics*

A compression test on the fabric samples was carried out by using the same testing machine along with a pair of compression plates with a diameter of 118 mm. The fabric samples were prepared with dimensions of 50 mm × 50 mm. The compression rate was 12 mm/min with a maximum compression stress of 60 kPa. Four specimens of each sample were tested. The compression energy of each sample was calculated as the integral of the compressive loading (WC) and unloading (WC′). All of the fabric samples were allowed to relax for one week after released from the knitting machine and stored in a standard environment (20 ± 2 °C, 65 ± 2% relative humidity) before testing. ANOVA was adopted to analyse the effect of the inlaid materials on the compression strain and compression energy. A Sidak post hoc test was used to analyse the effect between pairs. The alpha level was set at 0.05 for statistical significance.

## 3. Results and Discussion

### *3.1. Analysis of the Tensile Properties of the Inlaid Tubular Materials*

The plotted loading and unloading curves of the first and the fifth cycles of the tensile test of the three tubular samples are presented in Figure 3a,b. The force at 100% and 200% elongation of the tubes at the first and fifth loading cycles and the fifth unloading cycle, together with the Young's modulus measured from the first extension loading are shown in Figure 3c,d. Figure 3e shows the displacement–force curves obtained from the compression test. The three tubular samples show very different non-linear elastic behaviours. Silicone foam is a porous viscoelastic polymer foamed from silicone rubbers [32,33]. Silicone foam has the properties of silicone combined with foam properties, light weight, and good flexibility [34]. T1 is the most elastic and has the lowest Young's modulus amongst the three tubular samples. A relatively small force is required to extend T1 and the loss in elastic hysteresis is also small. T2 is solid rod form of silicone, has the highest Young's modulus, requires the largest force for extension, and shows a large hysteresis, especially in the first cycle of extension. T3 is hollow tube form of silicone and therefore has a lower weight and tensile strength than T2. T3 can be extended to the longest length at break. In comparison to T1, T3 requires a slightly higher force of 0.1 N to extend to a strain of 100% but 0.1 N less to extend to a strain of 200%. T2 has the highest compressive stiffness, followed by T3, whereas T1 is the softest material and most easily compressed. By inlaying the three tubular samples which have a different tensile strength, elasticity and stiffness into the spacer fabric, the effect of the mechanical properties of the inlaid materials on fabric compression properties can be identified.

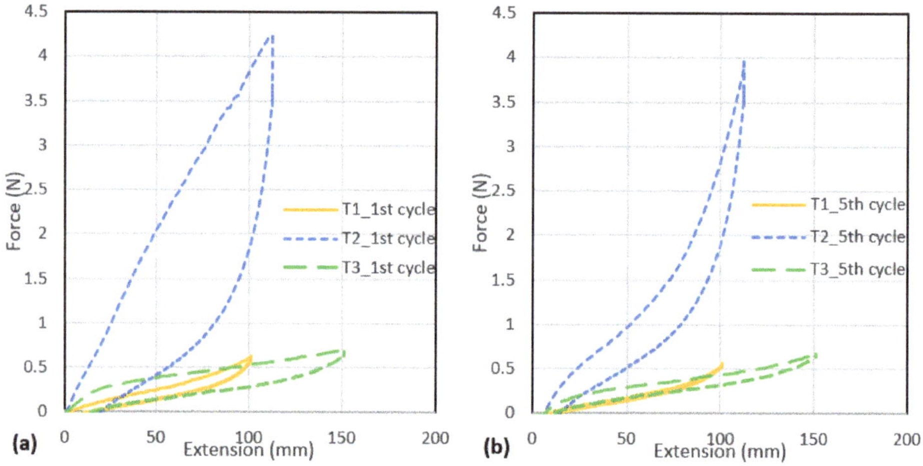


**Figure 3.** *Cont.*

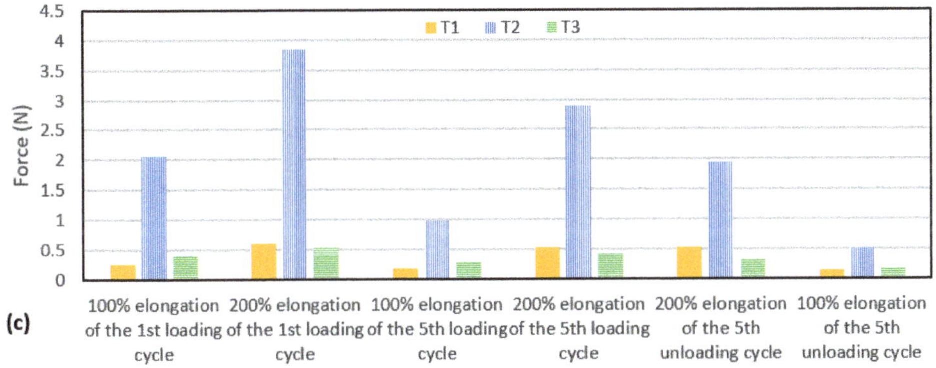


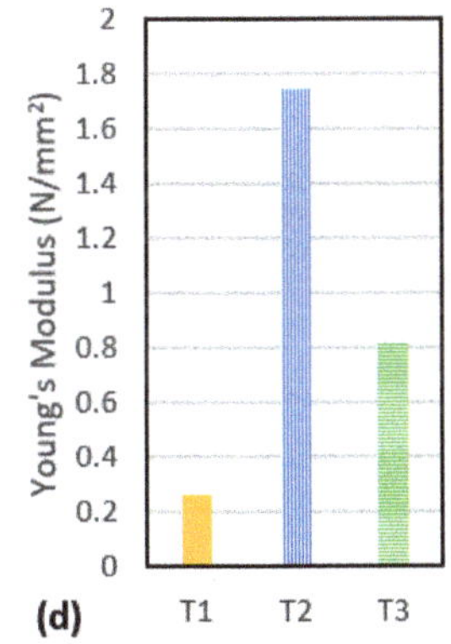


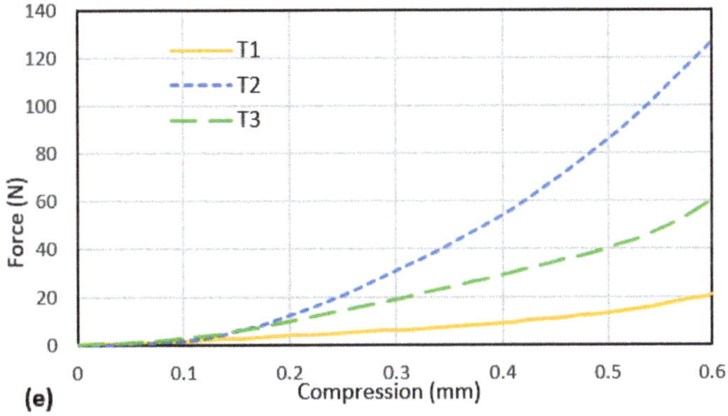


**Figure 3.** Tensile extension–force curves for (**a**) the first cycle of extension, and (**b**) the fifth cycle of extension; (**c**) force for 100% and 200% elongation; (**d**) Young's modulus, and (**e**) compression displacement–force curves of the three tubular samples.

### 3.2. *Effect of the Inlaid Tubes on the Compression Behaviour of Spacer Fabric*

The compression stress–strain curves, fabric strain at a stress of 60 kPa, and the compression energy of the four fabric samples are shown in Figure 4. At a compressive strain of 0 to 10%, the compression behaviour of the four fabrics is very similar because they are constructed with the same surface and connective structures that have the same materials. The initial stress up to 3.5 kPa compresses the loose surface layers and tightens the spacer structure. This shows that the initial softness of the spacer fabric is not affected by the presence of the inlaid tube in the connective layer.

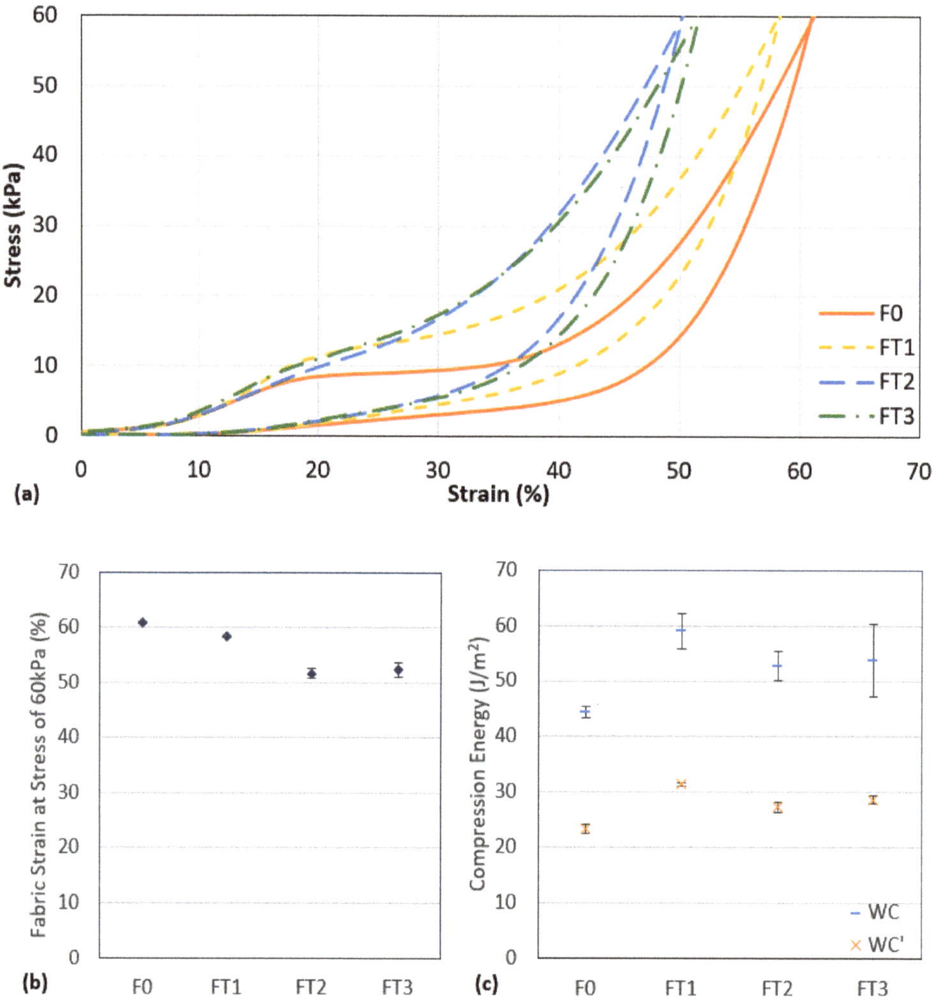


**Figure 4.** (**a**) Compression stress–strain curve, (**b**) fabric strain at a stress of 60 kPa, and (**c**) compression energy of four fabric samples.

When the compression stress is further increased, the monofilament yarns start to deform and buckle. F0 starts to collapse, thus showing a decrease in the slope of the stress—strain curve and entering a plateau stage at a stress of 8.5 kPa. In the plateau stage, the monofilament yarns can no longer hold the connective structure which leads to the shearing of the fabric layers and rotation of the yarns. The fabric can be easily compressed with a small increase in stress. FT1, FT2, and FT3 consist of inlaid tubes to give extra support to the structure and withstand some of the stress applied. As shown in Figure 4a, the inlaid spacer fabric can withstand a higher compression stress than the one without inlay. This supports that inlaid yarns decrease the deformation ability of knitted fabrics [35]. The different inlaid tubes have different Young's moduli and mechanical properties and thus different compression properties can be found for all three fabrics. FT1 reaches the plateau stage at 11.1 kPa of compression. The plateau stage of FT1 covers a smaller range of strain when compared with F0. This is because the inlaid silicone foam rods act as a buffer to absorb a certain amount of the compression energy applied to the connective layer. On the other hand, there is no prominent plateau stage for FT2 and FT3. The inlaid tubes T2 and T3 are relatively stiff and can withstand most of the compression stress that acts on the connective layer and exceed the energy absorption capacity of the monofilament yarns. Therefore, the plateau stage that typically appears in the compression curve of spacer fabric is not shown for FT2 and FT3.

Significant differences ($p < 0.05$) between the four samples on the fabric strain at a stress of 60 kPa, WC, and WC′ are found in the results of ANOVA. For the fabric strain at 60 kPa, F0 and F1 show significant difference with all the other samples while there is no significant difference between F2 and F3. At a stress of 60 kPa, F0 was compressed to the highest strain of 61%. With the silicone foam rod inlay, the strain at 60 kPa of stress for FT1 decreases by 58%. The fabric structure of FT2 and FT3 is supported by the relatively stiffer inlaid tubes and hence even smaller strains are shown at a stress of 60 kPa. Moreover, the spacer fabrics with inlay have a significantly higher WC than conventional spacer fabric with no inlay. This shows that the inlaid tubes could help to absorb more compression energy than regular spacer fabric. The inlaid structure can provide better support against impact forces when used as padding or cushioning materials. Although no significant difference ($p > 0.05$) was shown on the WC between the pairwise comparison of the three spacer fabrics with inlay, FT1 shows a significant difference with FT2 and FT3 on the WC′. The compression behaviour and the compression energy of the spacer fabric can be affected by the inlaid material used.

### 3.3. *Relationship between Properties of Inlaid Tubes and Spacer Fabric Properties*

The relationship between the elasticity of the inlaid tubes and the compression properties of the spacer fabric was further investigated. In Figure 5, the logarithmic relationships between the Young's modulus of the inlaid tube samples with the WC of the spacer fabric samples and the fabric strain at a stress of 60 kPa show a high coefficient of determination (R2 > 0.9). The inlaid tubular materials show a significant effect on the compression behaviour of the spacer fabric. Amongst the three types of inlaid spacer fabrics, FT1 shows the highest WC. By inlaying a softer and higher elastic material such as the silicone foam rods, the spacer fabric can absorb a larger amount of compression energy. T2 is however heavy and has high tensile strength. Therefore, FT2 is stiffer than FT1 and FT3, especially when the compressive strain is above 35%, where the monofilament yarns have buckled and the inlaid tubes mainly support the fabric against compression forces. The fabric compression behaviour of FT2 and FT3 is similar. The hollow silicone tube, T3, has a Young's modulus and compressive stiffness that ranges between those of T1 and T2. Therefore, FT3 has a slightly higher WC than FT2. As only three samples are studied, it is difficult to generalise the results to all the different types of inlay materials and inlaid spacer fabric. However, the correlation between tensile properties of the silicone foam rod, silicone rod, and silicone hollow tube used in this study and the corresponding inlaid spacer fabric is observed. The mechanical properties of the inlaid materials can affect the compression properties of the inlaid spacer fabric.

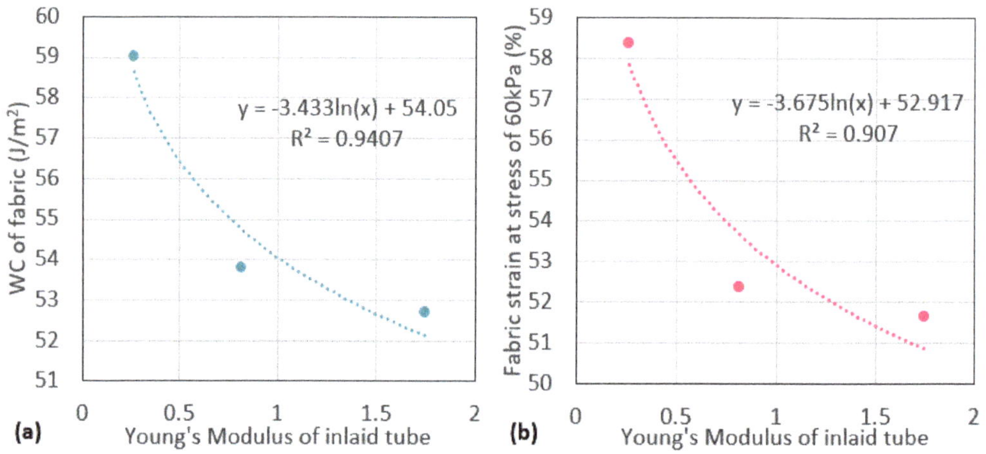


**Figure 5.** Relationship between the Young's modulus of the inlaid tube, (**a**) the compression energy of the fabric samples, and (**b**) the fabric strain at a stress of 60 kPa.

## 4. Conclusions

The effect of inlaying tubular materials in the connective layer of spacer fabric on compression reinforcement has been investigated. Three weft-knitted spacer fabric samples inlaid with different tubular materials and one conventional spacer fabric without inlaid material as the reference were fabricated. The mechanical properties of the tubular samples and the compression properties of the fabric samples and the relationship between them were evaluated. The following conclusions were made based on the findings:

- The compression behaviour of the spacer fabric at an initial compressive strain of 10% is not affected by the presence of inlaid tubes in the connective layer.
- The inlaid spacer fabrics require higher stress to enter the plateau stage than the conventional spacer fabric. When an inlaid material with higher tensile strength and compression strength is used, no obvious plateau stage can be found in the compression stress–strain curves of the fabric.
- The inlaid spacer fabrics not only have higher compression strength but can also absorb more compression energy than the conventional spacer fabric. The inlaying of elastic materials such as silicone foam or silicone rods effectively reinforces the spacer fabric.
- Different inlay materials with different Young's moduli and tensile behaviours can affect the compression energy and stiffness of the resultant fabrics. The spacer fabric inlaid with silicone foam rods, which have lower tensile strength and compression strength than silicone rods and silicone hollow tubes, can absorb more compression energy. On the other hand, the spacer fabric that is inlaid with silicone rods with a high tensile strength and compression strength has the highest compressive stiffness amongst the fabric samples.

A better understanding of the effect of different types of inlaid tubes on the compression properties of weft-knitted spacer fabric is provided. The findings can be used as a reference in the design and development of spacer fabrics to meet the requirements of various cushioning applications.

**Author Contributions:** A.Y. conceived, designed, performed experiments, analysed the results, wrote the manuscript, supervised, and acquired funding. S.S. gave advice, reviewed, and edited the paper writing. M.S. performed experiments and analysed the results. All authors have read and agreed to the published version of the manuscript.

**Funding:** This work was support by JSPS KAKENHI (Grant Number JP20K14638).

**Data Availability Statement:** Not applicable.

**Conflicts of Interest:** The authors declare no conflict of interest.

## References


1. Monie, F.; Vidil, T.; Grignard, B.; Cramail, H.; Detrembleur, C. Self-foaming polymers: Opportunities for the next generation of personal protective equipment. *Mater. Sci. Eng. R Rep.* **2021**, *145*, 100628. [CrossRef]
2. Jin, F.-L.; Zhao, M.; Park, M.; Park, S.-J. Recent Trends of Foaming in Polymer Processing: A Review. *Polymers* **2019**, *11*, 953. [CrossRef]
3. Yick, K.; Ng, S.; Wu, L. Innovations in the bra cup molding processes. In *Advances in Women's Intimate Apparel Technology*; Yu, W., Ed.; Woodhead Publishing: Cambridge, UK, 2016; pp. 69–87.
4. Liu, Y.; Hu, H. Compression property and air permeability of weft-knitted spacer fabrics. *J. Text. Inst.* **2011**, *102*, 366–372. [CrossRef]
5. Rajan, T.P.; Souza, L.D.; Ramakrishnan, G.; Zakriya, G.M. Comfort properties of functional warp-knitted polyester spacer fabrics for shoe insole applications. *J. Ind. Text.* **2016**, *45*, 1239–1251. [CrossRef]
6. Onal, L.; Yildirim, M. Comfort properties of functional three-dimensional knitted spacer fabrics for home-textile applica-tions. *Text. Res. J.* **2012**, *82*, 1751–1764. [CrossRef]
7. Yang, Y.; Hu, H. Application of Superabsorbent Spacer Fabrics as Exuding Wound Dressing. *Polymers* **2018**, *10*, 210. [CrossRef]
8. Chen, Q.; Shou, D.; Zheng, R.; Tang, K.-P.M.; Fu, B.; Zhang, X.; Ma, P. Moisture and Thermal Transport Properties of Different Polyester Warp-Knitted Spacer Fabric for Protective Application. *Autex Res. J.* **2020**, *1*, 182–191. [CrossRef]
9. Guo, Y.; Chen, L.; Qiang, S.; Qian, X.; Xue, T.; He, F. Comparing Properties of the Warp-knitted spacer fabric instead of sponge for automobile seat fabric. *J. Phys. Conf. Ser.* **2021**, *1948*, 012196. [CrossRef]
10. Yoneda, M.; Nakajima, C.; Inoue, T. Compression-recovery properties and compressive viscoelastic properties of polyurethane foam for cushioning material for home furnishings use. *J. Jpn. Res. Assoc. Text. End-Uses* **2014**, *55*, 61–68.
11. Obi, B.E. *Polymeric Foams Structure-Property-Performance*; Elsevier: Amsterdam, The Netherlands, 2018; pp. 299–331.
12. Chen, F.; Hu, H.; Liu, Y. Development of weft-knitted spacer fabrics with negative stiffness effect in a special range of compression displacement. *Text. Res. J.* **2015**, *85*, 1720–1731. [CrossRef]
13. Yu, A.; Sukigara, S.; Takeuchi, S. Effect of inlaid elastic yarns and inlay pattern on physical properties and compression behaviour of weft-knitted spacer fabric. *J. Ind. Text.* **2020**. [CrossRef]
14. Yu, A.; Sukigara, S.; Masuda, A. Investigation of vibration isolation behaviour of spacer fabrics with elastic inlay. *J. Text. Eng.* **2020**, *66*, 65–69. [CrossRef]
15. Asayesh, A.; Amini, M. The effect of fabric structure on the compression behavior of weft-knitted spacer fabrics for cushioning applications. *J. Text. Inst.* **2020**, *112*, 1568–1579. [CrossRef]
16. Liu, Y.; Hu, H. Finite element analysis of compression behaviour of 3D spacer fabric structure. *Int. J. Mech. Sci.* **2015**, *94–95*, 244–259. [CrossRef]
17. Chen, S.; Zhng, X.-p.; Chen, H.-x.; Gao, X.-p. An experimental study of the compression properties of polyurethane-based warp-knitted spacer fabric com-posites. *Autex Res. J.* **2017**, *17*, 199. [CrossRef]
18. Yu, S.; Dong, M.; Jiang, G.; Ma, P. Compressive characteristics of warp-knitted spacer fabrics with multi-layers. *Compos. Struct.* **2021**, *256*, 113016. [CrossRef]
19. Yip, J.; Ng, S.-P. Study of three-dimensional spacer fabrics: Physical and mechanical properties. *J. Mater. Process. Technol.* **2008**, *206*, 359–364. [CrossRef]
20. Yang, Y.; Hu, H. Spacer fabric-based exuding wound dressing—Part I: Structural design, fabrication and property evaluation of spacer fabrics. *Text. Res. J.* **2017**, *87*, 1469–1480. [CrossRef]
21. Chen, C.; Du, Z.; Yu, W.; Dias, T. Analysis of physical properties and structure design of weft-knitted spacer fabric with high porosity. *Text. Res. J.* **2018**, *88*, 59–68. [CrossRef]
22. Liu, Y.; Hu, H.; Zhao, L.; Long, H. Compression behavior of warp-knitted spacer fabrics for cushioning applications. *Text. Res. J.* **2012**, *82*, 11–20. [CrossRef]
23. Zhao, T.; Long, H.; Yang, T.; Liu, Y. Cushioning properties of weft-knitted spacer fabrics. *Text. Res. J.* **2018**, *88*, 1628–1640. [CrossRef]
24. Arumugam, V.; Mishra, R.; Militky, J.; Salacova, J. Investigation on thermo-physiological and compression characteristics of weft-knitted 3D spacer fabrics. *J. Text. Inst.* **2016**, *108*, 1095–1105. [CrossRef]
25. Lo, W.-T.; Wong, D.P.; Yick, K.-L.; Ng, S.P.; Yip, J. The biomechanical effects and perceived comfort of textile-fabricated insoles during straight line walking. *Prosthet. Orthot. Int.* **2018**, *42*, 153–162. [CrossRef] [PubMed]
26. Hamedi, M.; Salimi, P.; Jamshidi, N. Improving cushioning properties of a 3D weft knitted spacer fabric in a novel design with NiTi monofilaments. *J. Ind. Text.* **2018**, *49*, 1389–1410. [CrossRef]
27. Yu, A.; Sukigara, S.; Yick, K.-L.; Li, P.-L. Novel weft-knitted spacer structure with silicone tube inlay for enhancing mechanical behavior. *Mech. Adv. Mater. Struct.* **2020**, 1–12. [CrossRef]
28. Patterson, R.F. 9—Silicones. In *Handbook of Thermoset Plastics*, 2nd ed.; Goodman, S.H., Ed.; William Andrew Publishing: Westwood, NJ, USA, 1998; pp. 468–497.

29. Grande, J.B.; Fawcett, A.S.; McLaughlin, A.J.; Gonzaga, F.; Bender, T.P.; Brook, M.A. Anhydrous formation of foamed silicone elastomers using the Piers–Rubinsztajn reaction. *Polymers* **2012**, *53*, 3135–3142. [CrossRef]
30. Malla, R.B.; Shrestha, M.R.; Shaw, M.T.; Brijmohan, S.B. Temperature Aging, Compression Recovery, Creep, and Weathering of a Foam Silicone Sealant for Bridge Expansion Joints. *J. Mater. Civ. Eng.* **2011**, *23*, 287–297. [CrossRef]
31. Fink, J.K. 8—Silicones. In *Reactive Polymers: Fundamentals and Applications*, 3rd ed.; Fink, J.K., Ed.; William Andrew Publishing: Westwood, NJ, USA, 2018; pp. 303–323.
32. Landrock, A.H. *Handbook of Plastic Foams: Types, Properties, Manufacture and Applications*; Elsevier: Amsterdam, The Netherlands, 1995.
33. Yan, S.; Jia, D.; Yu, Y.; Wang, L. Influence of γ-irradiation on mechanical behaviors of poly methyl-vinyl silicone rubber foams at different tem-peratures. *Mech. Mater.* **2020**, *151*, 103639. [CrossRef]
34. Métivier, T.; Cassagnau, P. Foaming behavior of silicone/fluorosilicone blends. *Polymers* **2018**, *146*, 21–30. [CrossRef]
35. Dusserre, G.; Bernhart, G. Knitting processes for composites manufacture. In *Advances in Composites Manufacturing and Process Design*; Elsevier: Amsterdam, The Netherlands, 2015; pp. 27–53.

*polymers*

*Article*

# Prediction of Methylene Blue Removal by Nano $TiO_2$ Using Deep Neural Network

**Nesrine Amor *, Muhammad Tayyab Noman * and Michal Petru**

Department of Machinery Construction, Institute for Nanomaterials, Advanced Technologies and Innovation (CXI), Technical University of Liberec, Studentská 1402/2, 461 17 Liberec 1, Czech Republic; michal.petru@tul.cz
* Correspondence: nesrine.amor@tul.cz (N.A.); muhammad.tayyab.noman@tul.cz (M.T.N.)

**Abstract:** This paper deals with the prediction of methylene blue (MB) dye removal under the influence of titanium dioxide nanoparticles ($TiO_2$ NPs) through deep neural network (DNN). In the first step, $TiO_2$ NPs were prepared and their morphological properties were analysed by scanning electron microscopy. Later, the influence of as synthesized $TiO_2$ NPs was tested against MB dye removal and in the final step, DNN was used for the prediction. DNN is an efficient machine learning tools and widely used model for the prediction of highly complex problems. However, it has never been used for the prediction of MB dye removal. Therefore, this paper investigates the prediction accuracy of MB dye removal under the influence of $TiO_2$ NPs using DNN. Furthermore, the proposed DNN model was used to map out the complex input-output conditions for the prediction of optimal results. The amount of chemicals, i.e., amount of $TiO_2$ NPs, amount of ehylene glycol and reaction time were chosen as input variables and MB dye removal percentage was evaluated as a response. DNN model provides significantly high performance accuracy for the prediction of MB dye removal and can be used as a powerful tool for the prediction of other functional properties of nanocomposites.

**Keywords:** artificial neural network; titanium dioxide nanoparticles; methylene blue dye removal

**Citation:** Amor, N.; Noman, M.T.; Petru, M. Prediction of Methylene Blue Removal by Nano $TiO_2$ Using Deep Neural Network. *Polymers* **2021**, *13*, 3104. https://doi.org/10.3390/polym13183104

Academic Editor: Raquel Verdejo

Received: 29 August 2021
Accepted: 11 September 2021
Published: 15 September 2021



## 1. Introduction

Modern world witnesses the miracles of nanotechnology as it manipulates matter on molecular level with at least having one dimension less than 100 nm [1,2]. Researchers are applying nanomaterials i.e., nanoparticles, nanowires, nanorods, nanosheets, nanoflowers etc in medicines, optical instruments, energy devices, civil and building industry, aeronautics and electronics for better performance [3–6]. $TiO_2$ is a functional material mostly used as a photo catalyst in industrial applications [7–9]. $TiO_2$ nanomaterials have been investigated for photodegradation of organic pollutants i.e., tetracycline and MB [10,11], self-cleaning coatings, antimicrobial coatings, sensors and for other purposes by the academic researchers and industrial experts [12–14]. In an experimental study, Noman et al. worked with the synthesis and single step coating of $TiO_2$ NPs on cotton to develop photocatalytically active cotton composites for antimicrobial and self-cleaning applications. They used sonication method and reported the average particle size for their samples 4 nm [15]. On the other hand, DNN models have achieved human-level performance and have shown great success in different real-world applications, including computer vision [16], textile process, biomedical engineering [17], material engineering [18]. DNN is an efficient machine learning tool suitable for the prediction of output parameters from input variables where there is an unknown relationship exists between input and output variables [19–21]. In recent years, DNN has been widely used to predict various properties of textiles. Lui et al. developed a new strategy to predict the initial failure strength criterion of woven fabric reinforced composites based on micro-mechanical model by modifying DNN and mechanics of structure genome (MSG) [22]. MGS is used to perform initial failure analysis of a square pack microscale model that trained the samples to detect yarn failure criterion. The effectiveness of this strategy was confirmed by testing yarns of mesoscale plain weave fabrics and fiber reinforced composite materials to compute the initial failure strength constants. Khude et al. applied artificial neural network (ANN) and adaptive network-based fuzzy inference system (ANFIS) to predict antimicrobial properties of knitted fabrics made with silver fibres [23]. Both studies reported good results during training and testing of datasets. However, ANFIS showed better performance with small datasets. Altarazi et al. used multiple algorithms i.e., stochastic gradient descent (SGD), ANN, k-nearest neighbors (kNN), decision tree (DT), regression analysis, support vector machine (SVM), random forest (RF), logistic regression (LoR) and AdaBoost (AB), for the prediction and classification of tensile strength of polymeric films with various compositions [24]. The obtained results show that SVM algorithm has superior predictive ability. In addition, the experimental results show that classification ability of used algorithms was excellent for sorting films into conforming and non-conforming parts. Yang et al. identified knitted fabric pilling behavior by modifying ANN into deep principle components analysis-based neural networks (DPCANN) [25]. In DPCANN, principal components automatically track down the fabric properties before and after pilling test and then neural network was applied to evaluate pilling grades. The obtained results elucidate that DPCANN has above average classification efficiency for pilling behavior of knitted fabric.

Many other researchers worked with DNN in textiles. Li et al. proposed Fisher criterion-based deep learning algorithm for defects detection of patterned fabrics [26]. A Fisher criterion-based stacked denoising method has been used for fabric images to classify into defective and defectless categories. The experimental results showed that the accuracy of proposed approach in defects detection is excellent for patterned fabrics and more complex jacquard warp-knitted fabric. Ni et al. proposed a novel online algorithm that detects and predicts the coating thickness on textiles by hyperspectral images [27]. The proposed algorithm was based on two different optimization modules i.e., the first module is called extreme learning machine (ELM) classifier whereas, the later is called a group of stacked autoencoders.The ELM module optimized by a new optimizer known as grey wolf optimizer (GWO), to determine the number of neurons and weights to get more accuracy while detection and classification. The results explained that online detection performance significantly improved with a combination of the variable-weighted stacked autoencoders (VW-SAE) with GWO-ELM that provide 95.58% efficiency. Lazzús et al. used the combined ANN with particle swarm optimization (PSO) to predict the thermal properties [28]. The results demonstrated that the proposed model ANN-PSO provided better results than feedforward ANN. Malik et al. applied ANN to predict yarns crimp for woven fabrics. Simulation results showed a good prediction accuracy, especially for warp yarn [29]. Lu et al. used ANN and multiple linear regression (MLR) based on acoustic emission detection for the prediction of tensile strength of single wool fibers [30]. The coefficients of determination of ANN and MLR show that there is a high correlation between the predicted and measured values of strength of wool. However, ANN model has the higher accuracy and lower error prediction compared to MLR. In another study, Tadesse et al. proposed ANN and fuzzy logic (FL) to predict the tactile comfort of functional fabrics parameters [31]. FL has been performed to predict the predicted hand values (HVs) from finishing parameters; then, the total hand values (THVs) has been predicted from the HV using FL and ANN model. FL provided an efficient prediction of the HV with lower relative mean percentage error (RMPE) and root mean square error (RMSE). In addition, the fuzzy logic model (FLM) and ANN models showed an effective prediction performance of the THV with lower RMSE and RMPE values. Mishra used ANN models during the production of cotton fabric for the prediction of yarn strength utilization [32]. The selected input parameters were yarn counts, initial crimps, total number of yarns and yarn strengths in both longitudinal and transverse directions along with the weave float length. The experimental results showed that yarn strength utilization percentage increased with an increase in yarn number in both directions, however, a decrease in crimp percentage and float length was observed. El-Geiheini et al. worked with different types of yarns and used ANN and image processing tools for modeling and simulation of yarn tenacity and elongation [33]. They reported that the proposed techniques are suitable for the estimation of various yarn properties with minimum error. Erbil et al. used ANN and regression tools for tensile strength prediction of ternary blended open-end rotor yarns [34]. They performed stepwise multiple linear regression (MLR) method and trained their ANN algorithm with Levenberg–Marquardt backpropagation function. Furthermore, they performed a comparison of both models for prediction efficiency. The reported results of their experimental study demonstrated that ANN models give a better prediction output than MLR for both parameters i.e., breaking strength and elongation at break. Lui et al. developed a new strategy to predict the initial failure strength criterion of woven fabric reinforced composites based on micromechanical model by modifying deep learning neural network (DNN) and mechanics of structure genome (MSG) [22]. MGS is used to perform initial failure analysis of a square pack microscale model that trained the samples to detect yarn failure criterion. The effectiveness of this strategy was confirmed by testing yarns of mesoscale plain weave fabrics and fiber reinforced composite materials to compute the initial failure strength constants. The literature shows that the prediction of mechanical behavior of composites materials is a complex task due to variations in boundary conditions and structures [5,35]. Wang et al. presented the prediction of the tensile strength of ultrafine glass fiber felts by ANN [36]. The tensile strength are modelled based on the mean diameter of fibers, bulk density and resin content. Simulation results showed that ANN model provided a high prediction accuracy and lower mean relative errors. Unal et al. selected single jersey knitted fabrics for the evaluation of air permeability and combined ANN algorithm with regression methods for the prediction of bursting strength of used knit structures [37]. Implementation of results showed that both methods were able to predict precisely the properties of knitted fabrics. However, ANN had a slightly positive edge when used for prediction. Recently, Amor et al. used ANN and MLR for the prediction of functional properties of nano $TiO_2$ coated cotton composites and their results show that ANN outperformed MLR for prediction accuracy [38].

The discussed literature reveal that ANN is mostly used machine learning tool for textile industry [29,34,38,39]. DNN is a category of ANN model with multiple layers between input and output layers. DNN takes an edge of automatic learning process. Therefore, main contributions of this paper are to investigate the accuracy of DNN model for the prediction of MB dye removal under the influence of $TiO_2$ NPs and compare the results with MLR.

This study is organized as follows: Section 2 describes materials, synthesis of $TiO_2$ NPs and explanation of DNN model framework. Section 3 discusses the simulation, comparison and results for the prediction of MB dye removal. Section 4 summarizes the main findings and concludes the paper.

## 2. Materials and Methods

### 2.1. Materials

All chemicals and reagents for the synthesis of $TiO_2$ NPs i.e., titanium tetraisopropoxide (TTIP), ethylene glycol (EG) and for photodegradation study i.e., MB dye were received from Sigma-Aldrich (Prague, Czech Republic) and used without any further purification. The experimental design with different amount of TTIP and EG is presented in Table 1.

**Table 1.** Experimental design and variables for the synthesis of $TiO_2$ NPs.

| Sample | Amount of TTIP [mL] | Amount of EG [mL] | Sonication Time [h] |
|---|---|---|---|
| 1 | 10 | 8 | 4 |
| 2 | 1 | 2 | 0.5 |
| 3 | 1 | 2 | 4 |
| 4 | 10 | 4 | 1 |
| 5 | 10 | 5 | 0.5 |
| 6 | 5 | 5 | 2 |
| 7 | 1 | 8 | 4 |
| 8 | 5 | 5 | 0.5 |
| 9 | 7 | 5 | 1 |
| 10 | 10 | 2 | 0.5 |
| 11 | 7 | 5 | 2 |
| 12 | 7 | 8 | 1 |
| 13 | 5 | 2 | 2 |
| 14 | 1 | 4 | 1 |
| 15 | 7 | 5 | 2 |
| 16 | 5 | 5 | 1 |
| 17 | 10 | 2 | 4 |
| 18 | 5 | 5 | 2 |
| 19 | 7 | 5 | 4 |
| 20 | 1 | 8 | 0.5 |

### 2.2. *Synthesis of $TiO_2$ NPs*

Initially, TTIP and EG were added in a beaker containing 50 mL ethanol. The molar ratio of TTIP: EG was 2:1. The mixture was magnetically stirred at 500 rpm for 15 min. The mixture was then sonicated by ultrasonic probe (Bandelin Sonopuls HD 3200, 20 kHZ, 200 W, Berlin, Germany) for time intervals based on the experimental design. The temperature was maintained at 80 °C by using hot plate equipped with magnetic stirrer. The resulting nanoparticles were washed with ethanol to remove impurities and then centrifuged at 4000 rpm for 5 min to remove liquid from inside. The resulting nanoparticles were further dried in an oven at 100 °C for 2 h. The experimental setup is shown in Figure 1.

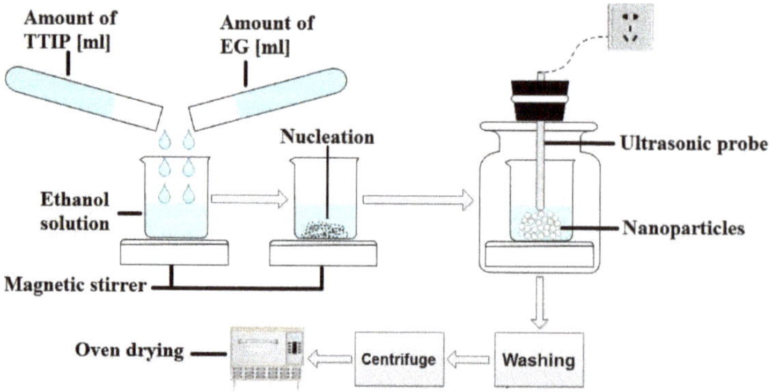


**Figure 1.** Graphical representation of experimental setup for the synthesis of $TiO_2$ NPs.

### 2.3. *MB Removal*

The photocatalytic activity of $TiO_2$ NPs was evaluated by the discoloration of MB solution under UV light. In this experiment, 1 g $L^{-1}$ $TiO_2$ NPs were mixed in 50 mL solution containing 100 mg $L^{-1}$ MB dye. The solution was stirred by a magnetic stirrer and placed in the dark for 40 min to reach an adsorption desorption equilibrium. The UV light source was a 500 W UV lamp with light intensity 30 W $m^{-2}$. MB residual concentration was calculated by spectrophotometer at 668 nm wavelength. The color removal efficiency ($CR\%$) was evaluated by the given Equation (1):

$$CR\% = \left[1 - \frac{C}{C_0}\right] * 100, \quad (1)$$

where $C_0$ and $C$ represents the initial and final concentration of MB in the solution respectively. The initial spectrum of dye solution without $TiO_2$ NPs was taken as standard sample. An aliquot was taken out by a syringe after a fix time interval to evaluate the results of MB removal.

### 2.4. Deep Neural Network

ANN model is widely used in the prediction of functional properties of composite structures [38,40]. Generally, ANN model contains three layers and ANN model with more than hidden layers is known as DNN model [18]. DNN is widely used to investigate the correlation between variables for critical problems [41]. Automatic creation and exploration of information from previous learning is an interesting feature of DNN [42]. DNN tunes the weights constantly until predicted and target values match. Back-propagation calculates the error between predicted and targeted and update weights to remove error after iterations [18,22]. The equation of DNN is given below:

$$y = f\left(\sum_i W_{ij} * X_i + b_j\right), \quad (2)$$

where, $y$ represents the output. $X_i$ represents the $i$th input variables. $W_{ij}$ represents the weight and $b_j$ is the bias. The weights and biases include the information that neuron recovers during training. $\varphi$ is the activation function mostly employed sigmoid activation function given in Equation (3) [43]:

$$f(x) = \frac{1}{1 + e^{\delta x}}, \quad (3)$$

where $\delta$ denotes the sigmoid function steepness parameter and $x$ is given by

$$x = \sum_i W_i * X_i, \quad (4)$$

A detail tutorial of DNN algorithms are presented by various researchers in their studies [44,45]. In the present work, the DNN model has been developed by a multi-layer feed-forward network. The hyperbolic tangent sigmoid function is used as the activation function for each layer. The amount of titanium precursor, amount of ehylene glycol and process time are taken as the input variables whereas the MB removal are outputs variable as described in Figure 2.

The DNN model has been trained by the Bayesian regularization backpropagation algorithm. 85% of the data was used for the training of the DNN model and 15% of the data was used for testing. After several tentative simulation, it was found that the best DNN architecture that gives the lowest relative error and highest correlation coefficient has a structure of 3-12-12-6-1, which represents three variables in the input layer, three hidden layers with 12, 12 and 6 neurons respectively, and one predicted output.

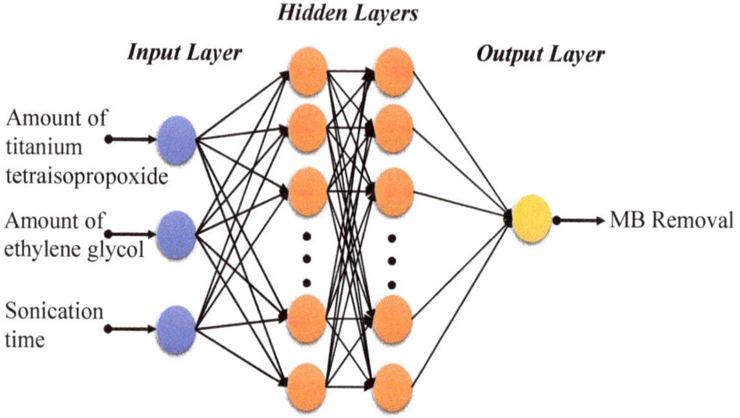


**Figure 2.** DNN model for the prediction of of the removal of methylene blue dye under the influence of nano $TiO_2$.

### 2.5. Evaluation of DNN Model

The performance and accuracy of the DNN model was evaluated using different method including mean absolute error ($MAE$), mean squared error ($MSE$), root mean squared error ($RMSE$), standard deviation ($SD$) of the error and coefficient of correlation ($R^2$), and they are expressed respectively as follows:

$$MAE = \frac{1}{n}\Sigma_{i=1}^{n}|(y_i - \hat{y}_i)|, \quad (5)$$

$$MSE = \frac{1}{n}\Sigma_{i=1}^{n}(y_i - \hat{y}_i)^2, \quad (6)$$

$$RMSE = \sqrt{\frac{1}{n}\Sigma_{i=1}^{n}(y_i - \hat{y}_i)^2}, \quad (7)$$

$$SD = \sqrt{\frac{1}{n-1}\sum_{i=1}^{n}((y-\hat{y})_i - \overline{(y-\hat{y})})^2}, \quad (8)$$

$$R^2 = \left(\frac{\Sigma_{i=1}^{n}(y_i - \bar{y})(\hat{y}_i - \bar{\hat{y}})}{\sqrt{\Sigma_{i=1}^{n}(y_i - \bar{y})^2}\sqrt{\Sigma_{i=1}^{n}(\hat{y}_i - \bar{\hat{y}})^2}}\right)^2. \quad (9)$$

where $y_i$ and $\hat{y}$ are the actual and predicted outputs, respectively. $\bar{y}$ is the mean of the actual variables and $\bar{\hat{y}}$ represents the mean of the predicted variables. $n$ is the number of samples.

In addition, statistical analysis (ANOVA) has been performed to test the statistical significance of input and output variables [38,46,47].

## 3. Results and Discussion

### 3.1. Scanning Electron Microscopy (SEM) Analysis

The results of SEM analysis for the synthesis of $TiO_2$ NPs are presented in Figure 3. It is observed from SEM results that the synthesized $TiO_2$ NPs are quasi spherical in shape and homogeneously distributed. The estimated size of the particles by image analysis was 20 nm. Moreover, for texture properties, the randomly selected samples were examined by Brunauer-Emmett Teller (BET surface area and pore size analyzer Quantachrome—NOVA 2200e (Boynton Beach, FL, USA)), Atomic Force Microscopy (AFM) system (Park System$^{TM}$, Suwon, South Korea) and Dynamic Light Scattering (DLS Malvern Pananalytical Zetasizer Ultra (Malvern, UK)) techniques and the obtained results are presented in Table 2.

**Table 2.** Textural and microstructural properties of $TiO_2$ NPs.

| Sample | Surface Area $[m^2/g]$ | Pore Volume $[cm^3/g]$ | Surface Roughness [nm] | Hydrodynamic Diameter [nm] |
|---|---|---|---|---|
| 1 | $187 \pm 4$ | $0.17 \pm 0.5$ | $5.24 \pm 2$ | $28 \pm 3$ |
| 6 | $163 \pm 2$ | $0.21 \pm 1$ | $2.11 \pm 1$ | $31 \pm 1$ |
| 9 | $149 \pm 5$ | $0.26 \pm 0.5$ | $1.38 \pm 1$ | $24 \pm 2$ |
| 16 | $201 \pm 4$ | $0.13 \pm 0.5$ | $3.81 \pm 2$ | $41 \pm 1$ |

In addition, XRD analysis was carried out to confirm the crystallite size and the purity of the crystalline phase. XRD results are illustrated and discussed (see the Supporting Information).

### 3.2. *Evaluation of MB Removal*

MB discoloration was investigated with 1 g $L^{-1}$ $TiO_2$ NPs and with 100 mg $L^{-1}$ initial concentration of MB. The results of MB removal explain that a complete discoloration of MB was done within 40 min under UV light. In order to confirm that this change was due to the presence of $TiO_2$ NPs and not by the poor light fastness of MB, dye solution was exposed to UV light without $TiO_2$ NPs. This solution didn't change its color even for longer irradiations time. Therefore, it is confirmed that $TiO_2$ NPs are highly photo active as their minimal quantity discolor MB solution in a short span of time.

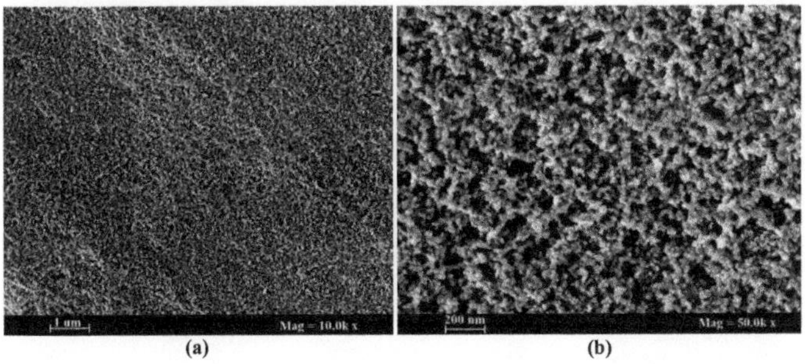

**Figure 3.** SEM images of as synthesized $TiO_2$ NPs (**a**) at magnification 10.0 k × and (**b**) at magnification 50.0 k ×.

### 3.3. *Analysis of the Proposed DNN Model*

We applied the DNN to predict the removal of methylene blue dye under the influence of titanium dioxide. After many trials, the best results that include the lower MSE and high prediction performance of the methylene blue removal obtained from a DNN model with five-layers, i.e., an input, three hidden, and an output layers, and the number of hidden layer nodes are 12, 12 and 6, where the network provides highly accurate results. The transfer functions used for hidden and output layers in this work is the type of tangent sigmoid (tansig) function. The best selection of a transfer function for input and output layers guarantee the accuracy of the predicted results. In addition, Bayesian regularization backpropagation was used to train the DNN. The setting of training parameters of the DNN are presented in Table 3. The obtained results with the DNN model were compared with MLR.

**Table 3.** Parameters and settings of training network.

| Parameters | Settings |
|---|---|
| Training function | trainbr |
| Transfer function of hidden layers | tansig, tansig, tansig |
| Transfer function of output layer | tansig |
| Epochs | 1000 |
| Input node | 3 |
| Hidden node | 12,12, 6 |
| Output node | 1 |
| Performance goal | 0.00001 |

Figure 4 illustrates the predicted values of MB dye removal using DNN and MLR. Figure 5 shows the absolute prediction error given by the difference between predicted and actual values using both MLR and DNN. We noticed that the values of prediction error were significantly lower for the DNN model as compared to MLR. It is clear from the Figure 5 that MLR model has one error burst at sample number 4 and has higher errors for the prediction of most values. Table 4 represents the computed $MSE$, $RMSE$, $MAE$, $SD$ and $R^2$ for both used models to predict the methylene blue removal. We observed that the proposed DNN model outperformed MLR with lower errors according to $MSE$, $RMSE$, $MAE$, $SD$, and high accuracy according to $R^2$.

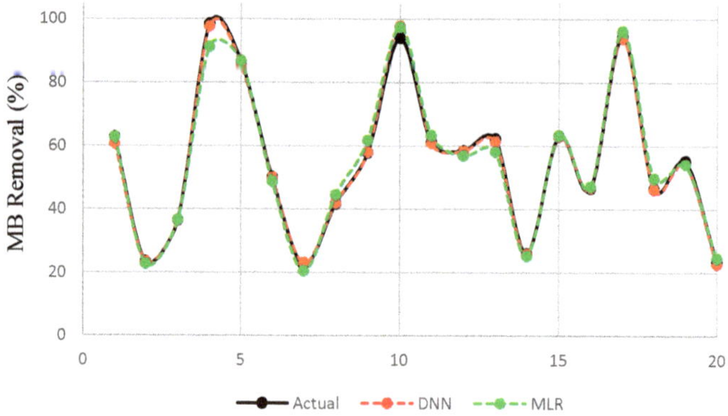


**Figure 4.** The predicted and actual values using DNN and MLR.

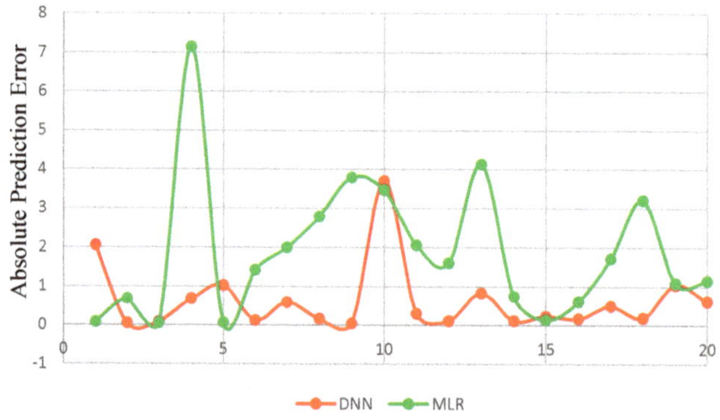


**Figure 5.** Results of absolute error for actual and predicted values using DNN and MLR models.

**Table 4.** The performance measures of DNN and MLR.

| Methods | *MSE* | *RMSE* | *MAE* | *SD* | $R^2$ |
|---|---|---|---|---|---|
| DNN (training) | 1.1186 | 1.0576 | 0.6254 | 1.1719 | 0.9997 |
| DNN (testing) | 1.09958 | 1.0532 | 0.6213 | 0.3558 | 0.9999 |
| MLR | 6.6044 | 2.5699 | 1.8933 | 2.6366 | 0.9882 |

The correlation between actual and predicted values using DNN for training and validation of all data sets is illustrated in Figure 6. It is observed from the results that the correlation coefficients ($R$-value) provide an excellent correlation between the predicted and actual values, where $R > 99\%$ which confirm the highly prediction accuracy of DNN. Figure 7 showed the correlation between the actual and predicted values using MLR model. We noticed that MLR model provide a good prediction accuracy for the MB removal, where $R = 98\%$. It is clear from the obtained results that both DNN and models verifies their accuracy and effectiveness in the prediction process. However, the performance of DNN model outperformed MLR with higher accuracy and lower errors.

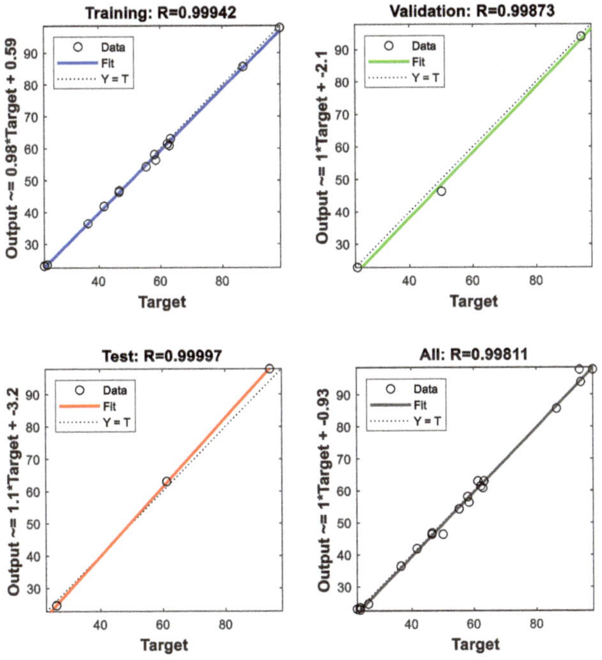


**Figure 6.** Correlation coefficient for experimental and predicted values by back-propagation DNN model.

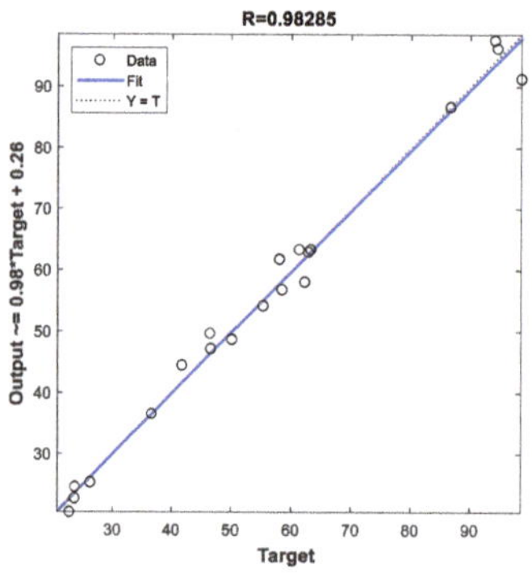


**Figure 7.** Correlation of actual and predicted values for all data sets by MLR model.

One-way ANOVA was used to check the robustness of the results obtained through DNN, MLR and experiments. ANOVA analysis helps to understand the relationship between predicted values of MB dye removal under the influence of nano $TiO_2$ with all process variables. Table 5 shows ANOVA results for MB dye removal obtained by DNN, MLR and experiments. The overall results show that DNN model is statistically more significant than experimental values and MLR values as DNN provides lowest *p*-value.

**Table 5.** Analysis report of DNN, MLR and experimental values for methylene blue removal.

| Methods | *p*-Value | F-Value |
|---|---|---|
| DNN | $1.73 \times 10^{-10}$ | 97.31 |
| MLR | $2.821 \times 10^{-9}$ | 66.97 |
| Experimental | $1.771 \times 10^{-9}$ | 71.33 |

## 4. Conclusions

In this paper, we introduced DNN model for the prediction of MB dye removal under the influence of $TiO_2$ NPs. In the fist phase, $TiO_2$ NPs were successfully synthesized by sonication. Scanning electron microscopy results showed quasi spherical shape of all prepared samples. The particles were homogeneous and successfully used in MB dye removal. The prediction of MB dye removal was performed by DNN model. In comparison, DNN showed more accurate results than MLR model. The obtained results for MSE, RMSE, MAE and SD elucidate that DNN model has lower error than MLR. The successful utilization of DNN model shows a non-linear behaviour for the prediction of MB dye removal. The obtained results confirm that DNN model can also be effectively used for the prediction of MB dye removal and for the removal of other organic pollutants in different industries.

**Supplementary Materials:** The following are available online at https://www.mdpi.com/article/10.3390/polym13183104/s1, Figure S1: XRD patterns of randomly selected samples i.e., Sample 1, Sample 6, Sample 9 and Sample 16 of $TiO_2$ NPs.

**Author Contributions:** N.A. and M.T.N. conceived, designed and performed experiments; analysed the results and wrote manuscript. M.P. analyzed the results, supervised and acquired funding. All of the authors participated in critical analysis and preparation of the manuscript. All authors have read and agreed to the published version of the manuscript.

**Funding:** This work was supported by the Ministry of Education, Youth and Sports of the Czech Republic and the European Union (European Structural and Investment Funds—Operational Programme Research, Development and Education) in the frames of the project "Modular platform for autonomous chassis of specialized electric vehicles for freight and equipment transportation", Reg. No. CZ.02.1.01/0.0/0.0/16_025/0007293.

**Institutional Review Board Statement:** Not applicable.

**Informed Consent Statement:** Not applicable.

**Data Availability Statement:** Not applicable.

**Conflicts of Interest:** The authors declare no conflict of interest.

## References


1. Ashraf, M.; Wiener, J.; Farooq, A.; Šašková, J.; Noman, M. Development of Maghemite Glass Fibre Nanocomposite for Adsorptive Removal of Methylene Blue. *Fibers Polym.* **2018**, *19*, 1735–1746. [CrossRef]
2. Waqas, H.; Farooq, U.; Shah, Z.; Kumam, P.; Shutaywi, M. Second-order slip effect on bio-convectional viscoelastic nanofluid flow through a stretching cylinder with swimming microorganisms and melting phenomenon. *Sci. Rep.* **2021**, *11*, 11208. [CrossRef] [PubMed]
3. Azeem, M.; Noman, M.T.; Wiener, J.; Petru, M.; Louda, P. Structural design of efficient fog collectors: A review. *Environ. Technol. Innov.* **2020**, *20*, 101169. [CrossRef]
4. Mansoor, T.; Hes, L.; Bajzik, V.; Noman, M.T. Novel method on thermal resistance prediction and thermo-physiological comfort of socks in a wet state. *Text. Res. J.* **2020**, *90*, 1987–2006. [CrossRef]
5. Noman, M.T.; Petru, M.; Militký, J.; Azeem, M.; Ashraf, M.A. One-Pot Sonochemical Synthesis of ZnO Nanoparticles for Photocatalytic Applications, Modelling and Optimization. *Materials* **2020**, *13*, 14. [CrossRef] [PubMed]
6. Mahmood, A.; Noman, M.T.; Pechočiaková, M.; Amor, N.; Petrů, M.; Abdelkader, M.; Militký, J.; Sozcu, S.; Hassan, S.Z.U. Geopolymers and Fiber-Reinforced Concrete Composites in Civil Engineering. *Polymers* **2021**, *13*, 2099. [CrossRef] [PubMed]
7. Wang, J.; Li, M.; Feng, J.; Yan, X.; Chen, H.; Han, R. Effects of $TiO_2$-NPs pretreatment on UV-B stress tolerance in Arabidopsis thaliana. *Chemosphere* **2021**, *281*, 130809. [CrossRef] [PubMed]
8. Pourhashem, S.; Duan, J.; Zhou, Z.; Ji, X.; Sun, J.; Dong, X.; Wang, L.; Guan, F.; Hou, B. Investigating the effects of chitosan solution and chitosan modified $TiO_2$ nanotubes on the corrosion protection performance of epoxy coatings. *Mater. Chem. Phys.* **2021**, *270*, 124751. [CrossRef]
9. Lee, K.H.; Chu, J.Y.; Kim, A.R.; Kim, H.G.; Yoo, D.J. Functionalized $TiO_2$ mediated organic-inorganic composite membranes based on quaternized poly(arylene ether ketone) with enhanced ionic conductivity and alkaline stability for alkaline fuel cells. *J. Membr. Sci.* **2021**, *634*, 119435. [CrossRef]
10. Yuan, N.; Cai, H.; Liu, T.; Huang, Q.; Zhang, X. Adsorptive removal of methylene blue from aqueous solution using coal fly ash-derived mesoporous silica material. *Adsorpt. Sci. Technol.* **2019**, *37*, 333–348. [CrossRef]
11. Zhang, X.; Yuan, N.; Li, Y.; Han, L.; Wang, Q. Fabrication of new MIL-53(Fe)@$TiO_2$ visible-light responsive adsorptive photocatalysts for efficient elimination of tetracycline. *Chem. Eng. J.* **2021**, *428*, 131077. [CrossRef]
12. Noman, M.T.; Petrů, M.; Amor, N.; Yang, T.; Mansoor, T. Thermophysiological comfort of sonochemically synthesized nano $TiO_2$ coated woven fabrics. *Sci. Rep.* **2020**, *10*, 17204. [CrossRef]
13. Noman, M.T.; Ashraf, M.A.; Jamshaid, H.; Ali, A. A Novel Green Stabilization of $TiO_2$ Nanoparticles onto Cotton. *Fibers Polym.* **2018**, *19*, 2268–2277. [CrossRef]
14. Noman, M.T.; Militky, J.; Wiener, J.; Saskova, J.; Ashraf, M.A.; Jamshaid, H.; Azeem, M. Sonochemical synthesis of highly crystalline photocatalyst for industrial applications. *Ultrasonics* **2018**, *83*, 203–213. doi: 10.1016/j.ultras.2017.06.012. [CrossRef]
15. Noman, M.T.; Wiener, J.; Saskova, J.; Ashraf, M.A.; Vikova, M.; Jamshaid, H.; Kejzlar, P. In-situ development of highly photocatalytic multifunctional nanocomposites by ultrasonic acoustic method. *Ultrason. Sonochem.* **2018**, *40*, 41–56. [CrossRef] [PubMed]
16. Chen, R.; Mihaylova, L.; Zhu, H.; Bouaynaya, N. A Deep Learning Framework for Joint Image Restoration and Recognition. *Circuits Syst. Signal Process.* **2020**, *39*, 1561–1580. [CrossRef]
17. Vahid, A.; Mückschel, M.; Stober, S.; Stock, A.; Beste, C. Applying deep learning to single-trial EEG data provides evidence for complementary theories on action control. *Commun. Biol.* **2020**, *3*, 112. [CrossRef] [PubMed]
18. Zazoum, B.; Triki, E.; Bachri, A. Modeling of Mechanical Properties of Clay-Reinforced Polymer Nanocomposites Using Deep Neural Network. *Materials* **2020**, *13*, 4266. [CrossRef]
19. Low, C.Y.; Park, J.; Teoh, A.B.J. Stacking-Based Deep Neural Network: Deep Analytic Network for Pattern Classification. *IEEE Trans. Cybern.* **2020**, *50*, 5021–5034. [CrossRef]
20. Ha, M.H.; Chen, O.T.C. Deep Neural Networks Using Capsule Networks and Skeleton-Based Attentions for Action Recognition. *IEEE Access* **2021**, *9*, 6164–6178. [CrossRef]
21. Amor, N.; Noman, M.T.; Petru, M. Classification of Textile Polymer Composites: Recent Trends and Challenges. *Polymers* **2021**, *13*, 2592. [CrossRef]
22. Liu, X.; Gasco, F.; Goodsell, J.; Yu, W. Initial failure strength prediction of woven composites using a new yarn failure criterion constructed by deep learning. *Compos. Struct.* **2019**, *230*, 111505. [CrossRef]

23. Khude, P.; Majumdar, A.; Butola, B.S. Modelling and prediction of antibacterial activity of knitted fabrics made from silver nanocomposite fibres using soft computing approaches. *Neural Comput. Appl.* **2019**, *32*, 9509–9519. [CrossRef]
24. Altarazi, S.; Allaf, R.; Alhindawi, F. Machine Learning Models for Predicting and Classifying the Tensile Strength of Polymeric Films Fabricated via Different Production Processes. *Materials* **2019**, *12*, 1475. [CrossRef]
25. Yang, C.S.; Lin, C.; Chen, W. Using deep principal components analysis-based neural networks for fabric pilling classification. *Electronics* **2019**, *8*, 474. [CrossRef]
26. Li, Y.; Zhao, W.; Pan, J. Deformable Patterned Fabric Defect Detection With Fisher Criterion-Based Deep Learning. *IEEE Trans. Autom. Sci. Eng.* **2017**, *14*, 1256–1264. [CrossRef]
27. Ni, C.; Li, Z.; Zhang, X.; Sun, X.; Huang, Y.; Zhao, L.; Zhu, T.; Wang, D. Online Sorting of the Film on Cotton Based on Deep Learning and Hyperspectral Imaging. *IEEE Access* **2020**, *8*, 93028–93038. [CrossRef]
28. Lazzús, J.A. Neural network-particle swarm modeling to predict thermal properties. *Math. Comput. Model.* **2013**, *57*, 2408–2418. doi: 10.1016/j.mcm.2012.01.003. [CrossRef]
29. Malik, S.A.; Gereke, T.; Farooq, A.; Aibibu, D.; Cherif, C. Prediction of yarn crimp in PES multifilament woven barrier fabrics using artificial neural network. *J. Text. Inst.* **2018**, *109*, 942–951. [CrossRef]
30. Lu, D.; Yu, W. Predicting the tensile strength of single wool fibers using artificial neural network and multiple linear regression models based on acoustic emission. *Text. Res. J.* **2021**, *91*, 533–542. [CrossRef]
31. Tadesse, M.G.; Loghin, E.; Pislaru, M.; Wang, L.; Chen, Y.; Nierstrasz, V.; Loghin, C. Prediction of the tactile comfort of fabrics from functional finishing parameters using fuzzy logic and artificial neural network models. *Text. Res. J.* **2019**, *89*, 4083–4094. [CrossRef]
32. Mishra, S. Prediction of Yarn Strength Utilization in Cotton Woven Fabrics using Artificial Neural Network. *J. Inst. Eng. Ser. E* **2015**, *96*, 151–157. [CrossRef]
33. El-Geiheini, A.; ElKateb, S.; Abd-Elhamied, M.R. Yarn Tensile Properties Modeling Using Artificial Intelligence. *Alex. Eng. J.* **2020**, *59*, 4435–4440. [CrossRef]
34. Erbil, Y.; Babaarslan, O.; Ilhan, İ. A comparative prediction for tensile properties of ternary blended open-end rotor yarns using regression and neural network models. *J. Text. Inst.* **2018**, *109*, 560–568. [CrossRef]
35. Noman, M.T.; Amor, N.; Petru, M.; Mahmood, A.; Kejzlar, P. Photocatalytic Behaviour of Zinc Oxide Nanostructures on Surface Activation of Polymeric Fibres. *Polymers* **2021**, *13*, 1227. [CrossRef]
36. Wang, F.; Chen, Z.; Wu, C.; Yang, Y.; Zhang, D.; Li, S. A model for predicting the tensile strength of ultrafine glass fiber felts with mathematics and artificial neural network. *J. Text. Inst.* **2021**, *112*, 783–791. [CrossRef]
37. Unal, P.; Üreyen, M.; Mecit, D. Predicting properties of single jersey fabrics using regression and artificial neural network models. *Fibers Polym.* **2012**, *13*, 87–95. [CrossRef]
38. Amor, N.; Noman, M.T.; Petru, M. Prediction of functional properties of nano $TiO_2$ coated cotton composites by artificial neural network. *Sci. Rep.* **2021**, *11*. [CrossRef]
39. Amor, N.; Noman, M.T.; Petrů, M.; Mahmood, A.; Ismail, A. Neural network-crow search model for the prediction of functional properties of nano $TiO_2$ coated cotton composites. *Sci. Rep.* **2021**, *11*, 1–13. [CrossRef]
40. Breuer, K.; Stommel, M. Prediction of Short Fiber Composite Properties by an Artificial Neural Network Trained on an RVE Database. *Fibers* **2021**, *9*, 8. [CrossRef]
41. Xie, Q.; Suvarna, M.; Li, J.; Zhu, X.; Cai, J.; Wang, X. Online prediction of mechanical properties of hot rolled steel plate using machine learning. *Mater. Des.* **2021**, *197*, 109201. [CrossRef]
42. Wang, Z.; Di Massimo, C.; Tham, M.T.; Julian Morris, A. A procedure for determining the topology of multilayer feedforward neural networks. *Neural Netw.* **1994**, *7*, 291–300. [CrossRef]
43. Rojas, R. *Neural Networks—A Systematic Introduction;* Springer: Berlin/Heidelberg, Germany, 1996.
44. Sze, V.; Chen, Y.H.; Yang, T.J.; Emer, J.S. Efficient Processing of Deep Neural Networks: A Tutorial and Survey. *Proc. IEEE* **2017**, *105*, 2295–2329. [CrossRef]
45. Chang, C.H. Deep and Shallow Architecture of Multilayer Neural Networks. *IEEE Trans. Neural Netw. Learn. Syst.* **2015**, *26*, 2477–2486. [CrossRef] [PubMed]
46. Noman, M.T.; Amor, N.; Petru, M. Synthesis and applications of ZnO nanostructures (ZONSs): A review. *Crit. Rev. Solid State Mater. Sci.* **2021**, *2*, 1–44. [CrossRef]
47. Noman, M.T.; Petrů, M.; Amor, N.; Louda, P. Thermophysiological comfort of zinc oxide nanoparticles coated woven fabrics. *Sci. Rep.* **2020**, *10*, 21080. [CrossRef] [PubMed]

polymers

MDPI

Article

# Preparation and Ballistic Performance of a Multi-Layer Armor System Composed of Kevlar/Polyurea Composites and Shear Thickening Fluid (STF)-Filled Paper Honeycomb Panels

**Chang-Pin Chang [1,2], Cheng-Hung Shih [3], Jhu-Lin You [1,2], Meng-Jey Youh [4,*], Yih-Ming Liu [1,2] and Ming-Der Ger [1,2,*]**

1 Department of Chemical & Materials Engineering, Chung Cheng Institute of Technology, National Defense University, Taoyuan 335, Taiwan; changpin24@gmail.com (C.-P.C.); yolin1014001@gmail.com (J.-L.Y.); takuluLiu@gmail.com (Y.-M.L.)
2 System Engineering and Technology Program, National Chiao Tung University, Hsinchu 300, Taiwan
3 Graduate School of Defense Science, Chung Cheng Institute of Technology, National Defense University, Taoyuan 335, Taiwan; s922326@gmail.com
4 Department of Mechanical Engineering, Ming Chi University of Technology, Taishan, New Taipei City 243, Taiwan
* Correspondence: mjyouh@mail.mcut.edu.tw (M.-J.Y.); mingderger@gmail.com (M.-D.G.); Tel.: +886-3-3891716 (M.-J.Y.); Fax: +886-3-3892494 (M.-J.Y.)

**Abstract:** In this study, the ballistic performance of armors composed of a polyurea elastomer/Kevlar fabric composite and a shear thickening fluid (STF) structure was investigated. The polyurea used was a reaction product of aromatic diphenylmethane isocyanate (A agent) and amine-terminated polyether resin (B agent). The A and B agents were diluted, mixed and brushed onto Kevlar fabric. After the reaction of A and B agents was complete, the polyurea/Kevlar composite was formed. STF structure was prepared through pouring the STF into a honeycomb paper panel. The ballistic tests were conducted with reference to NIJ 0101.06 Ballistic Test Specification Class II and Class IIIA, using 9 mm FMJ and 44 magnum bullets. The ballistic test results reveal that polyurea/Kevlar fabric composites offer better impact resistance than conventional Kevlar fabrics and a 2 mm STF structure could replace approximately 10 layers of Kevlar in a ballistic resistant layer. Our results also showed that a high-strength composite laminate using the best polyurea/Kevlar plates combined with the STF structure was more than 17% lighter and thinner than the conventional Kevlar laminate, indicating that the high-strength protective material developed in this study is superior to the traditional protective materials.

**Keywords:** multi-layer armor; ballistic performance; polyurea elastomers; shear thickening fluid

**Citation:** Chang, C.-P.; Shih, C.-H.; You, J.-L.; Youh, M.-J.; Liu, Y.-M.; Ger, M.-D. Preparation and Ballistic Performance of a Multi-Layer Armor System Composed of Kevlar/Polyurea Composites and Shear Thickening Fluid (STF)-Filled Paper Honeycomb Panels. *Polymers* **2021**, *13*, 3080. https://doi.org/10.3390/polym13183080

Academic Editors: Muhammad Tayyab Noman and Michal Petrů

Received: 15 August 2021
Accepted: 6 September 2021
Published: 13 September 2021



## 1. Introduction

With the development of technology and techniques, the structural design of armor systems is gradually moving towards a multi-layer armor system [1–3]. For protection purposes, the multi-layer armor system can be used as a simple way of increasing the number of layers of protection material to meet the need for increased impact resistance and to reduce the additional costs associated with engineering changes. However, this approach also tends to increase the thickness and weight of the garment, causing restrictions on movement and strain on the wearer, so the light weighting of protective equipment has been a continuous improvement effort by developers from all walks of life [4–6]. Bulletproof materials must have high strength and modulus to protect against shear damage and tensile deformation caused by projectiles during high-speed impacts [7–9]. The mass effect is an important factor affecting impact resistance, and in the pursuit of light weight, it also affects the protection capability of ballistic materials. Recently, many studies have proposed the use of composite materials to achieve the goal of improving impact resistance [10–12].

In addition, shear thickened fluids have been subjected to a series of tests in the field of impact resistant materials in recent years due to their special shear thickening properties.

Shear thickened fluids are intelligent materials that are non-Newtonian in nature and exhibit solid-like properties when the shear force or shear rate exceeds the critical shear stress or critical shear rate, resulting in a rapid increase in viscosity [13–18]. Due to this unique shear thickening rheological property, they facilitate the absorption of impact energy for protection in the event of an impact. However, in a study by Arora et al., it was noted that increased yarn-to-yarn friction was not necessarily beneficial in terms of absorbing impact energy, but rather reduced the protective capacity of these high-performance fabrics made from finer yarns shaped from specific fabrics [19]. The results of these studies have also highlighted a key point that when combining STF with other materials for impact-resistant protective gear applications, the mutual influence between the rheological properties of STF and the structural parameters of the combined material must be taken into account, in order to avoid stress concentration in the composite material during the impact process, which could lead to a weakness in the material and loss of the original protective effect, which plays a crucial role in determining its impact resistance.

Reviewing the literature and fundamental theories on the ballistic mechanism of high-performance Kevlar fiber fabric, the invasion process and damage mechanism of high-performance Kevlar fiber fabric subjected to high-velocity projectile impact are well understood [20–24]. A large number of studies have now shown that 2D woven fabrics can be coated with polymers, for example, to improve the fracture toughness of the structure [25–29]. In this study, an attempt was made to enhance the impact resistance of high-performance Kevlar fabric with polyurea polymers. From the literature on high-performance Kevlar fiber composites, it is clear that when adding reinforcement to high-performance Kevlar fiber fabrics, attention must be paid to the effect of the added content on their ballistic protection mechanism [4,30,31]. Therefore, in this study, polyurea elastomer/Kevlar composites were prepared using diluted tetrahydrofuran solution containing various ratios of polyurea polymer raw materials in order to obtain suitable composites.

The multi-layer armor system blunts the penetration and penetration capability of high-speed projectiles by means of a protective material at the front end of the armor stack, and then depletes the kinetic energy of the warhead by means of a bulletproof material at the rear end. In addition to the unique properties of the material (fibers) itself, a variety of mechanisms should be considered to enhance impact resistance [32–35]. For example, after selecting the appropriate material properties, a common way to improve protection is to add more layers in a different arrangement during the production of the laminate stack. However, even though increasing the number of layers in a composite material stack can improve protection, it can also affect the total weight and flexibility of the final protective equipment [36–39]. In order to achieve the goal of developing a soft armor, the multi-layer armor designed in the study uses fibers as the main structure at the front end and incorporates an STF structure at the rear end to enhance the efficiency of impact energy absorption.

This research follows our previous research on high energy absorbing nano-mesophase reinforced ballistic materials. Using the experience in developing ballistic materials, we further combine the characteristics of polymer composite materials and STF structures, with the aim to produce high-strength composite ballistic material with excellent protection for users. It has been reported that polymer matric composites or polymer-coated fabrics can change their ballistic performance [30,40]. However, the impact behavior and failure mode are affected by the polymer used [41,42]. Due to its effective energy absorption property, polyurea is commonly used as a protective coating on concrete or steel structures. However, studies on the ballistic performance of polyurea-coated fabric are rare [43]. For this purpose, high-velocity impact tests were conducted according to the NIJ 0101.06 Class IIIA standard [44] to determine the ballistic performance of neat Kevlar fabric and samples composed of Kevlar/polyurea composites and STF structures. The remarkable results from the comparison between the untreated Kevlar fabrics and panels with an adequate arrangement of Kevlar/polyurea composites and STF structure showed that a high-strength composite laminate using the best polyurea/Kevlar plates combined with the STF structure was lighter and thinner than the conventional Kevlar laminate, while maintaining the same ballistic resistance.

## 2. Experimental

### 2.1. Materials

In this study, nanoscale silica was mixed with polyethylene glycol (PEG) to prepare a shear thickening fluid. The silica was a nanoscale fumed silica powder and purchased from Kingmaker Chemical (ECHO, Miaoli, Taiwan). PEG was an oligomer produced by ECHO with a molecular weight of 200 g/mol. A wet dispersant, Disperbyk-111, was used to enhance the stability of silica particles.

### 2.2. Fabrication of STF-Filled Paper Honeycomb

When preparing STF, problems such as uneven mixing are often encountered. Therefore, in the preparation process, we initially used a planetary mixer to mix the solid dispersed particles with the liquid dispersing medium, and then used a three-roller mixer to roll the shear thickening fluid to disperse the solid particles in the shear thickening fluid well. In the first stage, the required amount of polyethylene glycol was weighed with an electronic scale and poured into the mixing tank of the planetary mixer, and the required amount of nanosilica (10–30 nm particle size) and additive (Disperbyk-111) was weighed and poured into the tank. The mixing tank was then placed in the planetary mixer and the suspension was stirred at speed of 2000 rpm for 10 min. After that, the suspension was defoamed at 2500 rpm for 5 min. The above steps were repeated until the silica content of the shear thickening liquid reached the required solid content. In the second stage, the shear thickening liquid prepared in the first stage was fed into and passed three roll mills five times with a gap of 150, 100, 50, 10 and 5 μm, separately, to improve the homogeneity of the STF. As shown in Figure 1, the STF-filled paper honeycomb structure was prepared through pouring the STF into a honeycomb paper panel (Asazawa Industrial Co., Ltd., Taoyuan, Taiwan) with a specified size and an NY vacuum bag, provided by Futian Packaging Co., Ltd, Taipei, Taiwan was used to seal the STF. The purpose of using honeycomb paper spacers was to maintain the shear thickened liquid in a liquid state. By controlling the size of the honeycomb spacer, the amount of shear thickened liquid can be fixed and the thickness as well as weight of the target plate can be controlled.

### 2.3. Preparation of Polyurea Elastomer/Kevlar Plates

The preparation process is displayed in Figure 2. The sample layers were finally investigated according to the subsequent experimental design.

### 2.4. Rheological Test

Rheological measurements of STF suspensions were performed at 25 °C using a stress-controlled rheometer (model: HAAKE RS600, Thermo Fisher Scientific, Newington, CT, USA). For rheometer testing, a tapered plate (No. C20/2 Ti) with a flat fixture was used for measurement at room temperature (25 °C). The diameter of the tapered plate was 20 mm, the angle of the outer taper was 2 degrees and the distance between measurement positions (gap) was set at 0.1 mm.

### 2.5. Stab Tests

A home-made drop hammer tester (Figure 3) was used to perform the puncture resistance test. The test was carried out with a fixed weight of 176.3 g and a knife weighing 102.6 g, for a total weight of 278.9 g and a height of 142 cm. The puncture effect was assessed by impacting the piercing knife on a high energy absorbing nano-media reinforced ballistic material on the test bench using the free fall principle with a certain amount of energy.

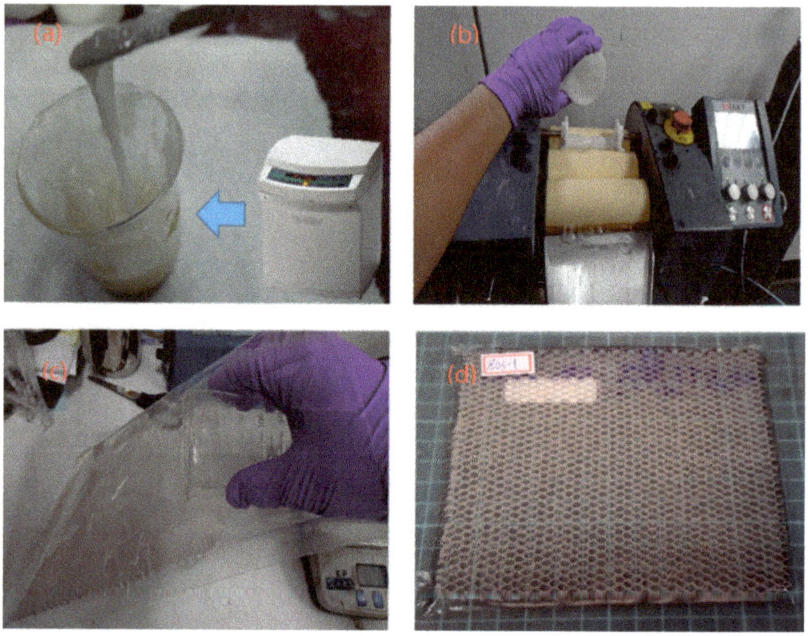


**Figure 1.** Fabrication of STF-filled paper honeycomb. (**a**) planetary mixer (**b**) three-roller mixer (**c**) Filled with good flowing STF into honeycomb paper (**d**) Schematic diagram of STF structure.

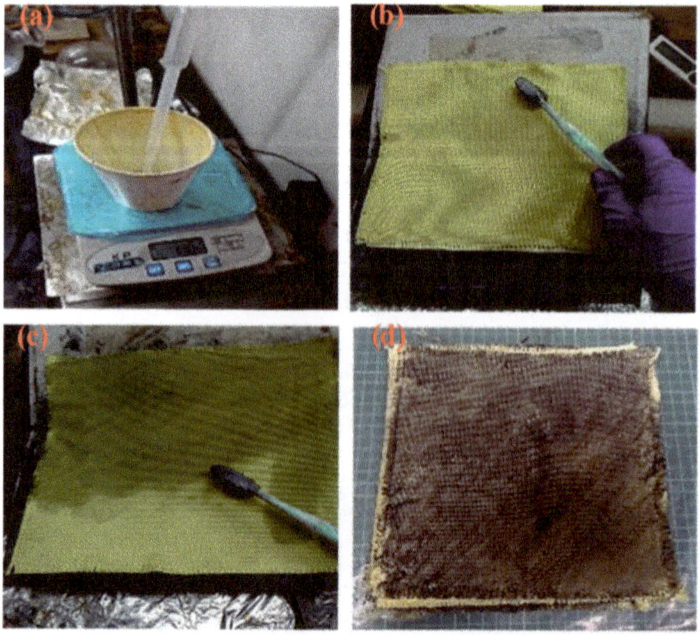


**Figure 2.** Preparation of polyurea elastomers/Kevlar plates (**a**) Preparation of polyurea elastomers (**b**) lay the cut Kevlar fabric flat on the carrier (**c**) painting polyurea elastomers (**d**) polyurea elastomers/Kevlar plates schematic diagram.

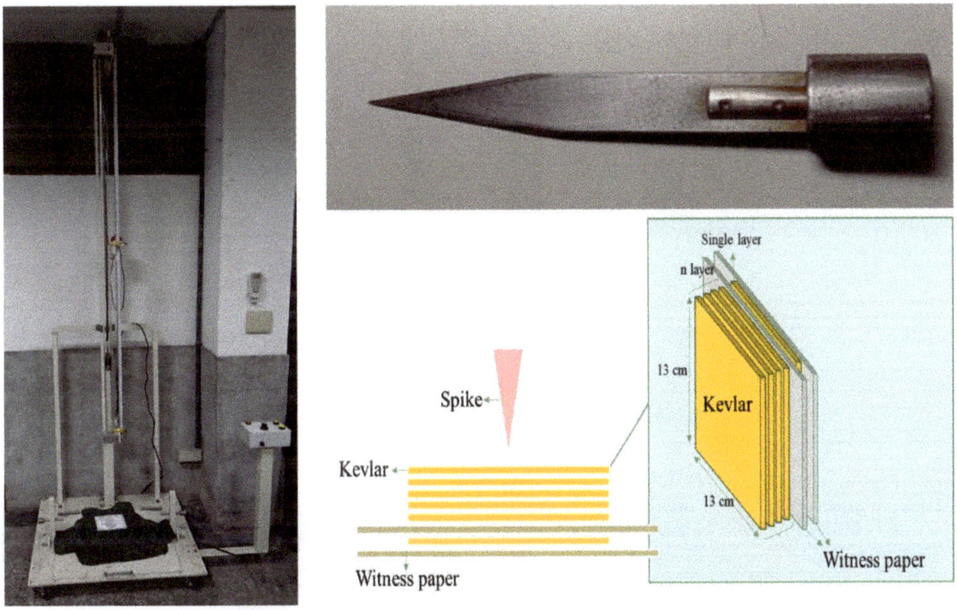


**Figure 3.** Schematic diagram of the stab test setup.

*2.6. Ballistic Impact Testing*

The ballistic impact tests were conducted with reference to NIJ 0101.06 Ballistic Test Specification Class II and Class IIIA [44], using 9 mm FMJ and 44 magnum bullets for live ballistic testing. Ballistic tests were conducted with the projectile flying in a steady direction through the inner rifling of the accuracy barrel to the target, with the muzzle 5 m from the target and the target 6 m from the wooden retaining wall. A schematic diagram of the ballistic test is shown in Figure 4. The first set of light gates was set at the muzzle from the target to measure the initial velocity of the projectile, and the second set of light gates was set from the target to the wooden retaining wall to measure the final velocity of the projectile, and the depth of the rear mud depression was the standard for impact resistance. Although a bulletproof vest is generally resistant to penetration by bullets, it can also cause bodily injury due to the force of the impact, so the depth of indentation in the ballistic test has always been a key criterion for a bulletproof vest. In this study, the back face signature (BFS), as defined by the NIJ 0101.06 standard, was used to determine the level of protection of the protective material and to compare the protection performance of the samples in terms of the depth of depression. In addition, during the ballistic test, we used Photron's FASTCAM SA1.1 high-speed video camera to capture the dynamic changes in the STF structure during the impact of the bullet on the bullet-resistant laminated sample.

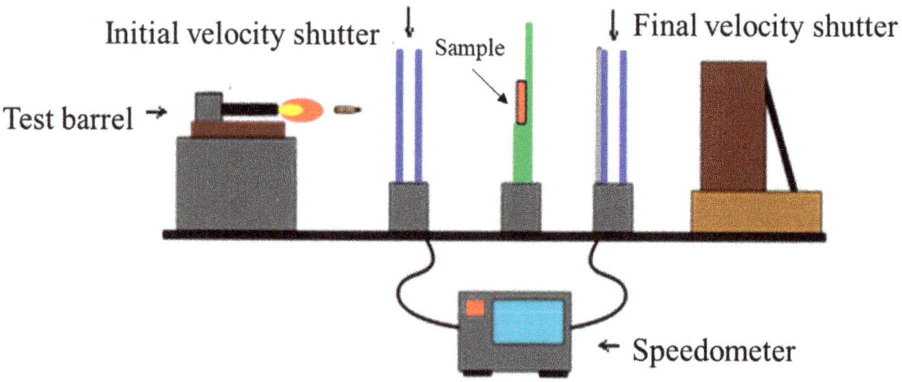


**Figure 4.** Schematic diagram of the ballistic test setup.

## 3. Results and Discussion

### *3.1. Rheological Analysis of the STF*

By measuring the rheological properties and analyzing the rheology, we could further identify the physical changes in the flow and deformation of the shear thickened liquid materials prepared for the study. The solid content of STF is an important parameter affecting the rheological properties of STF. According to the literature, as the solid content of STF increases, the shear thickening properties of STF will reach the maximum viscosity more rapidly with the increase in shear rate, and STF with a higher solid content has a higher maximum viscosity value [45]. By controlling the solid content of the STF, the same can be obtained for STF materials with different maximum viscosities. Figure 5a shows the rheological curves obtained from the planetary mixer and the triple-roller blender after processing different STF solid contents by rheological analysis. From Figure 5a, it can be seen that as the solid content of STF increases, the maximum viscosity obtained by the STF material under shear also tends to increase. In order to achieve the level of ballistic protection and high-velocity impact resistance required for practical protection applications, it is essential that the solid particles in the STF should be evenly dispersed in the solvent. Another important point to note is the reversibility of STF. In protective applications, STF differs from other materials in that it hardens on impact, but slowly returns to its fluid state after the impact force has dissipated. This special phenomenon also allows the use of STF in bulletproof vests to avoid "secondary damage" during the bulletproofing process compared to rigid vests. Therefore, in addition to the use of a triple-roller mixer to improve the dispersibility of the STF, the addition of an interfacial activator to the system to promote the homogeneous dispersion of the particles in the suspension was used to further increase the dispersibility, reversibility and stability of the prepared STF system. The main function of adding an interfacial activator is to reduce the interfacial tension between $SiO_2$ and PEG (solid–liquid) in the system and to help prevent the coalescence of $SiO_2$ particles. The addition of a dispersant has the effect of maintaining the stability of the dispersion in the normal fluid state of the STF, and also helps with the reversibility of the STF after it has been impacted and formed a solid state, reducing the time required to return to the original fluid state. Some of the test results on the effect of the added dispersant content on the rheological properties of the STF are shown in Figure 5b. The results of Figure 5b show that the rheological curve of STF gradually shifts to the left and the critical shear rate gradually decreases with the increase in the added dispersant. The trend of the experimental results obtained is consistent with the results in the literature on the effect of different molecular weights of liquid media on STF [46]. The addition of dispersant increases the molecular weight of the liquid medium in the STF, and as the molecular weight of the liquid medium increases, the viscosity of the liquid medium also increases. However, as the dispersant affects the coalescence of the solid particles in the STF, the observed shear thickening curve tends to flatten out as the dispersant content in the STF increases. In order to reduce the effect of dispersant on the rheological properties of STF and to increase the dispersion effect of STF, a dispersant amount of 3 g was used in this study for subsequent experiments.

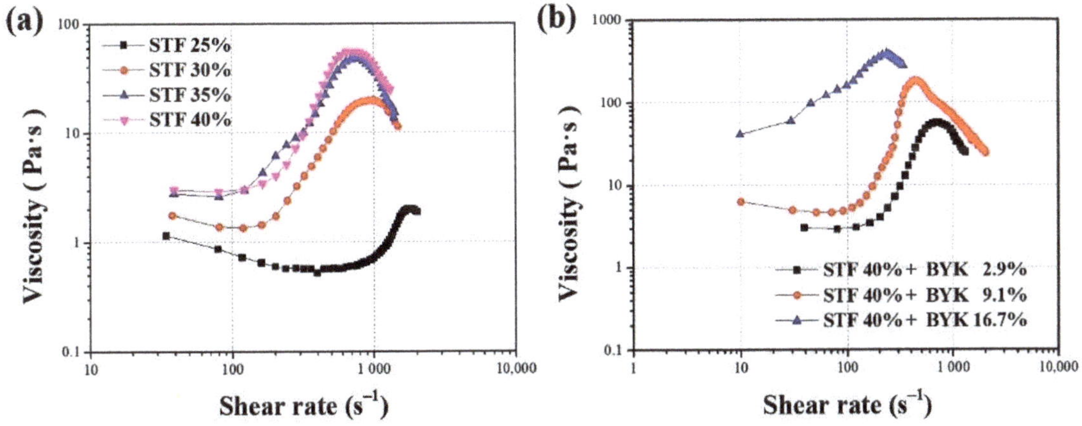


**Figure 5.** Rheology graph for as-prepared STF sample (**a**) STF rheological curves with different solid contents (**b**) The rheological curve of STF with 40% solid content and different content of dispersant.

### 3.2. Ballistic Performance of Kevlar/STF-Filled Paper Honeycomb Plates

The rheological properties of STFs required to resist high- or low-velocity impacts are very different, so STFs with different rheological properties should be used for different types of protection in order to achieve efficient protection capabilities. Our previous studies have shown that STF with a high critical shear rate has better protection properties [47]. A shear thickening fluid with a solid content of 40 wt% was prepared using nano-grade silica and poured into honeycomb paper of various thicknesses to continue the research on weight reduction and thinning of ballistic materials to find a balance between weight, thickness and protection. The test specimens were 40 wt% solids shear thickening fluid in honeycomb paper of 2 mm, 3 mm and 4 mm in thickness, sealed with PE bags and combined with Kevlar. Ballistic testing has been performed on different laminating sequences of composite panels composed of one STF structure layer and nineteen layers of Kevlar fabric which were obtained by simply placing the STF structure at different positions in the composite panels in our previous study [47] and showed that the STF structure placed at the rear position can significantly contribute to the increase in impact resistance. Thus, the test specimens used in the following study were prepared by putting Kevlar laminate in the front end and the STF structure in the rear end. Ballistic testing was conducted using National 9 mm pistol ammunition and 44 Magnum pistol ammunition in accordance with NIJ 0101.06 Class IIA. As can be seen from the test results in Table 1, compared to the control 29-layer Kevlar sample (Std. 29), the A-1 sample prepared in the first stage of research was already lighter and thinner than the control 29-layer Kevlar sample, and a comparison of the sludge depression depth results between the two indicated that the STF/Kevlar composite (No. A-1) prepared was more impact resistant. This result also shows that the amount of shear thickening fluid in the sample is directly proportional to the impact resistance, and that shear thickening fluid is indeed effective in absorbing impact energy. Based on the results of these experiments, the design of a bulletproof vest can be further developed in accordance with the product specifications to find the optimum weight and thickness of the bullet-resistant structure. In addition, for the ballistic test upgraded to NIJ 0101.06 Level IIIA, the test results are shown in Table 1, for samples

A-4 and A-5. The test results show that a stack of 37 layers of Kevlar fabric and an STF structure positioned at the second layer is needed to satisfy the requirement of NIJ 0101.06 Level IIIA.

**Table 1.** Results for ballistic test of Kevlar/STF-filled paper honeycomb plates.

| Plate ID | Composition | Layers | Honeycomb Structure Thickness (mm) | Areal Density (g/cm²) | BFS (mm) | NIJ 0101.06 Test Level |
|---|---|---|---|---|---|---|
| Std. 29 | 29 plies of Kevlar fabric | 29 | - | 1.29 | 34.01 | IIA |
| A-1 | 18 plies of Kevlar fabric and one ply of STF-filled paper honeycomb structure | 19 | 4 | 1.27 | 33.84 | IIA |
| A-2 | | 19 | 3 | 1.16 | 34.32 | IIA |
| A-3 | | 19 | 2 | 1.08 | 41.37 | IIA |
| A-4 | 37 plies of Kevlar fabric and one ply of STF-filled paper honeycomb structure | 38 | 4 | 2.08 | 39.45 | IIIA |
| A-5 | | 38 | 2 | 1.85 | 43.71 | IIIA |

### 3.3. Ballistic Performance of the Polyurea Elastomer/Kevlar Plates

From reviewing the literature and fundamental theories on the ballistic protection mechanism of high-performance Kevlar fiber fabric, the process and damage mechanism of high-performance Kevlar fiber fabric exposed to high-velocity projectile impact are well understood. Using the results of previous experiments on the impact resistance of high-performance fiber composites [48], Kevlar high-performance fiber fabric coated with polyurea was used to enhance the impact resistance. Then, a comparison of the impact resistance between the Kevlar fabric coated with polyurea and STF-impregnated Kevlar fiber composites reported in the literature was made. From the literature on Kevlar fiber composites, it is clear that when reinforcements are added into Kevlar fiber fabrics, attention must be paid to the effect of the amount of reinforcement added on their ballistic protection mechanism. The test results (Table 2) show that, when polyurea elastomers diluted with THF are brushed onto Kevlar, the performance of the Kevlar fabric is improved and the number of layers of the Kevlar stack is reduced, and Class II ballistic testing shows that there is one level above the original Class IIA in ballistic performance. When a projectile impacts a fabric, it is considered that the projectile kinetic energy is dissipated through a combination of mechanisms such as tension in primary yarns, deformation of fabric, energy dissipated through frictional slips (yarn/yarn and projectile/yarn), yarn breakage and yarn pull-out from the fabric [49]. The polyurea coating helps primary yarns to transfer the impact load into the secondary yarns and it also increases the friction between yarns. Consequently, the Kevlar/polyurea composites achieved a better ballistic performance compared to the pristine Kevlar fabric. The results from Table 2 for No. B-3 and No. B-4 tests showed that they were approximately the same, so we chose the lighter sample parameters and used a 1:5 THF dilution ratio for the following tests.

### 3.4. Stab Resistance of the Polyurea Elastomer/Kevlar Plates

Commercially available Kevlar fabric is woven in an orthogonal plain weave, which makes it difficult to protect against the impact of sharp projectiles. The puncture resistance of the polyurea elastomer/Kevlar composite was verified by simulating a bayonet puncture attack using a drop hammer puncture tester and comparing it with the Kevlar fabric as a control group. The results are presented in Table 3. As can be seen from Table 3 for Test No. D-1, a combined stack of 21 layers of Kevlar fabric is required to protect against the impact of a hammer piercing without penetration by a bayonet. However, a polyurea elastomer/Kevlar composite laminated construction (No. D-2) requires only 12 layers to resist the impact of a falling bayonet puncture without penetration when tested with the same standard drop hammer puncture test.

**Table 2.** Results for ballistic test of polyurea elastomer/Kevlar plates.

| Plate ID | Composition | Layers | Dilution Ratio (Polyurea/THF) | Areal Density (g/cm$^2$) | BFS (mm) | NIJ 0101.06 Test Level |
|---|---|---|---|---|---|---|
| Std. 19 | 19 plies of Kevlar fabric | 19 | - | 0.87 | 43.9 | II |
| B-1 | 19 plies of polyurea elastomer/ Kevlar plates | 17 | 1:10 | 0.79 | 55.3 | II |
| B-2 | | 17 | 1:5 | 0.83 | 36.2 | II |
| B-3 | | 15 | 1:5 | 0.75 | 45.2 | II |
| B-4 | | 15 | 1:3 | 0.91 | 46.4 | II |
| B-5 | | 15 | 1:1 | 0.97 | 39.9 | II |

**Table 3.** Results for stab tests.

| Plate ID | Composition | Not Penetrated Limit (Layers) | Weight (g) | Thickness (mm) |
|---|---|---|---|---|
| D-1 | Neat Kevlar fabrics | 21 | 150.5 | 11.17 |
| D-2 | Polyurea elastomers/Kevlar | 12 | 93.7 | 7.15 |

### *3.5. Ballistic Test Results of the Multi-Layer Armor System*

For stacked structures, in order to enhance the protection performance of the ballistic stack, the number of stacked layers is generally increased in order to increase the surface density of the stacked structure to achieve the purpose of enhancing the protection performance. However, by increasing the number of layers, the weight of the protective equipment is increased and the load on the person wearing it is increased [50]. The variety of high-strength composites produced in this study provides more scope for design discretion in optimizing the impact resistance of the laminated structure due to the diverse properties of the composites. In order to achieve the goal of lightweighting of protective equipment, the experiments in this section were carried out to verify the design patterns of different functional composite materials and boundary conditions (target size and thickness of the STF structure) in order to obtain the optimal design pattern of the high-strength composite material stack.

To summarize the results of this study and to consider the need for a soft ballistic material, the final high-strength composite layer will be designed using the best impact-resistant high-performance polyurea elastomer/Kevlar fiber composite, combined with a flexible STF structure. The ballistic test was conducted under the conditions of NIJ 0101.06 Class IIIA (two levels higher than Class IIA), and the size of the sample prepared for the test was 13 cm × 13 cm, a smaller surface area than that of a normal bulletproof vest. A domestic soft bulletproof vest with 29 layers of Kevlar fabric was used for comparison. Although the sample had the same number of layers (29), it was not able to protect against Class IIIA projectiles. This could be attributed to the size boundary effect caused by its smaller dimensional area. Under Class IIIA test conditions, a 47-ply Kevlar fabric (No. Std. 47) was selected as the control group to compare the impact resistance of the experimental groups, which met the 44 mm depth of sludge depression. From the analysis of the above experimental results, a lighter polyurea elastomer/Kevlar laminate was used to replace the

Kevlar fabric laminate structure at the front of the original sample, and a shear thickening fluid structure was added to form a high-strength composite ballistic laminate sample. Table 4 displays the results of ballistic tests which were conducted using the Class IIIA standard of NIJ 0101.06. It can be seen from Table 4 that the depth of sludge depressions for samples E-1 and E-6 is 38.01 mm and 43.38 mm, respectively. Both met the Class IIIA standard of NIJ 0101.06. The lighter and thinner sample, No. E-6, is 17.91% thinner and 17.68% lighter than the control group, No. Std. 47, indicating that the reinforced high-strength composite material is lighter and thinner and meets the Class IIIA standard of NIJ 0101.06. To optimize the high-strength composite laminates, samples E-1 and E-6 were prepared using the best data from this study and their ballistic tests were performed according to the Class IIIA standard of NIJ 0101.06. The depths of sludge depressions for these two samples were 40.01 mm and 43.38 mm, respectively, which met the requirements of the standard of less than 44.0 mm. The error value is approximately ±2.0 mm, which is in line with the standard error value set by the American Judicial Association.

**Table 4.** The influence of boundary effects (target area size, surface density and thickness parameters).

| Plate ID | Composition | Layers | Thickness (mm) | Areal Density ($g/cm^2$) | BFS (mm) | NIJ 0101.06 Test Level |
|---|---|---|---|---|---|---|
| Std. 47 | 47 plies of Kevlar fabric | 47 | 24.91 | 2.11 | 43.79 | IIIA |
| E-1 | Polyurea elastomer/ Kevlar plates (37-ply) + STF structure (40%, 2 mm, 1-ply) | 38 | 23.22 | 1.98 | 38.01 | IIIA |
| E-2 | Polyurea elastomers Kevlar plates (35-ply) + STF structure (40%, 2 mm, 1-ply) | 36 | 22.74 | 1.94 | 38.32 | IIIA |
| E-3 | Polyurea elastomer/ Kevlar plates (32-ply) + STF structure (40%, 2 mm, 1-ply) | 33 | 20.31 | 1.77 | 40.01 | IIIA |
| E-4 | Polyurea elastomer/ Kevlar plates (32-ply) + STF structure (40%, 4 mm, 1-ply) | 33 | 23.14 | 2.03 | 38.56 | IIIA |
| E-5 | Polyurea elastomer/ Kevlar plates (32-ply) + STF structure (40%, 3 mm, 1-ply) | 33 | 22.12 | 1.91 | 37.58 | IIIA |
| E-6 | Polyurea elastomer/ Kevlar plates (32-ply) + STF structure (40%, 2 mm, 1-ply) | 33 | 20.45 | 1.74 | 43.38 | IIIA |

## 4. Conclusions

In this study, we introduced a new approach for developing a high-strength composite laminate which is lighter and thinner than the conventional Kevlar laminate and can still meet the Class IIIA standard of NIJ 010106 for bulletproof vests. The results of this study show that relying only on shear thickening fluids to improve the protection of the elastic cascade is limited. On the other hand, it generally will increase the weight and thickness of the sample. In this study, a polyurea/Kevlar fabric composite was prepared and utilized to develop a protective material that is comfortable to wear and meets the requirements of reduced wear load. A close comparison in ballistic test results reveals that polyurea/Kevlar fabric composites offer better impact resistance than conventional Kevlar fabric. In order to further improve the protection performance of the body armor and to understand the correct application of STF, this study used STF structures to replace some of the layers of fabric in soft body armor. The results of the NIJ 0101.06 Class IIIA ballistic test showed that a 2 mm STF structure could replace approximately 10 layers of Kevlar in a ballistic-resistant layer. Finally, a high-strength composite laminate (13 cm × 13 cm) using the best polyurea/Kevlar plates combined with the STF structure was more than 17% lighter and thinner than the conventional Kevlar laminate. Further work needs to be carried out to obtain the optimal results. However, our research provides a promising way to fabricate protective materials that can be applied in the future.

**Author Contributions:** Conceptualization, C.-P.C. and M.-D.G.; Data curation, C.-H.S. and J.-L.Y.; Methodology, C.-P.C.; Project administration, M.-D.G.; Resources, Y.-M.L.; Supervision, M.-D.G.; Validation, M.-J.Y.; Writing – original draft, C.-H.S.; Writing – review & editing, C.-P.C. and M.-J.Y. All authors have read and agreed to the published version of the manuscript.

**Funding:** This work is sponsored by the Ministry of Science and Technology of Taiwan under Grant No. MOST 109-2221-E-13 -007-MY2 and No. MOST 107-2623-E-606-001-D.

**Data Availability Statement:** The data presented in this study are available on request from the corresponding authors.

**Conflicts of Interest:** The authors declare no conflict of interest.

## References

1. Deju, Z.; Aditya, V.; Barzin, M.; Subramaniam, D.R. Finite Element Modeling of Ballistic Impact on Multi-Layer Kevlar 49 Fabrics. *Compos. B Eng.* **2014**, *56*, 254–262.
2. Yang, C.C.; Ngo, T.; Tran, P. Influences of Weaving Architectures on the Impact resistance of Multi-Layer Fabrics. *Mater. Des.* **2015**, *85*, 282–295. [CrossRef]
3. Liu, J.; Long, Y.; Ji, C.; Liu, Q.; Zhong, M.; Zhou, Y. Influence of Layer Number and Air Gap on the Ballistic Performance of Multi-Layered Targets Subjected to High Velocity Impact by Copper EFP. *Int. J. Impact Eng.* **2018**, *112*, 52–65. [CrossRef]
4. Kang, T.J.; Kim, C. Energy-Absorption Mechanisms in Kevlar Multiaxial Warp-Knit Fabric Composites Under Impact Loading. *Compos. Sci. Technol.* **2000**, *60*, 773–784. [CrossRef]
5. Min, S.; Chen, X.; Chai, Y.; Lowe, T. Effect of Reinforcement Continuity on the Ballistic Performance of Composites Reinforced with Multiply Plain Weave Fabric. *Compos. B Eng.* **2016**, *90*, 30–36. [CrossRef]
6. Gürgen, S.; Kuşhan, M.C. The Ballistic Performance of Aramid Based Fabrics Impregnated with Multi-Phase Shear Thickening Fluids. *Polym. Test.* **2017**, *64*, 296–306. [CrossRef]
7. Sockalingam, S.; Chowdhury, S.C.; Gillespie, J.W.; Keefe, M. Recent Advances in Modeling and Experiments of Kevlar Ballistic Fibrils, Fibers, Yarns and Flexible Woven Textile Fabrics—A Review. *Text. Res. J.* **2016**, *87*, 511–524. [CrossRef]
8. Fábio, O.B.; Fernanda, S.L.; Sergio, N.M.; Édio, P.L. Effect of the Impact Geometry in the Ballistic Trauma Absorption of a Ceramic Multilayered Armor System. *J. Mater. Res. Technol.* **2018**, *7*, 554–560.
9. Vemu, V.P.; Sowjanya, T. A Review on Reinforcement of Basalt and Aramid (Kevlar 129). *Mater. Today Proc.* **2018**, *5*, 5993–5998.
10. Tabiei, A.; Nilakantan, G. Ballistic Impact of Dry Woven Fabric Composites: A Review. *Appl. Mech. Rev.* **2008**, *61*, 1–13. [CrossRef]
11. Singh, T.J.; Samanta, S. Characterization of Kevlar Fiber and Its Composites: A Review. *Mater. Today Proc.* **2015**, *2*, 1381–1387. [CrossRef]
12. Clifton, S.; Thimmappa, B.H.S.; Selvam, R.; Shivamurthy, B. Polymer Nanocomposites for High-Velocity Impact Applications-a Review. *Compos. Commun.* **2020**, *17*, 72–86. [CrossRef]
13. Ding, J.; Tracey, P.; Li, W.; Peng, G.; Whitten, P.G.; Wallace, G.G. Review on Shear Thickening Fluids and Applications. *Text. Light Ind. Sci. Technol.* **2013**, *2*, 161–173.
14. Majumdar, A.; Butola, B.S.; Srivastava, A. Optimal Designing of Soft Body Armour Materials Using Shear Thickening Fluid. *Mater. Des.* **2013**, *46*, 191–198. [CrossRef]
15. Fahool, M.; Sabet, A.R. Parametric Study of Energy Absorption Mechanism in Twaron Fabric Impregnated with a Shear Thickening Fluid. *Int. J. Impact Eng.* **2016**, *90*, 61–71. [CrossRef]
16. Gurgen, S.; Kushan, M.C.; Li, W.H. Shear Thickening Fluids in Protective Applications: A Review. *Prog. Polym. Sci.* **2017**, *75*, 48–72. [CrossRef]
17. Mawkhlieng, U.; Majumdar, A. Designing of Hybrid Soft Body Armour Using High-Performance Unidirectional and Woven Fabrics Impregnated with Shear Thickening Fluid. *Compos. Struct.* **2020**, *253*, 112776–112785. [CrossRef]
18. Zhang, Q.; Qin, Z.; Yan, R.; Wei, S.; Zhang, W.; Lu, S.; Jia, L. Processing Technology and Ballistic-Resistant Mechanism of Shear Thickening Fluid/High-Performance Fiber-Reinforced Composites: A Review. *Compos. Struct.* **2021**, *266*, 113806–113818. [CrossRef]
19. Arora, S.; Majumdar, A.; Butola, B.S. Structure Induced Effectiveness of Shear Thickening Fluid for Modulating Impact Resistance of UHMWPE Fabrics. *Compos. Struct.* **2019**, *210*, 41–48. [CrossRef]
20. Guoqi, Z.; Goldsmith, W.; Dharan, C.K.H. Penetration of Laminated Kevlar by Projectiles-I. Experimental Investigation. *Int. J. Solids Struct.* **1992**, *29*, 399–420. [CrossRef]
21. Yang, H.H. *Kevlar Aramid Fiber*; John Wiley & Sons: Chichester, UK, 1993.
22. Dobb, M.G.; Robson, R.M.; Roberts, A.H. The Ultraviolet Sensitivity of Kevlar 149 and Technora Fibres. *J. Mater. Sci.* **1993**, *28*, 785–788. [CrossRef]

23. Chowdhury, S.C.; Gillespie, J.W. A Molecular Dynamics Study of the Effects of Hydrogen Bonds on Mechanical Properties of Kevlar® Crystal. *Comput. Mater. Sci.* **2018**, *148*, 286–300. [CrossRef]
24. Li, D.; Wang, R.; Liu, X.; Zhang, S.; Fang, S.; Yan, R. Effect of Dispersing Media and Temperature on Inter-Yarn Frictional Properties of Kevlar Fabrics Impregnated with Shear Thickening Fluid. *Compos. Struct.* **2020**, *249*, 112557–112565. [CrossRef]
25. Cunniff, P.M. An Analysis of the System Effects in Woven Fabrics under Ballistic Impact. *Text. Res. J.* **1992**, *62*, 495–509. [CrossRef]
26. Naik, N.K.; Reddy; K.S. Delaminated Woven Fabric Composite Plates Under Transverse Quasi-Static Loading: Experimental Studies. *J. Reinf. Plast. Compos.* **2002**, *21*, 869–877. [CrossRef]
27. Sadegh, A.M.; Cavallaro, P.V. Mechanics of Energy Absorbability in Plain-Woven Fabrics: An Analytical Approach. *J. Eng. Fibers Fabr.* **2012**, *7*, 10–25. [CrossRef]
28. Shanazari, H.; Liaghat, G.; Hadavinia, H.; Aboutorabi, A. Analytical Investigation of High-Velocity Impact on Hybrid Unidirectional / Woven Composite Panels. *J. Thermoplast. Compos. Mater.* **2017**, *30*, 545–563. [CrossRef]
29. Haris, A.; Lee, H.P.; Tay, T.E.; Tan, V.B.C. Shear Thickening Fluid Impregnated Ballistic Fabric Composites for Shock Wave Mitigation. *Int. J. Impact Eng.* **2015**, *80*, 143–151. [CrossRef]
30. Kim, Y.; Kumar, S.; Park, Y.; Kwon, H.; Kim, C.G. High-Velocity Impact onto a High-Frictional Fabric Treated with Adhesive Spray Coating and Shear Thickening Fluid Impregnation. *Compos. B Eng.* **2020**, *185*, 107742. [CrossRef]
31. Mukesh, B.; Abhijit, M.; Bhupendra, S.B.; Sanchi, A.; Debarati, B. Ballistic Performance and Failure Modes of Woven and Unidirectional Fabric Based Soft Armour Panels. *Compos. Struct.* **2021**, *255*, 112941–112951.
32. Billon, H.H.; Robinson, D.J. Models for the Ballistic Impact of Fabric Armour. *Int. J. Impact Eng.* **2001**, *25*, 411–422. [CrossRef]
33. Chen, X.; Zhu, F.; Wells, G. An Analytical Model for Ballistic Impact on Textile Based Body Armour. *Compos. B Eng.* **2013**, *45*, 1508–1514. [CrossRef]
34. Zhang, D.; Sun, Y.; Chen, L. Influence of Fabric Structure and Thickness on the Ballistic Impact Behavior of Ultrahigh Molecular Weight Polyethylene Composite Laminate. *Mater. Des.* **2014**, *54*, 315–322. [CrossRef]
35. Pacek, D.; Rutkowski, J. The Composite Structure for Human Body Impact Protection. *Compos. Struct.* **2021**, *265*, 113763–113776. [CrossRef]
36. Morye, S.S.; Hine, P.J.; Duckett, R.A.; Carr, D.J.; Ward, I.M. Modelling of the Energy Absorption by Polymer Composites upon Ballistic Impact. *Compos. Sci. Technol.* **2000**, *60*, 2631–2642. [CrossRef]
37. Hog, P.J. *Composites for Ballistic Applications*; Department of Materials Queen Mary, University of London: London, UK, 2003; pp. 1–11.
38. Goode, T.; Shoemaker, G.; Schultz, S.; Peters, K.; Pankowa, M. Soft Body Armor Time-Dependent Back Face Deformation (BFD) with Ballistics Gel Backing. *Compos. Struct.* **2019**, *220*, 687–698. [CrossRef]
39. Nagaraj, M.H.; Reiner, J.; Vaziri, R.; Carrera, E.; Petrolo, M. Compressive Damage Modeling of Fiber-Reinforced Composite Laminates Using 2D Higher-Order Layer-Wise Models. *Compos. B Eng.* **2021**, *215*, 108753–108766. [CrossRef]
40. Vieille, B.; Casado, V.M.; Bouvet, C. About the impact behavior of woven-ply carbon fiber-reinforced thermoplastic-and thermosetting-composites: A comparative study. *Compos. Struct.* **2013**, *101*, 9–21. [CrossRef]
41. Zangana, S.; Epaarachchi, J.; Ferdous, W.; Leng, J.; Schubel, P. Behaviour of continuous fibre composite sandwich core under low-velocity impact. *Thin-Walled Struct.* **2021**, *158*, 107157. [CrossRef]
42. Ferdous, W.; Manalo, A.; Yu, P.; Salih, C.; Abousnina, R.; Heyer, T.; Schubel, P. Tensile Fatigue Behavior of Polyester and Vinyl Ester Based GFRP Laminates-A Comparative Evaluation. *Polymers* **2021**, *27*, 386. [CrossRef] [PubMed]
43. Feaga, M. The effect of projectile strike velocity on the performance of polyurea coated RHA plates under ballistic impact. Master's Thesis, Lehigh University, Bethlehem, PA, USA, 2007.
44. United States National Institute of Justice Standard (NIJ). *Ballistic Resistance of Body Armor NIJ*; Standard 0101.06; NIJ: Washington, DC, USA, 2008.
45. Tan, V.B.C.; Tay, T.E.; Teo, W.K. Strengthening Fabric Armour with Silica Colloidal Suspensions. *Int. J. Solids Struct.* **2005**, *42*, 1561–1576. [CrossRef]
46. Qin, J.; Zhang, G.; Shi, X. Study of a Shear Thickening Fluid: The Suspensions of Monodisperse Polystyrene Microspheres in Polyethylene Glycol. *J. Dispers. Sci. Technol.* **2017**, *38*, 935–942. [CrossRef]
47. Shih, C.H.; Chang, C.P.; Liu, Y.M.; Chen, Y.L.; Ger, M.D. Ballistic Performance of Shear Thickening Fluids (STFs) Filled Paper Honeycomb Panel: Effects of Laminating Sequence and Rheological Property of STFs. *Appl. Compos. Mater.* **2021**, *28*, 201–218. [CrossRef]
48. Shih, C.H.; Huang, Y.R.; Chang, C.P.; Liu, Y.M.; Chen, Y.L.; Ger, M.D. Design and Ballistic Performance of Hybrid Plates Manufactured from Aramid Composites for Developing Multilayered Armor Systems. *Compos. Struct.* **2021**. under review.
49. Carr, D. Failure mechanisms of yarns subjected to ballistic impact. *J. Mater. Sci Lett.* **1999**, *18*, 585–588. [CrossRef]
50. Zhang, T.G.; Satapathy, S.S.; Walsh, S.M. *Effect of Boundary Conditions on the Back Face Deformations of Flat UHMWPE Panels*; Army Research Laboratory: Adelphi, MD, USA, 2014.

polymers

*Review*

# Classification of Textile Polymer Composites: Recent Trends and Challenges

**Nesrine Amor, Muhammad Tayyab Noman * and Michal Petru**

Department of Machinery Construction, Institute for Nanomaterials, Advanced Technologies and Innovation (CXI), Technical University of Liberec, 461 17 Liberec, Czech Republic; nesrine.amor@tul.cz (N.A.); michal.petru@tul.cz (M.P.)
* Correspondence: muhammad.tayyab.noman@tul.cz

**Abstract:** Polymer based textile composites have gained much attention in recent years and gradually transformed the growth of industries especially automobiles, construction, aerospace and composites. The inclusion of natural polymeric fibres as reinforcement in carbon fibre reinforced composites manufacturing delineates an economic way, enhances their surface, structural and mechanical properties by providing better bonding conditions. Almost all textile-based products are associated with quality, price and consumer's satisfaction. Therefore, classification of textiles products and fibre reinforced polymer composites is a challenging task. This paper focuses on the classification of various problems in textile processes and fibre reinforced polymer composites by artificial neural networks, genetic algorithm and fuzzy logic. Moreover, their limitations associated with state-of-the-art processes and some relatively new and sequential classification methods are also proposed and discussed in detail in this paper.

**Keywords:** classification; fiber reinforced polymer composites; artificial neural network; fuzzy logic; Sequential Monte Carlo methods

**Citation:** Amor, N.; Noman, M.T.; Petru, M. Classification of Textile Polymer Composites: Recent Trends and Challenges. *Polymers* **2021**, *13*, 2592. https://doi.org/10.3390/polym13162592

Academic Editor: Barbara Simončič

Received: 3 July 2021
Accepted: 2 August 2021
Published: 4 August 2021



## 1. Introduction

Classification of textiles and polymer based nanocomposites by computer added programs is relatively a new approach that develops the simulations of human brain in the form of algorithms to solve complex problems. Machine learning is a subcategory of artificial intelligence that provides the solution of various issues i.e., grading, classification, defects detection, quality control, prediction and process optimization, through advanced tools such as image processing, soft computing and computer vision algorithms. The adaptation of machine learning in textiles has aroused in recent years [1]. Durable, sustainable and quality products are produced with the help of machine learning algorithms with minimal effort. These algorithms are essential parts of modern artificial intelligence systems and researchers have been significantly used these systems for the betterment of textiles. Therefore, machine learning based automatic fabric defects detection system are integrated in modern textile machines to evaluate fiber grading, yarn quality and fabric performance.

Due to their economic benefits, textiles and polymers based composites have received tremendous attention and researchers used them in various fields due to their excellent mechanical, electrical and interfacial properties [2–6]. In terms of current research, the interfacial performance of fibre/resin for composites was observed to be sensitive to the actual service environment. The potential fibre/resin debonding may occur. In addition, the fatigue resistance is a key advantage of textile-based composites compared to the steel materials, that expands the application circle of composites in automobiles, aerospace, oil extraction industry, civil and building industry [7]. A group of researchers worked with the interfacial, mechanical and thermal properties of fibre reinforced composites and reported interesting results [8,9]. Li et al. worked with interfacial shear strength of pultruded rod made of carbon/glass. They investigated the effect of hydraulic pressure and water immersion on interfacial properties. The results revealed that the hydraulic pressure had positive impact on the interfacial performance of the carbon/glass composite and conversely, water immersion reduced the interfacial strength of carbon/glass rod [10]. In a recent study, Li et al. expanded their work on carbon/glass pultruded rod and investigated the interfacial, thermal and mechanical properties at elevated temperature. They reported that at elevated temperature, interfacial shear strength decreased with time for hybrid fibre reinforced composites. Longer exposure led to more degradation and plasticizing as well as hydrolysis observed due to the diffusion of water molecules [11]. In another study, Li et al. worked with the mechanical properties and life service evolution of unidirectional hybrid carbon/glass pultruded rod under harsh and elevated conditions. The results showed that fibre/resin debonding occurred under longer exposure and overall mechanical, thermal and interfacial properties decreased [12].

The use of machine learning in textiles, especially for classification, has shown its potential exponentially in the current era [13,14]. Zimmerling et al. reported the application of Gaussian regression algorithm to improve the geometrical shapes of fiber reinforced textile composites. This method significantly improved the assessment criterion for fibre reinforced plastics components. Based on the efficiency, they suggested machine learning as an economical tool than finite element method, for the evaluation of textile processes [15]. In another work, Seçkin et al. reported a production fault in gloves industry. They used time-series data for the simulation and forecasting of this problem and classified it with different machine learning algorithms [16]. Ribeiro et al. proposed an automatic method to predict different properties of woven fabrics based on design and finishing features [17]. Due to the complexities of their micro-structures and boundary conditions, the classification of overall characteristics of textiles and polymer composites is still a challenging task even for machine learning. Therefore, a highly efficient and accurate approach is required that can predict the microscopic structural performance under different geometries. Apart from the above discussion, hardware utilization and technical issues are two other major constrains during the application of machine learning in textiles. However, to address these problems, various machine learning and computer vision-based applications are reported in the literature as deterministic and non-deterministic models. Mathematical models, empirical models and computer aided models i.e., finite element method (FEM), are deterministic models. However, genetic algorithm (GA), artificial neural network (ANN), chaos theory (CT) and fuzzy logic (FL) are non-deterministic approaches. Figure 1 shows the difference between machine learning approaches and traditional deterministic engineering models.

In recent years, considerable research efforts have been made to the development of machine learning tools for classification, prediction and defects detection. Although, the prediction and defects detection for textiles and polymer composites have been reviewed, the comprehensive review on the classification of fiber reinforced polymer composites is still missing. Therefore, our motivation is to provide a detail description about the used algorithms for fiber reinforced composites. This may help our readers to find out best possible algorithm for their future research endeavours. We categorized the applications of machine learning and computer vision algorithms into four classes based on standard textile manufacturing processes i.e., spinning, weaving, finishing and fiber reinforced polymer composites. In each of these four classes, we elucidated the recently reported work for defects detection, identification, classification and prediction by image segmentation-based approaches, color-based approaches, texture-based defect detection and by deep learning. In addition, we reviewed the limitations of existing state of the art methods and proposed a possible future research direction in textile composites using the sequential Monte Carlo methods.

After the introduction, this paper is organized in a following sequence: Section 2 focuses the utilization of machine learning in spinning, weaving, non-woven and textile finishing applications. The following Section 3 provides the limitations of widely used approaches. Section 4 explains the possibilities of future challenges for sequential Monte

Carlo methods in textiles and polymer-based composites and the final Section 5 summarizes the paper with few suggestions to overcome the future challenges. Figure 2 explains the scheme that elucidates the methodology of the proposed paper. The authors believe that the approach delineates here opens up a new gateway for researchers to choose the best suitable machine learning tool in order to work with textile substrates and composites.

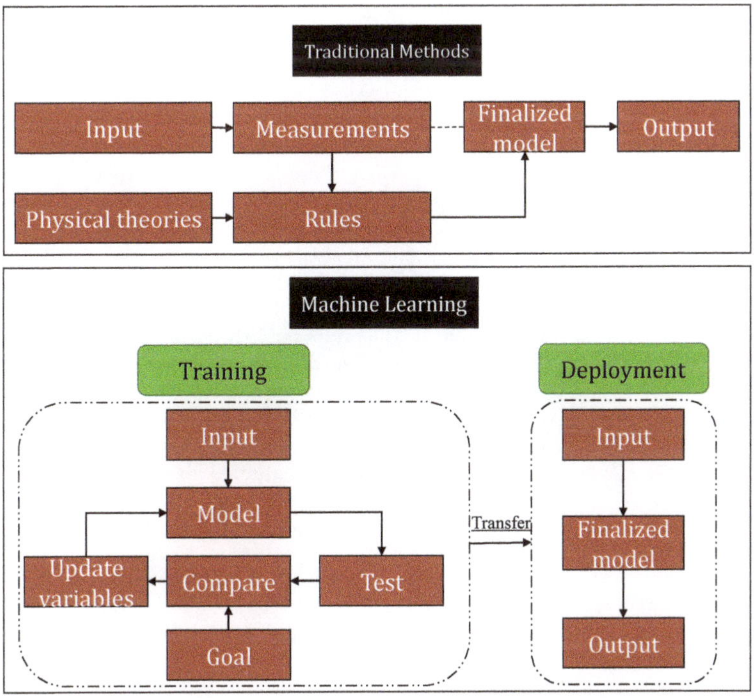


**Figure 1.** Comparison of machine learning approaches with traditional deterministic models.

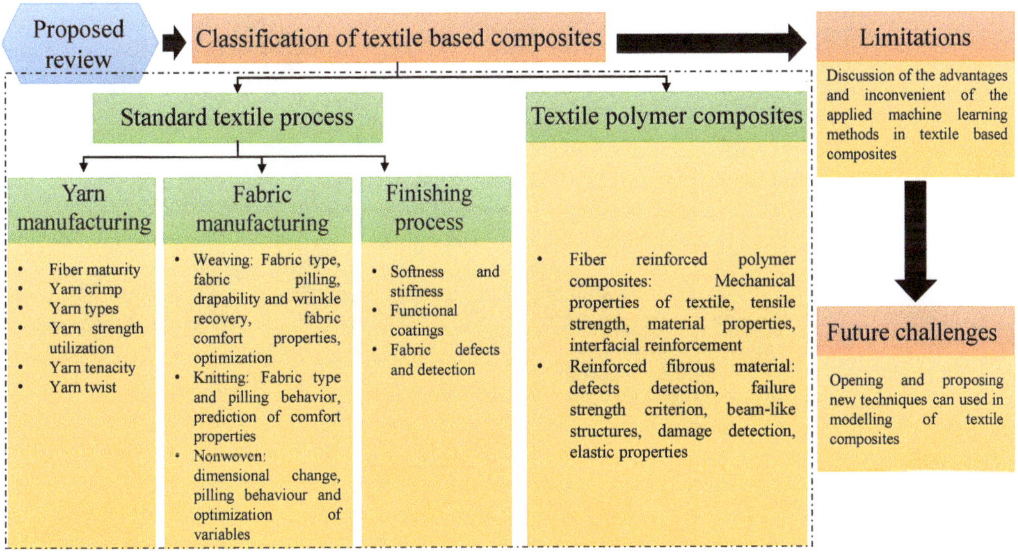


**Figure 2.** A schematic illustration of this study.

## 2. Classification Based on Textile Processes

The necessity to process raw data and explore valuable information from it has become essential in every field of science, engineering, business and medicine. In textiles, even when a simple product e.g., a t-shirt is considered, bulk amount of data is generated from raw materials, quality parameters and machine settings. The data could be nonlinear and multivariable depend on the relationship between fiber properties and yarn properties or between fabric performance and machine settings. Moreover, improvements and innovations in textiles with the introduction of technical textiles have occurred with exceptional performance expectations at extreme conditions e.g., against sun, cold, impact, knife, bullet and microorganisms [18,19]. Therefore, the demand for data processing and discovery of valuable information from this data is continuous in textile industry. Many traditional models e.g., statistical and mathematical, have been reported in numerous studies to process textile data. However, these classical models remain incapable of discovering the complex relationship between the variables. To solve this challenge, machine learning models are implemented in almost all areas of textile engineering. In textile processes, the trend of reported literature elucidates that the researchers tried to establish a critical relationships among essential parameters of fibers, yarns and fabrics. Machine learning algorithms are robust and powerful tools for modeling and solving complex and nonlinear applications. In this context, Majumdar published a book on soft computing in textile engineering [20]. This book compiled various research studies based on ANN and FL approaches during yarn modelling, fabric manufacturing, garments modelling by FL, composites modelling for quasi-static mechanical properties, viscoelastic behaviour and fatigue behaviour using ANN, textile quality evaluation using image processing, ANN and FL approaches.

ANN has been widely studied for textile data since last decade and helped the researcher to get better efficiency for fiber classification, defect detection, prediction and modeling of yarn, fabric, color matching, color separation and their coordinates conversion [21]. Vassiliadis et al. introduced a comprehensive overview of ANN applications in fabric manufacturing [22]. ANN had been successfully utilised for fibrous properties (classification of fibers, color grading, selection of cotton bales, identification of control parameter), yarns parameters (detection of faults, prediction of tensile properties and shrinkage) and for fabrics properties (defects detection, prediction of thermophysiological, sensorial and comfort properties, bursting of woven and knitted fabrics).

The literature discussed above provides an overview of machine learning till 2011. Therefore, we will present a review of recent advanced works in this field. In this section, we categorized the textile applications using machine learning algorithms into four classes based on standard textile-based processes i.e., spinning, weaving, finishing and fiber reinforced polymer composites.

### *2.1. Classification Based on Yarn's Production*

Yarn is generally considered as a primary element for the manufacturing of high quality textiles and in recent years, numerous studies were conducted on the classification, modeling and prediction of essential yarn parameters as given below:

#### 2.1.1. Fibre Maturity

Fibre maturity is an important and significantly crucial parameter especially when the researchers deal with yarns properties and desire excellent final product. In general, fibre maturity is considered as a functional and primary building block of any good textile product. Therefore, many researcher worked with the prediction of fibre maturity. Farook et al. proposed that ANN algorithms are excellent prediction tools in order to predict cotton fibre maturity [23]. They selected various fibre characteristics as an input variables and analysed fibre maturity as an output variable. The simulation results showed that ANN predicted cotton fibre maturity was higher when compared with the experimental values. However, there was no optimal result in this application.

#### 2.1.2. Yarn Crimp

Malek et al. evaluated the performance of ANN for the prediction of yarn crimp in woven barrier fabrics [24]. They performed two experiments to predict yarn crimp. The purpose of both experiments was to predict the crimp in warp yarn and weft yarn respectively. The input variables were weave style, density of warp and weft yarns, fibre and filament fineness, shed time and loom speed. However, the only exception in the second experiment was the replacement of yarn fineness with filament fineness. ANN results showed good results for the prediction of yarn crimp with the exception of two small deviations between actual and predicted output. In a different work, Majumdar et al. introduced mathematical, statistical and ANNs models to predict breaking force at elongation of ring-spun cotton yarns [25]. The inputs for these three models were yarn count and cotton fiber properties. ANN model provided a better prediction performance compared to the statistical and mathematical models. Later, they reported the implementation of a hybrid neuro-fuzzy system for the prediction of yarn strength [26]. The results were compared with standard ANN and regression models for prediction accuracy. ANN showed better prediction results than others.

#### 2.1.3. Yarn Types

Before the production of yarn, the prediction of quality parameters is important to overcome production faults. In an experimental study, Almetwally et al. used ANN and linear regression for the prediction of core spun yarn strength, elongation and rupture [27]. The results showed that ANN models provided significantly accurate prediction for yarn strength. Recently, Doran et al. reported the utilization of ANN and support vector machine (SVM) methods to avoid faulty fabric production [28]. In addition, they used statistical tools i.e., analysis of variance (ANOVA) and principal component analysis (PCA) to overcome input dimensions. The test results showed that both ANN and SVM methods provided effective predictions for yarn quality characteristics. However, SVM showed slightly better results than ANN for mean absolute percentage error (MAPE) and coefficient of correlation (R).

#### 2.1.4. Yarn Tenacity

Dashti et al. worked with yarn tenacity through ANN and produced a decision support system by applying GA [29]. Experimental results showed that ANN offered an accurate prediction for yarn tenacity with less than 3.5% error. In addition, GA was applied to obtain optimal input parameters for yarn production. The obtained tenacity was greater than the desired tenacity, therefore, a reduction in production cost was observed. The implementation of this strategy was useful to find good input conditions in order to achieve desired tenacity.

#### 2.1.5. Yarn Strength Utilization

Mishra used ANN models during the production of cotton fabric for the prediction of yarn strength utilization [30]. The selected input parameters were yarn counts, initial crimps, total number of yarns and yarn strengths in longitudinal and transverse directions along with the weave float length. The experimental results showed that yarn strength utilization percentage increased with an increase in yarn number in both directions. However, a decrease in crimp percentage and float length was observed. Mozafary et al. proposed a combined approach, where they used K-means algorithm for data clustering and ANNs for defects detection i.e., yarn unevenness [31]. The feedforward ANN and Levenberg–Marquardt training function of back propagation were applied in this method and the effectiveness was demonstrated by a comparative analysis with standard ANN results. In an experimental study, Malik et al. applied a back propagation ANN to analyse the prediction efficiency of used model for tensile properties of even and uneven yarns extracted from polyester-cotton blend [32]. The selected parameters were twist multiplier, cot hardness and Break draft ratio. The reported results of linear regression and ANN for tensile properties were compared with standard methods. El-Geiheini et al. worked with different types of yarns and used ANN and image processing tools for modeling and simulation of yarn tenacity and elongation [33]. They reported that the proposed techniques were suitable for the estimation of various yarn properties with minimum error. In another study, Erbil et al. used ANN and regression tools for tensile strength prediction of ternary blended open-end rotor yarns [34]. They applied multiple linear regression (MLR) and trained ANN algorithm with Levenberg–Marquardt backpropagation function. Furthermore, they compared both models for prediction efficiency. The results demonstrated that ANN models gave better prediction output than MLR for both parameters i.e., breaking strength and elongation at break.

#### 2.1.6. Yarn Twist

Yarn twist i.e., S twist, Z twist etc., is another important and noteworthy parameter in order to estimate end product's performance. Therefore, different researcher worked with this variable and reported interesting results. Azimi et al. used ANN in order to predict twist type for textured yarns [35]. They investigated the effects of heater temperature, texturing speed and the effects of twist type on yarns crimp stability for hybrid yarns. The testing results demonstrated that ANN models were excellent for the prediction of yarn properties under selected variables.

### 2.2. *Classification Based on Fabric Manufacturing*

In this part, we classified the use of machine learning tools into three categories based on fabric manufacturing methods i.e., weaving, knitted and non-woven.

#### 2.2.1. Weaving

The process of weaving is based on interlacing of yarns in warp and weft directions. However, with time, textile woven structures have become more and more complex by the addition of diagonal yarns in interlacing. Therefore, prediction of woven textiles is now a complex task that requires accumulated empirical knowledge about various parameters of woven textiles. Some of those variables are listed below where researchers performed machine learning algorithms to gain better performance and utilization of woven textiles.

##### Fabric Type

Woven structures are the mostly used structures not only in textile production but also in composites [36–41]. Ribeiro et al. proposed an automated machine learning method to predict the physical properties of woven fabrics based on finishing features and textile design. They investigated nine different properties including pilling, abrasion and elasticity and reported improved prediction results for all properties with low prediction error. They applied Cross-Industry Standard Process for Data Mining (CRISP-DM) iterations, where every iteration was based on the verification of standard input parameters. At CRISP-DM stage, an automated machine learning (AutoML) algorithm was performed to choose optimal regression model among six different machine learning algorithms. The results demonstrated that significantly better output was achieved by the selected codes for fixed sequence of yarns and fabric finishing treatment [17]. In an experimental study, Hussain et al. proposed a novel machine learning algorithm depends on transfer learning and data augmentation in order to recognize and classify the complex patterns of woven textiles. The texture and pattern of textiles are considered as essential factors to design and produce high-quality fabrics. The proposed algorithm worked with residual network through which textures of woven fabric were extracted and auto classified as an end-to-end manner. They suggested that the reported results would be effective even all of the fabric properties are altered [42].

Fabric Pilling, Drapability and Wrinkle Recovery

Fabric pilling, drapability and wrinkle recovery are the aesthetic properties of textiles and considered as performance indicator of textile fabrics for quality evaluation. Eldessouki et al. applied adaptive neuro-fuzzy models (ANFIS) for the evaluation of pilling resistance of woven fabrics. In this proposed approach, they classified the selected samples on the basis of texture patterns and compared their results with standard method of pilling resistance for correlation [43]. Xiao et al. predicted cotton-polyester fabric pilling with ANN (Back-propagation approach) and then used GA for optimization of their results. The optimized results revealed that GA algorithms were better in terms of root mean square error (RMSE), MAPE and mean absolute error (MAE) compared to ANN [44]. Drapability is one of the most important aesthetic properties that plays crucial role in providing graceful effects to textile fabrics. Drapability depends on experience and skills of humans and is judged subjectively. It renders the complexities during drape comparisons particularly when judged by different persons. Taieb et al. used ANN for the prediction of fabric drape ability under low stress. They reported that the application of ANN for the prediction of aesthetic properties including drapability is a promising one and is physical factors played crucial role during the prediction of fabric drape ability [45]. Hussain et al. compared ANN with adaptive neuro-fuzzy inference system (ANFIS) during the evaluation of fabrics wrinkle recovery [46]. They found that for both types of algorithms, the input conditions suitable for better wrinkle recovery were linear densities of both warp and weft sides. However, the suitable output variables were crease recovery angles of warp and weft yarns. The results demonstrated that simulation performed by ANN produced slightly better results than ANFIS with significant accuracy percentage. However, ANFIS process was more useful while drawing surface plots among variables. ANN algorithms do not have this feature.

Fabric Comfort Properties

Comfort evaluation is a noteworthy parameter in terms of fabric overall performance [47–51]. Majority of machine learning algorithms were applied on fabric's thermophysiological and sensorial comfort. Malik et al. used ANN algorithm to predict woven fabrics thermophysiological property i.e., air permeability with respect to fabric construction, raw materials involved during production and process variables [52]. ANN algorithm was trained with feedforward neural function under a hybrid back propagation method composed of Bayesian regularization and Levenberg-Marquardt function. Simulation results showed that the proposed model provided promising results on test data with lower MAE. In addition, Malik et al. employed another ANN algorithm to show a relationship between loom parameters, used material and construction of fabric in terms of porosity, mean pore flow, mean pore size with air permeability [53]. The experimental result showed that ANN algorithms were excellent for the prediction of comfort properties with minimal error. Wong et al. applied a hybrid approach that combined ANN and fuzzy-logic for overall prediction of clothing comfort by considering physical properties as input variables [54]. Simulation results provided maximum correlation coefficient.

Optimization

Optimization of process variables in order to reduce cost and improve production efficiency is another important factor in textiles. Therefore, many studies were carried out to investigate the process of optimization. Xu et al. combined differential evolution and Kriging surrogate algorithms to study the optimization of enzyme washing and production cost incurred on indigo dyed cotton [55]. They selected Taguchi L16 orthogonal array algorithm for optimization and applied it in their study. temperature of bath and concentration of enzymes were chosen as input variables and enzymatic washing as output response. Kriging model was used to analyse the relationship between variables and results revealed that the applied method was significantly efficient for the optimization of overall cost and can be further utilized in the analysis of mean square error, absolute error and relative error.

#### 2.2.2. Knitting

Knitting is another important production method in which prediction and modelling of knitting parameters are significantly complex tasks due to diversified variables i.e., knitted structures, knitting machine variables and selected yarn attributes etc. Several researchers have used different soft computing and machine learning algorithms to predict knitted fabric's comfort properties, spirality, pilling, bursting strength as well as other aesthetic and physical properties.

##### Fabric Type and Pilling Behaviour

Pilling is a serious fault in textile production especially in knit wear. Therefore, machine learning is a useful tool to forecast pilling behaviour of knitted fabric. Unal et al. selected single jersey knitted fabrics for the evaluation of air permeability and combined ANN algorithm with regression methods for the prediction of bursting strength of knit structures [56]. Implementation of results showed that both methods were able to predict the properties of knitted fabrics. However, ANN had a slightly positive edge when used for prediction. Yang et al. identified knitted fabric pilling behaviour by modifying ANN into deep principle components analysis-based neural networks (DPCANNs) [57]. In DPCANNs, principle components automatically tracked down the fabric initial and after pilling test properties and then neural network was applied to evaluate pilling grades. The obtained results revealed that DPCANNs had above average classification efficiency for pilling behaviour of knitted fabric. Another important work using ANN was performed by Kayseri et al. where pilling tendency was predicted by selecting fabric cover factor as an input parameter [58]. They observed that by changing cover factor, fabric pilling was controlled to a greater extent. In this study, they concluded that pilling behaviour was the outcome of pilling grade, mean pilling height as well as covered pilling area. They reported that used algorithms had very good prediction power in determining fabric pilling behaviour.

##### Prediction of Comfort Properties

Fayla et al. applied ANN algorithm on knitted fabrics to predict thermal conductivity [59]. They selected yarn conductivity, porosity, fabric weight and air permeability as input conditions. The results revealed that ANN algorithm predicted the thermal conductivity with significantly high correlation coefficient. Majumdar used ANN to predict the thermal conductivity of cotton, bamboo and their blended yarns. The input variables were bamboo fiber proportion, linear density of yarn, thickness of fabric and areal density. The correlation coefficient of this study very high [60]. Knanat et al. used ANN for the prediction of thermal resistance of wet knitted fabrics [61]. Here, they used two different ANN networks for the prediction of thermal resistance. In the first network, the input variables were moisture content, yarn, fiber and fabric parameters. However, in the second network, input variables were yarn, fiber and fabric parameters, and the output response was thermal resistance under varying moisture level. The results from both networks showed efficient prediction of thermal resistance. Mitra et al. used ANN for the prediction of thermal resistance of handloomed cotton fabric [62]. The input fabric parameters were picks per inch (PPI), ends per inch (EPI), weft and warp count. The results revealed that used ANN algorithm achieved good prediction efficiency for thermal resistance under low MAE values. In addition, EPI, warp count and weft count were major contributors for the evaluation of thermal resistance.

#### 2.2.3. Nonwoven

Machine learning algorithms have gained tremendous importance during the last few years to enhance productivity of nonwoven textiles by predicting various important parameters i.e., dimensional change, pilling behaviour and optimization of variables. Wang et al. measured the pilling of nonwoven fabrics using wavelet analysis [63]. The results demonstrated that wavelet analysis was quite similar to traditional method for pilling evaluation. Kalkanci et al. estimated fabric shrinkage by applying ANN algorithm inside relaxation methods [64]. Thermofixing, sanforizing, drying and washing were important processes applied on fabrics during finishing . Dimensional changes were predicted at the end of finishing processes by ANN. Two-layer feedforward perceptron function was used for ANN algorithm to evaluate the width of dimensional change. The experimental results showed that ANN gave better prediction results for dimensional change. Abhijit et al. applied a combination of GA and ANN as a hybrid algorithm to predict the comfort performance and the range of ultraviolet protection factor (UPF) [65]. ANN was applied as a prediction tool and GA was utilised as an optimization tool. For experimental purpose, a set of four samples were selected for the evaluation of functional properties. The proposed ANN–GA method was carried out until the required results were achieved. The results achieved by this method were in good agreement with the standard methods.

### 2.3. *Classification Based on Finishing Processes*

#### 2.3.1. Handle Modifications (Softness and Stiffness)

Modification of textile end product by applying softeners and stiffeners is necessary to improve the aesthetic properties. The selection of these materials attracts the scientists and researchers to build and train special machine learning algorithms, special mathematical models and soft computing tools. Farooq et al. used ANN to predict the shade change of dyed knitted fabrics after finishing application [66]. The inputs were the shade percentage, dye color and finishing concentrations. The output was delta values with respect to standard samples. Simulations results showed that ANN provided high prediction accuracy for shade change that occurred during finishing with minimal value of error between actual and predicted values.

#### 2.3.2. Functional Coatings

Malik et al. used ANN for the prediction of antimicrobial performance of chitosan/AgCl-$TiO_2$ coated fabrics. The input variables were curing time and concentration of colloids [67]. Samples were developed with different blends of selected colloid under different curing time. Feedforward ANN was trained under a hybrid combination of Bayesian regularization and Levenberg Marqaurdt algorithms. The testing results had an acceptable MAE during network training. Furferi et al. introduced a novel ANN algorithm for the prediction of coating process on textile fabrics [68]. Testing results demonstrated the significance of ANN model for coating mechanism. Ni et al. proposed a novel online algorithm that detected and predicted the coating thickness of textiles by hyperspectral images [69]. The proposed algorithm was based on two different optimization modules i.e., the first module was called extreme learning machine (ELM) classifier whereas the second one was called a group of stacked autoencoders. The lateral module was designed to take data from hyperspectral images. However, ELM module optimized by a new optimizer known as grey wolf optimizer (GWO). GWO was used to determine the number of neurons and weights to get more accuracy during classification. The results explained that online detection performance significantly improved with a combination of VW-SAET with GWO-ELM that provided 95.58% efficiency.

#### 2.3.3. Fabric Defects and Detection

Fabrics are occasionally the end product of any textile manufacturing process and fabric defects inspection is very important in terms of post manufacturing processes i.e., marketing, merchandising and branding. In a simple term, Fabric defects detection is a crucial process applied to control the quality of textile production. Machine learning algorithms have also played their role in this detection/inspection process. The most famous machine learning tools used in defects detection are ANN and image processing algorithms that have been applied for defects detection and grading of woven, knitted and nonwoven textiles. Hanbay et al. presented a literature review about the methods used for the detection of fabric defects and explained that detection methods had several types including structural, hybrid, spectral, model-based and statistical [70]. Czimmermann et al. presented a detail review based on automatically detection of fabric faults and fabric defect [71]. Rasheed et al. reported a comprehensive study on faults detection methods of textiles [72]. The widely used detection methods are based on image segmentation, color coordinates, frequency domain, texture-based, image morphology operations and deep learning. Eldessouki et al. applied a defects detection method composed of a hybrid combination of sepctral (Fourier transform) and (spatial) statistical functions that detected the fabric defects from images [73]. They applied component analysis to overcome input characteristics of selected datasets. The use of PCA in this application increased the classification rate. Liu et al. proposed an algorithm composed of low-rank decomposition and multi-scale convolution neural networks for defects detection [74]. Convolution neural networks were applied to extract multiple characteristics of defects from images for the improvement of image characterization ability to deal with complex textures. However, low-rank decomposition tool was established to analyze matrix characteristics for background (low-rank part) and for (salient defects). Furthermore, the salient defects map produced by sparse matrix was further diversified under threshold to localize the defected area of fabric. The test results showed that extracted features by neural network were accurate enough to analyse fabric texture than traditional standard methods i.e., local binary pattern and histogram of the oriented gradient.

Many other researchers utilised machine learning algorithms for defects detection. Sezer et al. applied independent component analysis (ICA) for defects detection at block level using a sample image [75]. They reported that this method provided satisfactory results for plain weave fabrics. However, for twill and texture weave patterns, this method is not generalized yet. Yapi et al. proposed redundant contourlet transform (RCT) method for defects detection [76]. A finite mixture of generalized Gaussians (MoGG) was used for modeling RCT coefficients that constituted statistical signatures to differentiate the defected fabric from defect-free fabric. The proposed approach was based on three steps: (1) detection of basic pattern for image decomposition and signature calculation, (2) discrimination between defected and defect-free fabric through Bayes classifier (BC) based on labeled fabric samples, and (3) detection of defects during image inspection by testing local patches. Experimental results revealed that the used approach achieved good results compared to ICA, local binary patterns (LBPs) and slope difference distribution (SDD). Li et al. proposed Fisher criterion-based deep learning algorithm for defects detection of patterned fabrics [77]. A Fisher criterion-based stacked denoising method was used for fabric images to classify into defective and defect free categories. The experimental results showed that the accuracy of proposed method was excellent for patterned fabrics and more complex jacquard warp-knitted fabric. Han et al. proposed the stacked convolutional autoencoders for defect detection [78]. The autoencoders were trained through synthetic defected data and non-defected data by using expert-based knowledge of defect characteristics, where, input was used as a defected image produced artificially and output was the corresponding clean image. Jeffrey Kuo et al. detected the following four defects in embroidery textile patterns i.e., stitch missing, joint defect, yarn floating knit and unregistered defect recognition [79]. The results demonstrated that the applied procedure was more effective than back propagation for detects detection as it took less time to train the network. Huang et al. used machine learning tools and image analysis for pilling assessment of fleece [80]. The applied methods were discrete Fourier transform, Gaussian filtering and Daubechies wavelet, for the extraction of important features of image information i.e., pilling area, pilling density and number of pilling points. ANN and SVM were used to classify the textile grade. Experimental results showed that the use of Fourier-Gaussian method improved the efficiency of classification for ANN and SVM. Table 1 elucidates a comparison of related work for defects detection in textile processes.

**Table 1.** A comparison of previously performed work for defects detection in textile processes.

| Proposed Models | Purposes | Methods | Major Findings | Authors |
|---|---|---|---|---|
| Gabor filters and pulse coupled neural network (PCNN) | Fabric defect detection for of warp knitting fabrics | Enhanced the image contrast using Gabor filters and they applied PCNN for segmentation purpose | Results of the experiments have demonstrated that the proposed PCNN with Gabor has higher detection accuracy (98.6%) | Li et al. [81] |
| Convolutional neural networks (CNN) | Automatic quality control for fiber placement manufacturing | A pixel-by-pixel classification has been created for the defects of the whole part scan | Simulation results showed that the proposed strategy failed to achieve satisfactory results due to their small training dataset (confront with over-fit problem) | Sacco et al. [82] |
| Fuzzy ARTMAP neural network | Evaluation of yarn surface qualities based on the extracted features | Wavelet texture analysis, attention-driven fault detection, and statistical measurement are used to extract the characteristic features of yarn surface appearance from images. and a fuzzy ARTMAP neural network is employed to classify and grade yarn surface qualities based on the extracted features. | The experimental results showed that the fuzzy ARTMAP achieved superior results to classify yarn surfaces compared to ANN and SVM | Liang et al. [83] |
| CNN | Fabric defect detection | Mobile-Unet is used to improve the performance of CNN | Experimental results showed that the detection speed and the segmentation accuracy in the proposed method achieve powerful performance compared to SegNet and U-net | Jing et al. [84] |
| CNN | Fabric defect detection and classification system | (1) Prototyped an advanced image acquiring model using National Instruments NI Vision; (2) Train the CNN using standard textile fabrics. (3) Testing fabrics are examined by the trained CNN. | The experiment work produced good accuracy in defect detection compared to the Bayesian classifier and SVM methods. In addition, it provided better processing and classification on defective pattern variation in patterned fabric | Jeyaraj et al. [85] |
| CNN | Fabric texture defects classification | Compressive sampling theorem is used to compress and augment the data in small sample sizes | The classification results of the proposed model achieved higher accuracy 97.9%compared to KNN, ANN and SVM | Wei et al. [86] |
| Deep convolutional generative adversarial network | Localize the surface defects for woven fabrics | A new encoder block was used to reconstruct query image with normal texture and no defect | The experiments results showed that the proposed approach is not sensitive to image blurring or illumination changes. In addition, it has high flexibility and high detection accuracy for different types of texture structures and defects. | Hu et al. [87] |

### 2.4. *Classification Based on Textile Polymer Composites*

Composites are the most promising class of versatile and durable materials of modern age. Machine learning algorithms reduce time, cost and effort to search optimal conditions for selected variables of composite structures. Therefore, machine learning is an essential and effective tool for a comprehensive evaluation of composites. Machine learning is used to solve complex numerical and applied problems in composites. In general, the fabrication of fiber reinforced composites is considered more challenging than other anisotropic structures. Sapuan et al. presented a book on ANN applications for composite materials [88]. They reported the use of ANN for numerous tasks such as defects detection in composites and polymeric structures, localization of carbon fiber–reinforced plastics and perspex plates, prediction of mechanical behavior, aging cycles evaluation, fatigue life prediction and prediction of composites life under loading. Muzel et al. presented a comprehensive review on the applications of finite element method for composite materials, failure criteria, material properties and types of elements in aeronautics, aerospace, naval, automotive, energy, sports, civil, manufacturing and electronics [89]. Dixit et al. introduced a review on modelling approaches for the prediction of mechanical properties of textile based composites using finite element method [90]. However, there are many important parameters need to investigate for the development of new algorithms for composite materials. Therefore, In this study, we will discuss these variables in detail and propose new methods to develop machine learning algorithms for textiles and composite structures. Schimmack et al. used Extended Kalman Filter (EKF) algorithm as a virtual sensor for temperature detection, composed of metal-polymer fibre based heater structure [91]. The main purpose of this algorithm was to control temperature in case of overheating or in any other emergency condition. The results revealed the accuracy of proposed approach. In another study, Gonzalez et al. used CNN for the identification of flow disturbances of dissimilar materials in composites production [92]. Specifically, CNN was applied to detect the position, size and permeability of any embedded material on the surface of mould. In CNN, the region of dissimilar material was selected as an input variable in order to recognise disturbance flow. Altarazi et al. applied multiple algorithms at a time to predict and classify tensile strength of polymeric films of different compositions. The used algorithms were stochastic gradient descent (SGD), ANN, k-nearest neighbors (kNN), decision tree (DT), regression analysis, SVM, random forest (RF), logistic regression (LoR) and AdaBoost (AB) [93]. Experimental results demonstrated that SVM algorithm showed better prediction results. In addition, the results revealed that the classification ability of used algorithms was excellent for sorting films into conforming and non-conforming parts. Balcioglu et al. compared finite element analysis with machine learning algorithms (DT, KNN, RF, SVR) for fracture analysis of polymer composites [94]. Fracture behavior of laminated composites reinforced with pure carbon, glass and carbon/glass composition were tested and compared with standard samples. The RF algorithm showed the best result with lower MSE values compared to other algorithms.

#### 2.4.1. Fiber Reinforced Polymer Composites

The use of natural fibers as a reinforcement in polymer composites has gained commercial success in terms of durable, economical and environmentally friendly materials. Khan et al. investigated the mechanical properties of cross-ply laminated fibre-reinforced polymer composites. They developed model for the prediction of mechanical properties using ANN [95]. The composite samples were developed by altering glass fibre layers with carbon fibre layers and polyphenylene sulphide with high-density polyethylene. The fibers were used as reinforcement and polyphenylene sulphide was used as a polymer matrix. Mechanical properties i.e., hardness, flexural modulus, impact and rupture strength were investigated for both directions. The input variables for ANN model were material type, matrix layers, composition and number of reinforcement. Simulation results showed that ANN predicted the mechanical properties with low MAE. In a study, Boon et al. provided a literature review on recent advances in optimization and design of fiber-reinforced polymer composites [96]. They stated that the best approach to provide accurate results was deep learning (DL). He et al. proposed a delamination detection approach for the detection of location, size and interfacial bonding of delamination in fiber-reinforced polymer composites. This method was based on frequency changes in multiple modes [97]. They employed a combination of different algorithms i.e., support vector machine, extreme learning machine and back propagation neural network for the detection of delamination parameters. Experimental results showed that SVM algorithms provided excellent prediction and classification performance as compared to other two algorithms.

Carbon fiber reinforced polymer composites (CFRP) are the most durable and promising modern age composite materials. By applying machine learning algorithms, researchers significantly reduced cost, efforts and time to determine optimal design points and process variables to develop CFRP structures. Mathematical modeling together with machine learning algorithms provide comprehensive analysis of CFRP structures. Matsuzaki et al. proposed an approach for state estimation and material properties of thermoset CFRP by using data assimilation [98]. Thermosetting simulation based on a non-linear state-space model that utilise ensemble Kalman filter (EnKF) for the estimation of state using data assimilation. This method estimated the degree of curing and the distribution of temperature model with thermal conductivity distribution. Simulation results showed that EnKF was successful in the estimation of the state of thermal conductivity distribution and model parameters. However, the estimation of thermal conductivity in complex distributions is still a challenging task. After the effectiveness of EnKF to estimate various CFRP thermoset molding attributes, they applied EnKF for the estimation of internal temperature during curing [99]. In this application, they selected three samples with altered thermal conductivity. The experimental results validated the efficiency of this approach using these types of specimens. Figure 3 shows a typical problem-solving method under machine learning algorithms validated for numerous types of fiber reinforced polymer composites including CFRP, glass fiber reinforced polymer composites (GFRP), basalt fiber reinforced polymer composites (BFRP) and aramid fiber reinforced polymer composites (AFRP).

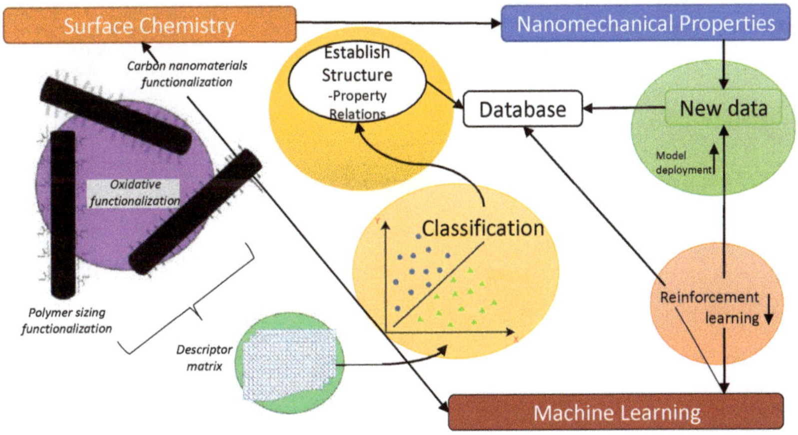


**Figure 3.** Summary of machine learning procedure validated for fiber reinforced polymer composites including CFRP, GFRP, BFRP and AFRP etc. [100].

Gonzalez introduced different mathematical models to detect nonlinear flexural deformation of CFRP based on stiffness level in compression and polymer matrices under different strength [101]. The study further presented modeling of different properties of fiber reinforced composite beams [102]. The proposed mathematical model described nonlinear elastic three-point bending of isotropic and reinforced beams under stiffness and strength levels. The obtained results revealed that nonlinear properties of reinforced materials and polymer matrices carefully investigated when designing real structures. Zhang et al. predicted the delaminations through Gaussian process regression (GPR) algorithm for CFRP composites during drilling [103]. Taguchi and GPR approaches explained that more data set were required for the extraction of optimal variables from fewer experimental trials. Konstantopoulos et al. used nanoindentation mapping data with machine learning algorithms to identify interfacial reinforcement [100]. Normalization and k-means clustering were applied to process data by filtering out from epoxy matrix. The used processs was trained by ANN, support vector machines and classification trees. The intrinsic modifications at the interface of CFRP proved that machine learning algorithms effectively patterned data and best fit can be obtained through SVM. Qi et al. employed the decision tree (regression tree) model to establish a relationship between variables properties and macroscopic variables of composite materials [104]. Here, representative volume element (RVE) algorithms for single-layer and multi-layer CFRP were established by a cross-scale FEM and periodic boundary conditions were loaded in order to verified the obtained results. Table 2 illustrates a comparison of related work in fiber reinforced polymer composites.

**Table 2.** Comparative study of related work in fiber reinforced polymer composites.

| Proposed Models | Purposes | Methods | Major Findings | Authors |
|---|---|---|---|---|
| Deep neural network (DNN) with finite-element method | Estimation of the stress distributions of the aorta | DNN model was constructed and trained, where the input is the results obtained by finite-element analysis method and the output was the aortic wall stress distributions | Simulations results showed that the proposed model was able to predict the stress distributions with lower error and accurate surrogate of finite-element analysis for stress analysis | Liang et al. [105] |
| Logical analysis of data (LAD) | Process control technique applied to the routing process for CFRP | Monitoring and evaluating the quality of the machined parts in CFRP by controlling some machining features and parameters | Experimental work showed that the proposed LAD outperformed ANN in both accuracy of controlling and monitoring variables | Shaban et al. [106] |
| Neural network regression | Damage location detection of the CFRP composite plate | Process the signals obtained by acoustic emission sensors in CFRP composite | Experiments are applied on the composite structure showed that the proposed approach provided a good result in the estimation of localization of damage signals comparing to the actual sources | Zhao et al. [107] |
| ANN | Prediction of damage progression and fatigue life in laser induced graphene interlayered fiberglass composites | Investigation of the potential of exploiting the piezoresistive properties of laser induced graphene interlayered fiberglass composites | Simulation results showed that piezoresistive laser induced graphene interlayers provided high prediction accuracy of fatigue life in multifunctional composite structures | Nasser et al. [108] |
| ANN, SVM and extreme learning machine | Assessment of delamination damage in fiber-reinforced polymer composite beams | Machine learning algorithms have been adopted as inverse algorithms to evaluate the delamination parameters | Experimental results demonstrated that the SVM provided the best prediction accuracy compared to ANN and extreme learning machine algorithms for delamination damage in fiber-reinforced polymer composites | He et al. [97] |
| Generative kernel principal component thermography and spectral normalized generative adversarial network | Defect detection in carbon fiber reinforced polymer composites | Extraction of nonlinear features from thermographic data and producing a number of informative thermographic data to improve the defect detection | Testing results showed that the proposed approach improved the detection accuracy of subsurface defects in CFRP | Liu et al. [109] |
| Bayesian regularized neural network | Weld quality classification for ultrasonic welding of CFRP | Proposing a feature selection methodology that combines new clustering overlap analysis with Fisher's ratio to improve the classification results | Simulations results in this application showed that the Bayesian regularized neural network have higher robustness and classification accuracy compared to SVM and kNN | Sun et al. [110] |

### 2.4.2. Prediction and Estimation of Reinforced Fibrous Material

Schimmack et al. applied a prediction approach based on wavelet for defects detection of any variable in a fiber reinforced polymer composite [111]. The applied algorithm was based on variance estimation for the local Lipschitz constant of any received signal over time. In addition, a modified recursive least squares (RLS) approach was applied to evaluated the various attributes of conductive multifilament fibers used as reinforcement during production process. The results proposed that RLS algorithms were useful for the estimation of time-varying sinusoidal disturbances as well as for inductance. Lui et al. developed a new strategy to predict the initial failure strength criterion of woven fabric reinforced composites based on micromechanical model by modifying deep learning neural network (DNN) and mechanics of structure genome (MSG) [112]. MGS is used to perform initial failure analysis of a square pack microscale model that trained the samples to detect yarn failure criterion. The effectiveness of this strategy was confirmed by testing yarns of mesoscale plain weave fabrics and fiber reinforced composite materials to compute the initial failure strength constants. Soman et al. used a novel algorithm based on Kalman Filter (KF) for load estimation in beam-like structures under complex loading [113]. Simulation results using experimental data showed that the used algorithm is efficient for classification and monitoring strains in continuous welded rails. In addition, Soman et al. used Kalman filter based neutral axis (NA) tracking algorithm for damage detection in composites structures under varying axial loading [114]. The proposed scheme was applied on a composite beam instrumented with fiber optic strain sensors. The change in neutral axis location is utilized to detect delamination in beams. Simulations results showed that the proposed formulation of KF for NA tracking provided more powerful use of NA location in various applications. Hallal et al. introduced a review on analytical modeling of elastic properties of textile composites [115]. Balokas et al. proposed FEM based multiscale prediction algorithm combined with ANN for the prediction of elastic properties of textile composites under different sources of aleatory uncertainty [116]. The results of sensitivity analysis showed that the proposed algorithm provided good prediction results for elastic yarn properties. Jiang et al. proposed an approach to predict elastic modulus of fiber braided composites with uncertainties using vibration test data [117]. Reference FEM was used for simulation of uncertain elastic parameters that reflected the dynamic characteristics of a braided composites. Statistical analysis of uncertain parameters revealed that uncertainties in elastic modulus can be identified by using modal data.

## 3. Limitations of the Proposed Techniques

The textile industry is benefited from machine learning tools by using them for different applications like prediction, classification, performance simulation, structural features modelling and image analysis etc. Table 3 summarizes mostly used machine learning techniques by various researcher in textile based applications.

**Table 3.** Techniques used in textiles and fiber reinforced polymer composites.

| Classes | Techniques | Applications |
|---|---|---|
| Yarn manufacturing | ANN, FL, GA and SVM | Prediction properties |
| Fabric manufacturing | DL, ANN, FL, DE, GA, SVM, DT, K-nearst, KNN, RF and SVR | Prediction, classification and recognition, optimization, identification and estimation |
| Finishing processes | ANN and image processing analysis | Prediction, defect detection |
| Textile based composites | ANN, DL, KF, EKF, EnKF, FEA, DT, KNN, RT and SVR | Prediction, defect detection, control, classification, estimation, tracking |

In general, textile processes are mostly non-linear in nature. Therefore, it is difficult to obtain analytic models for the technical design of fabrics due to the difficulties and complex structure imposed by the raw materials. Therefore, most researchers applied ANN in textiles during the confrontation nonlinear and multiparameter problems, without an analytical solution. Furthermore, the use of ANN in textile data prediction, detection, identification and classification problems covers fabrics, fibers, yarns, color, wet processing and garments. In addition, ANN has shown its potential as a successful tool for the prediction of different textiles, fiber reinforced composites and in the evaluation of structural properties of polymer composites. The most common ANN type used in textile industry is multilayer perceptron that represents a class of feedforward ANN. A feedforward network consists of single hidden layer and sigmoid activation function is used extensively to solve textile processing problems. However, the limitations of ANN are: it is not applicable outside the data range for which it is trained. In simple, the durability of ANN is limited to the selected range of data. In addition, ANN cannot answer the relationship between input and output variables i.e., ANN cannot predict why the selected input variables result in a significant increase or decrease in output variables and vice versa. Textile processes parameters prediction in a hybrid situation is a complex task for ANN because of highly variable nature of natural fibers, spinning processes, functional materials and fabric end use requirements. The shortcoming of ANN is the implicit nature of ANN models. Rather than developing an explicit analytic expression i.e., linear or nonlinear, of input and output variables, ANN processes the variables in order to gain iterative knowledge and store it in the system. In such a case, ANN subsequently simulates the system and predicts the results.

Fuzzy logic has been applied in various fields of textiles including the prediction of melt-spun yarn count and tensile strength of fibers, classification of colored cotton into different classes, automatic recognition of fabric weave pattern and intelligent diagnosis system for fabric inspection. However, the use of fuzzy logic in a highly complex system may become an obstacle to the verification of system reliability. In addition, validation and verification of a fuzzy knowledge-based system need extensive testing with hardware. Genetic algorithms are widely used to solve various problems in textile processing right from fiber production to garment design and manufacturing. However, the major limitations of GA are: it cannot guarantee to find an optimal solution and it is time consuming. In addition, the solution quality deteriorates with the increase of the dimension of the problem. The neuro-fuzzy hybrid model was applied in several cases for the prediction of fiber, yarn and fabric properties. The prediction reliability of this hybrid model had outperformed the conventional multiple regression model and the ANN model. Supervised learning techniques such as SVM, SVR, DT, k-nearst, KNN and RT were used for classification, identification and prediction properties. However, the training for these algorithms requires a lot of computation time. Recently, deep learning has been used in some textile applications and has shown its performance in identification, defect detection and prediction. Like every method, DL has some limitations. It requires large training data to provide better performance than other methods. However, it has high computational cost to train complex data models.

Kalman filter, ensemble Kalman filter, extended Kalman filter were used to estimate and track the state of materials and their properties in fiber reinforced polymer composites. Kalman filter provides optimal solution only when the state is linear with Gaussian model. For nonlinear and non Gaussian state-space models, optimal estimation problems do not typically admit analytic solutions. Therefore, a numerical method is needed to approximate the state as EKF, EnKF and PFs. EKF relies on the linearization of nonlinear state and observations, and this may result in an erroneous estimation of the state, and in a highly nonlinear case, the filter may diverge. EnKF works better with the Gaussian model, and the accuracy of its estimation depends on number of samples. These algorithms are not effective when the model is highly non-Gaussian and/or nonlinear. The particle filters (PFs), also called Sequential Monte Carlo (SMC), are able to proceed better in these situations. PFs is a sequential Monte Carlo method to estimate the posterior density of the state in a sequential manner, and does not make any assumptions about the linearity of the system model [118].

## 4. Future Challenges

The main reason of machine learning algorithms shortcoming is to not fully utilization of valuable and persistent information about the dynamics and physiology of system. In majority of cases, the model structure of textile composites is related to physical information that may not incorporated explicitly into machine learning algorithms. Physiologically stable models, in contrast, for textiles and composites carry all available information related to system and underlying properties, can be developed using state-space framework. In general, state-space hidden Markov models allow extremely flexible frameworks for simulation and modeling of discrete time data. In a linear system with additive Gaussian noise, an optimal estimation is provided using the Kalman filter [119]. However, the textile application state-space models are highly nonlinear and may be non-Gaussian.

Sequential Monte Carlo (SMC) is a famous and reliable class of numerical methods to evaluate optimal designs related problems in nonlinear non-Gaussian systems [118]. SMC is a powerful sampling tool that works with a set of random weighted samples in order to predict the optimal solution. These samples are technically known as *particles* that are utilized during the approximation of state density and statistics of interest [118]. Given enough particles, the SMC will always perform better than the EKF or EnKF, albeit at the expense of computational requirements [119,120]. In addition, it provides the SMC converges almost to the optimal solution [121]. Figure 4 illustrates a general schematic layout of state estimation with data assimilation using SMC.

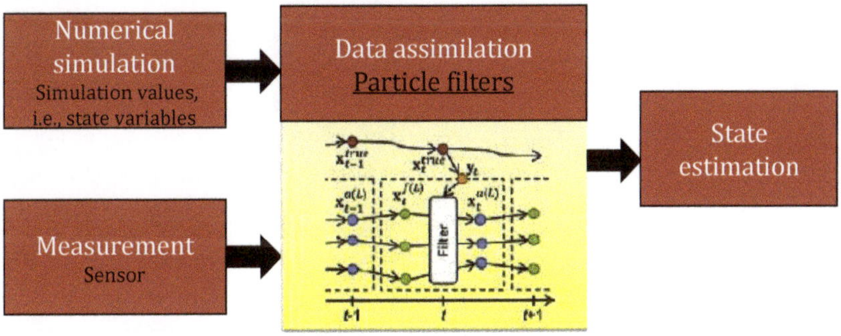


**Figure 4.** A general schematic layout of the state estimation method with data assimilation.

### 4.1. Classical Sequential Monte Carlo

We consider the following constrained discrete state-space model, where model representation consists of a dynamical process that captures temporal evolution of system state. As a result, the measurement model explains the relationship between the system state and the system output.

$$x_{k+1} = f_k(x_k) + u_k, \tag{1}$$

$$y_k = h_k(x_k) + v_k, \tag{2}$$

where $\boldsymbol{x}_k$ is the state transition vector and $\boldsymbol{y}_k$ is the measurement vector. $f_k$ and $h_k$ are possibly nonlinear state transition and measurement functions, respectively. $\boldsymbol{u}_k$ and $\boldsymbol{v}_k$ are the process and measurement zero-mean white noise sequences with known probability density functions (pdfs) $Q_k$ and $R_k$, respectively.

In the Bayesian framework, the optimal inference of the state $\boldsymbol{x}_k$ using the measurement history $\boldsymbol{y}_{1:k} = [\boldsymbol{y}_1, \ldots, \boldsymbol{y}_k]$ relies on the posterior density $p(\boldsymbol{x}_k|\boldsymbol{y}_{1:k})$. Using Bayes' rule, the posterior density can be computed recursively from the following prediction and update steps:

$$p(x_k|y_{1:k-1}) = \int p(x_{k-1}|y_{1:k-1})\, p(x_k|x_{k-1})\, dx_{k-1}, \tag{3}$$

$$p(x_k|y_{1:k}) = \frac{p(y_k|x_k)\, p(x_k|y_{1:k-1})}{\int p(y_k|x_k)\, p(x_k|y_{1:k-1})\, dx_k}. \quad (4)$$

In fact, the Equations (3) and (4) represent only a conceptual solution in the nonlinear case, because the integrals defined are intractable.

Sequential Monte Carlo approximate the posterior density of the unknown state using a set of $N$ particles and their associated weights $\{x_k^{(i)}, w_k^{(i)}\}_{i=1}^N$:

$$p^N(x_k|y_{1:k}) = \sum_{i=1}^{N} w_k^{(i)} \delta(x_k - x_k^{(i)}), \quad (5)$$

where $\delta$ represents the dirac delta function. In the ideal case, the particles required to be generated from the true posterior $p(x_k|y_{1:k})$, which is unknown. Thereby, another distribution named *proposal distribution* $q(x_k|x_{k-1}, y_k)$ is used to generate the particles [118]. The importance weight of each particle $x_k^{(i)}$ is computed using:

$$\tilde{w}_k^{(i)} = w_{k-1}^{(i)} \frac{p(y_k|x_k^{(i)})p(x_k^{(i)}|x_{k-1}^{(i)})}{q(x_k^{(i)}|x_{k-1}^{(i)}, y_k)}, \quad (6)$$

where the normalized weights are given by $w_k^{(i)} = \tilde{w}_k^{(i)} / \sum_{j=1}^{N} w_k^{(j)}$.

The conditional mean estimate of the state is then given by:

$$\widehat{x}_k = \mathbb{E}[x_k|y_{1:k}] \approx \sum_{i=1}^{N} w_k^{(i)} x_k^{(i)}. \quad (7)$$

The weights of the particles may perish and thus require resampling [118]. The particles are resampled according to their weights, i.e., removing particles with very small weights and duplicating particles with large weights. Thus, equal weights ($\frac{1}{N}$) are assigned to all selected $N$ particles. The detailed steps of sequential Monte Carlo are presented in Algorithm 1.

**Algorithm 1** Classical sequential Monte Carlo

**Initialization**
**for** i = 1, 2, $\cdots$, N **do**
  Generate $x_0^{(i)} \sim \mathcal{N}(x_0^{(j)}, R_k)$.
  Compute the initial weights using Equation (6) and normalize.
**end for**
**Estimation**
**for** k = 1, 2, $\cdots$, T **do**
  **for** i = 1, 2, $\cdots$, N **do**
    Generate sample $x_k^{(i)}$ from the system dynamics model (1).
    Compute weight using: $\tilde{w}_k^{(i)} = \tilde{w}_{k-1}^{(i)} p(y_k|x_k^{(i)})$.
  **end for**
  Normalize particle weights $w_k^{(i)} = \tilde{w}_k^{(i)} / \sum_{i=1}^{N} \tilde{w}_k^{(i)}$.
  Resample $\{x_k^{(i)}, \frac{1}{N}\}_{i=1}^N$.
  Compute the weighted mean $\widehat{x}_k = \sum_{i=1}^{N} \frac{1}{N} x_k^{(i)}$.
**end for**

### 4.2. Constrained Sequential Monte Carlo

Due to physical laws, kinematic constraints, mathematical properties such as target speed restrictions and road networks, technological limitations, geometric considerations, material balance, bounds on actuators and plants and maximum transmission capacity, various dynamical systems are limited within restricted regions [122–124]. Generally, these constraints may not indulged in state-space model without a major increase to avoid model complexities [119,124–126]. Nevertheless, it is not straightforward to take into account the physiological and modeling constraints on the state with SMC, due to the complex nature of computations in SMC. The current trend in constrained sequential Monte Carlo simply imposes the constraints on all particles of the SMC. These approaches are: (1) The acceptance/rejection approach, which enforces the constraints by simply rejecting the particles violating them [127,128]; (2) Constrained importance distribution, which imposes the constraints on all particles or equivalently sample from a constrained importance distribution [129–133]. The issue of how to impose the constraints -onto prior particles (particles before resampling), posterior particles (particles after resampling) or the estimated unconstrained conditional mean estimate- remains still open [129,134]. But these approach underlies the fundamental assumption that constraints on the conditional mean estimate (given in Equation (9)) can be effectively substituted by the same constraints on all particles. However, this is not true in general. It has been referred to these approaches as Point-wise Density Truncation (PoDeT) methods [135]. It was recently shown that such schemes result in incorrect estimate or irrelevant constraints altogether [135,136].

We consider a general constraint of the form [135,136]:

$$a_k \leq \phi_k(\widehat{x}_k) \leq b_k, \tag{8}$$

$\phi_k$ indicates the constraint function at time $k$. It is important to affirm that the constraint must only be satisfied by the state estimate provided by the conditional mean, defined as follows:

$$a_k \leq \phi_k(\widehat{x}_k) = \phi_k(E[x_k|Y_k]) \approx \phi_k\left(\sum_{i=1}^{N} w_k^{(i)} x_k^{(i)}\right) \leq b_k. \tag{9}$$

Recently, Amor et al. derived the optimal bounds of PoDeT [135]. They revealed that error estimation was bounded by the area of state posterior density that had not included constraining interval. Specifically, if most of the density lies within the interval, i.e., the density is well-localized in the constraining interval, then the PoDeT estimation error will be small. However, if a high probability region lies outside of the interval, i.e., the density is not well-localized in the constraining interval, then the PoDeT estimation error will be large [135]. Therefore, Amor et al. proposed a new algorithm referred as "Inductive Mean Density Truncation" (IMeDeT), which inductively samples particles that are guaranteed to satisfy the constraint on the mean of the unknown state [137]. The details the steps of IMeDeT algorithm are presented in Algorithm 2. They evaluated the robustness of the proposed algorithm on the dynamic brain source localization problem using EEG data. In addition, Amor et al. introduced a novel constrained particle filter algorithm called as "mean density truncation" (MiND) and established its convergence properties [136,138]. MiND is based on the principle of minimal perturbation strategy such that the constrained posterior density is "close"to the unconstrained posterior density. Specifically, they imposed the constraint on the mean of the unknown state by perturbing the unconstrained posterior density using only one particle. The details the steps of MiND algorithm are introduced in Algorithm 3. To assess the performance of the proposed algorithm, they applied MiND to solve the problem of movement identification for forearm prosthetic control using the non-negative synergy activation coefficients. The proposed algorithm provided an accurate result with error rates significantly lower than the state-of-the-art in the literature.

Many real-world applications in textile engineering and polymer composites [139–141], may take benefits from this research, i.e., constrained state estimation for nonlinear and non-Gaussian dynamical systems. The main objective of this paper is to emphasize the use of sequential Monte Carlo methods as well as their constrained formulation (IMeDeT and MiND) for the development and modelling of many applications in textile engineering based on, prediction, estimation, controlling, defect detection (See examples in Figure 5), identification and classification (See example in Figure 6) etc.

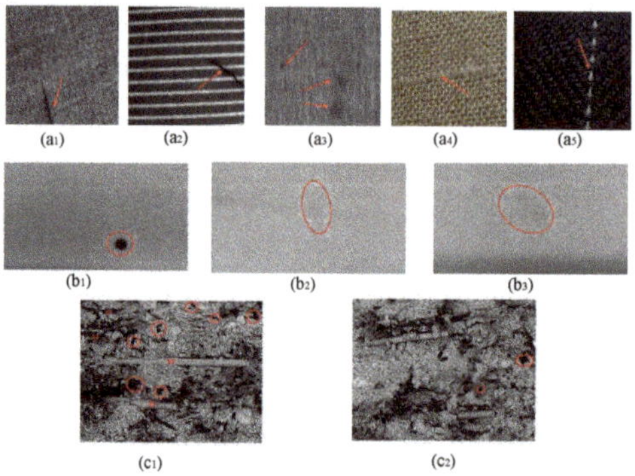


**Figure 5.** Examples of: Defective fabric samples with different patterned textures (from **a1–a5**), Different types of defects in cotton fabric (from **b1–b3**). Defect with polymer composite (**c1**,**c2**).

---

**Algorithm 2** Inductive Mean Density Truncation (IMeDeT)

---

**Initialization**
Denote by $C_k$ the constraint region::$C_k = \{x_k : a_k \leq \hat{x}_k \leq b_k\}$.
**for** j = 1, 2, $\cdots$, N **do**
  Generate $x_0^{(j)} \sim \mathcal{N}(x_0^{(j)}, R_k)$.
  Compute the initial weights using Equation (6) and normalize.
**end for**
**Unconstrained estimation**
**for** $k = 1, 2, \ldots, T$ **do**
  **for** $j = 1, 2, \ldots, N$ **do**
    Generate sample $x_k^{(i)}$ from the system dynamics model (1).
    Calculate the weights $w_k^{(j)}$ of $x_k^{(j)}$ using Equation (6); then , normalize the weights.
    **Constrained estimation**
    **for** i=1,2,. . . ,j **do**
      **if** $\sum_{i=1}^{j} w_k^{(i)} x_k^{(i)} \in C_k$ **then**
        Go to the next step.
      **else**
        Find a particle $x_k^{(j)}$ such that $\sum_{i=1}^{N} w_k^{(i)} x_k^{(i)} \in C_k$.
      **end if**
    **end for**
  **end for**
  Compute the constrained weighted mean $\hat{x}_k = \sum_{i=1}^{N} w_k^{(i)} x_k^{(i)}$.
**end for**

---

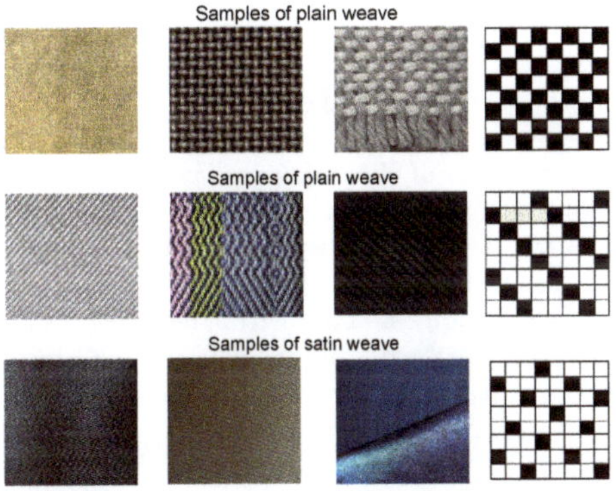


**Figure 6.** Example of identification and classification fabrics weave samples based on patterns.

**Algorithm 3** Mean Density Truncation (MiND)

<u>**Initialization**</u>
The same as initializing IMeDeT.
**for** k = 1, 2, · · · , T **do**
  <u>**Unconstrained estimation**</u>
  **for** j = 1, 2, · · · , N **do**
    Generate sample $x_k^{(j)}$ from the system dynamics model (1).
    Compute weight using Equation (6).
  **end for**
  Normalize particle weights $w_k^{(i)} = \tilde{w}_k^{(i)} / \sum_{j=1}^{N} \tilde{w}_k^{(j)}$.
  Resample $\{x_k^{(i)}, \frac{1}{N}\}_{i=1}^{N}$.
  Compute the weighted mean $\hat{x}_k = \sum_{i=1}^{N} \frac{1}{N} x_k^{(i)}$.
  <u>**Constrained estimation**</u>
  **if** $\hat{x}_k \notin C_k$ **then**
    Remove the furthest particle $x_k^{(i)}$.
    Add a new particle $x_k^N$ using $a_k \le \frac{1}{N}\sum_{i=1}^{N-1} x_k^{(i)} + \frac{1}{N} x_k^{(N)} \le b_k$ and $a_k' \le x_k^{(N)} \le b_k'$ where $a_k' = Na_k - \sum_{i=1}^{N-1} x_k^{(i)}$ and $b_k' = Nb_k - \sum_{i=1}^{N-1} x_k^{(i)}$.
    Compute the constrained weighted mean $\hat{x}_k = \sum_{i=1}^{N} \frac{1}{N} x_k^{(i)}$.
  **end if**
**end for**

## 5. Future Direction and Summary

This study focuses on machine learning classification methods specifically designed for textiles and polymer composites. It elucidates how classification methods are applied in fiber reinforced composites to deal with problems. Based on discussed literature, this study clearly explains that machine learning classification receives significant consideration in textiles and composites industries. SVM and ANN are widely used classification methods as they provide better prediction accuracy. In addition, this study provides the classification of carbon fiber reinforced composites and the inclusion of polymeric fibers in composites formation. It elaborates recently used advanced machine learning algorithms for textile processes and carbon fiber reinforced composites. It provides critical and in-depth information regarding the algorithms applied during yarn production, fabric manufacturing and textile finishing processes. Drawbacks and limitations of each method are discussed in detail. This study proposes gateway and opens new avenues not only for researcher community but also for the readership of the journal. In addition, we suggest the use of sequential Monte Carlo methods, i.e., particle filters for control, monitoring, prediction, and identification, in textiles and composites.

For future work, some novel algorithms e.g., golden eagle optimiser are suggested to study the performance of fiber reinforced polymer composites, besides the classification method. Golden eagle optimiser is a recent method and only one or two studies have applied it so far. This can be beneficial to discover complex relationships and useful patterns between textiles and fiber reinforced polymer composites.

In textiles and composites industries, researchers mostly used single classifier. However, combining multiple algorithms may provide a more accurate and semantic vision for the classification of textile processes. Therefore, researchers should start to use hybrid models in order to achieve better results.

**Author Contributions:** N.A. and M.T.N. conceived, designed and performed experiments; analysed the results and wrote manuscript. M.P. analyzed the results, supervised and acquired funding. All of the authors participated in critical analysis and preparation of the manuscript. All authors have read and agreed to the published version of the manuscript.

**Funding:** This work was supported by the Ministry of Education, Youth and Sports of the Czech Republic and the European Union (European Structural and Investment Funds—Operational Programme Research, Development and Education) in the frames of the project "Modular platform for autonomous chassis of specialized electric vehicles for freight and equipment transportation", Reg. No. CZ.02.1.01/0.0/0.0/16_025/0007293.

**Institutional Review Board Statement:** Not applicable.

**Informed Consent Statement:** Not applicable.

**Data Availability Statement:** Not applicable.

**Conflicts of Interest:** The authors declare no conflict of interest.

## References

1. Giri, C.; Jain, S.; Zeng, X.; Bruniaux, P. A Detailed Review of Artificial Intelligence Applied in the Fashion and Apparel Industry. *IEEE Access* **2019**, *7*, 95376–95396. [CrossRef]
2. Noman, M.T.; Wiener, J.; Saskova, J.; Ashraf, M.A.; Vikova, M.; Jamshaid, H.; Kejzlar, P. In-situ development of highly photocatalytic multifunctional nanocomposites by ultrasonic acoustic method. *Ultrason. Sonochem.* **2018**, *40*, 41–56. [CrossRef]
3. Noman, M.T.; Petrů, M. Functional Properties of Sonochemically Synthesized Zinc Oxide Nanoparticles and Cotton Composites. *Nanomaterials* **2020**, *10*, 1661. [CrossRef]
4. Behera, P.; Noman, M.T.; Petrů, M. Enhanced Mechanical Properties of Eucalyptus-Basalt-Based Hybrid-Reinforced Cement Composites. *Polymers* **2020**, *12*, 2837. [CrossRef]
5. Ashraf, M.; Wiener, J.; Farooq, A.; Šašková, J.; Noman, M. Development of Maghemite Glass Fibre Nanocomposite for Adsorptive Removal of Methylene Blue. *Fibers Polym.* **2018**, *19*, 1735–1746. [CrossRef]
6. Jamshaid, H.; Mishra, R.; Militký, J.; Pechočiaková, M.; Noman, M. Mechanical, thermal and interfacial properties of green composites from basalt and hybrid woven fabrics. *Fibers Polym.* **2016**, *17*, 1675–1686. [CrossRef]
7. Mahmood, A.; Noman, M.T.; Pechočiaková, M.; Amor, N.; Petrů, M.; Abdelkader, M.; Militký, J.; Sozcu, S.; Hassan, S.Z.U. Geopolymers and Fiber-Reinforced Concrete Composites in Civil Engineering. *Polymers* **2021**, *13*, 2099. [CrossRef]
8. Sarr, M.M.; Inoue, H.; Kosaka, T. Study on the improvement of interfacial strength between glass fiber and matrix resin by grafting cellulose nanofibers. *Compos. Sci. Technol.* **2021**, *211*, 108853. [CrossRef]
9. Liu, L.; Wang, X.; Wu, Z.; Keller, T. Tension-tension fatigue behavior of ductile adhesively-bonded FRP joints. *Compos. Struct.* **2021**, *268*, 113925. [CrossRef]
10. Li, C.; Xian, G.; Li, H. Influence of immersion in water under hydraulic pressure on the interfacial shear strength of a unidirectional carbon/glass hybrid rod. *Polym. Test.* **2018**, *72*, 164–171. [CrossRef]
11. Li, C.; Yin, X.; Liu, Y.; Guo, R.; Xian, G. Long-term service evaluation of a pultruded carbon/glass hybrid rod exposed to elevated temperature, hydraulic pressure and fatigue load coupling. *Int. J. Fatigue* **2020**, *134*, 105480. [CrossRef]
12. Li, C.; Yin, X.; Wang, Y.; Zhang, L.; Zhang, Z.; Liu, Y.; Xian, G. Mechanical property evolution and service life prediction of pultruded carbon/glass hybrid rod exposed in harsh oil-well condition. *Compos. Struct.* **2020**, *246*, 112418. [CrossRef]
13. Amor, N.; Noman, M.T.; Petrů, M. Prediction of functional properties of nano $TiO_2$ coated cotton composites by artificial neural network. *Sci. Rep.* **2021**, *11*, 12235. [CrossRef]

14. Amor, N.; Noman, M.T.; Petrů, M.; Mahmood, A.; Ismail, A. Neural network-crow search model for the prediction of functional properties of nano $TiO_2$ coated cotton composites. *Sci. Rep.* **2021**, *11*, 13649. [CrossRef]
15. Zimmerling, C.; Dörr, D.; Henning, F.; Kärger, L. A machine learning assisted approach for textile formability assessment and design improvement of composite components. *Compos. Part A Appl. Sci. Manuf.* **2019**, *124*, 105459. [CrossRef]
16. Seçkin, M.; Çağdaş Seçkin, A.; Coşkun, A. Production fault simulation and forecasting from time series data with machine learning in glove textile industry. *J. Eng. Fibers Fabr.* **2019**, *14*, 1558925019883462. [CrossRef]
17. Ribeiro, R.; Pilastri, A.; Moura, C.; Rodrigues, F.; Rocha, R.; Morgado, J.; Cortez, P. Predicting Physical Properties of Woven Fabrics via Automated Machine Learning and Textile Design and Finishing Features. In *Artificial Intelligence Applications and Innovations*; Maglogiannis, I., Iliadis, L., Pimenidis, E., Eds.; Springer International Publishing: Cham, Switzerland, 2020; pp. 244–255._21. [CrossRef]
18. Noman, M.T.; Petru, M.; Louda, P.; Kejzlar, P. Woven Textiles Coated with Zinc Oxide Nanoparticles and Their Thermophysiological Comfort Properties. *J. Nat. Fibers* **2021**, 1–13. [CrossRef]
19. Noman, M.T.; Amor, N.; Petru, M.; Mahmood, A.; Kejzlar, P. Photocatalytic Behaviour of Zinc Oxide Nanostructures on Surface Activation of Polymeric Fibres. *Polymers* **2021**, *13*, 1227. [CrossRef] [PubMed]
20. Majumdar, A. *Soft Computing in Textile Engineering*; Woodhead Publishing: Sawston, UK, 2011; pp. i–iii. [CrossRef]
21. Mohammad, A.T.; Mahboubeh, M. Artificial Neural Network Prosperities in Textile Applications. In *Artificial Neural Networks Industrial and Control Engineering Applications*; Suzuki, K., Ed.; IntechOpen: Rijeka, Croatia, 2011; pp. 35–64. [CrossRef]
22. Vassiliadis, S.; Rangoussi, M.; Cay, A.; Provatidis, C. Artificial Neural Networks and Their Applications in the Engineering of Fabrics. In *Woven Fabric Engineering*; Dubrovski, P.D., Ed.; Intechopen: London, UK, 2010; pp. 111–134. [CrossRef]
23. Farooq, A.; Sarwar, M.I.; Ashraf, M.A.; Iqbal, D.; Hussain, A.; Malik, S. Predicting Cotton Fibre Maturity by Using Artificial Neural Network. *Autex Res. J.* **2018**, *18*, 429–433. [CrossRef]
24. Malik, S.A.; Gereke, T.; Farooq, A.; Aibibu, D.; Cherif, C. Prediction of yarn crimp in PES multifilament woven barrier fabrics using artificial neural network. *J. Text. Inst.* **2018**, *109*, 942–951. [CrossRef]
25. Majumdar, P.K.; Majumdar, A. Predicting the Breaking Elongation of Ring Spun Cotton Yarns Using Mathematical, Statistical, and Artificial Neural Network Models. *Text. Res. J.* **2004**, *74*, 652–655. [CrossRef]
26. Majumdar, A.; Majumdar, P.K.; Sarkar, B. Application of an adaptive neuro-fuzzy system for the prediction of cotton yarn strength from HVI fibre properties. *J. Text. Inst.* **2005**, *96*, 55–60. [CrossRef]
27. Almetwally, A.A.; Idrees, H.M.; Hebeish, A.A. Predicting the tensile properties of cotton/spandex core-spun yarns using artificial neural network and linear regression models. *J. Text. Inst.* **2014**, *105*, 1221–1229. [CrossRef]
28. Doran, E.C.; Sahin, C. The prediction of quality characteristics of cotton/elastane core yarn using artificial neural networks and support vector machines. *Text. Res. J.* **2020**, *90*, 1558–1580. [CrossRef]
29. Dashti, M.; Derhami, V.; Ekhtiyari, E. Yarn tenacity modeling using artificial neural networks and development of a decision support system based on genetic algorithms. *J. AI Data Min.* **2014**, *2*, 73–78. [CrossRef]
30. Mishra, S. Prediction of Yarn Strength Utilization in Cotton Woven Fabrics using Artificial Neural Network. *J. Inst. Eng. Ser. E* **2015**, *96*, 151–157. [CrossRef]
31. Mozafary, V.; Payvandy, P. Application of data mining technique in predicting worsted spun yarn quality. *J. Text. Inst.* **2014**, *105*, 100–108. [CrossRef]
32. Malik, S.A.; Farooq, A.; Gereke, T.; Cherif, C. Prediction of Blended Yarn Evenness and Tensile Properties by Using Artificial Neural Network and Multiple Linear Regression. *Autex Res. J.* **2016**, *16*, 43–50. [CrossRef]
33. El-Geiheini, A.; ElKateb, S.; Abd-Elhamied, M.R. Yarn Tensile Properties Modeling Using Artificial Intelligence. *Alex. Eng. J.* **2020**, *59*, 4435–4440. [CrossRef]
34. Erbil, Y.; Babaarslan, O.; İlhami, I. A comparative prediction for tensile properties of ternary blended open-end rotor yarns using regression and neural network models. *J. Text. Inst.* **2018**, *109*, 560–568. [CrossRef]
35. Azimi, B.; Tehran, M.A.; Mojtahedi, M.R.M. Prediction of False Twist Textured Yarn Properties by Artificial Neural Network Methodology. *J. Eng. Fibers Fabr.* **2013**, *8*, 97–101. [CrossRef]
36. Noman, M.; Ashraf, M.A.; Jamshaid, H.; Ali, A. A Novel Green Stabilization of $TiO_2$ Nanoparticles onto Cotton. *Fibers Polym.* **2018**, *19*, 2268–2277. [CrossRef]
37. Azeem, M.; Noman, M.T.; Wiener, J.; Petru, M.; Louda, P. Structural design of efficient fog collectors: A review. *Environ. Technol. Innov.* **2020**, *20*, 1–17. [CrossRef]
38. Yang, T.; Hu, L.; Xiong, X.; Petrů, M.; Noman, M.T.; Mishra, R.; Militký, J. Sound Absorption Properties of Natural Fibers: A Review. *Sustainability* **2020**, *12*, 8477. [CrossRef]
39. Noman, M.; Ashraf, M.A.; Ali, A. Synthesis and applications of nano-$TiO_2$: A review. *Environ. Sci. Pollut. Res.* **2018**, *26*, 3262–3291. [CrossRef]
40. Noman, M.T.; Petru, M.; Militký, J.; Azeem, M.; Ashraf, M.A. One-Pot Sonochemical Synthesis of ZnO Nanoparticles for Photocatalytic Applications, Modelling and Optimization. *Materials* **2020**, *13*, 14. [CrossRef] [PubMed]
41. Noman, M.; Amor, N.; Petru, M. Synthesis and applications of ZnO nanostructures (ZONSs): A review. *Crit. Rev. Solid State Mater. Sci.* **2021**, *2*, 1–44. [CrossRef]
42. Hussain, M.; Khan, B.; Wang, Z.; Ding, S. Woven Fabric Pattern Recognition and Classification Based on Deep Convolutional Neural Networks. *Electronics* **2020**, *9*, 1048. [CrossRef]

43. Eldessouki, M.; Hassan, M. Adaptive neuro-fuzzy system for quantitative evaluation of woven fabrics' pilling resistance. *Expert Syst. Appl.* **2015**, *42*, 2098–2113. [CrossRef]
44. Xiao, Q.; Wang, R.; Zhang, S.; Li, D.; Sun, H.; Wang, L. Prediction of pilling of polyester–cotton blended woven fabric using artificial neural network models. *J. Eng. Fibers Fabr.* **2020**, *15*, 1558925019900152. [CrossRef]
45. Taieb, A.H.; Mshali, S.; Sakli, F. Predicting Fabric Drapability Property by Using an Artificial Neural Network. *J. Eng. Fibers Fabr.* **2018**, *13*, 87–93. [CrossRef]
46. Hussain, T.; Malik, Z.A.; Arshad, Z.; Nazir, A. Comparison of artificial neural network and adaptive neuro-fuzzy inference system for predicting the wrinkle recovery of woven fabrics. *J. Text. Inst.* **2015**, *106*, 934–938. [CrossRef]
47. Noman, M.T.; Petru, M. Effect of Sonication and Nano $TiO_2$ on Thermophysiological Comfort Properties of Woven Fabrics. *ACS Omega* **2020**, *5*, 11481–11490. [CrossRef]
48. Mansoor, T.; Hes, L.; Bajzik, V.; Noman, M.T. Novel method on thermal resistance prediction and thermo-physiological comfort of socks in a wet state. *Text. Res. J.* **2020**, *90*, 1987–2006. [CrossRef]
49. Noman, M.; Petrů, M.; Amor, N.; Yang, T.; Mansoor, T. Thermophysiological comfort of sonochemically synthesized nano $TiO_2$ coated woven fabrics. *Sci. Rep.* **2020**, *10*, 17204. [CrossRef]
50. Ali, A.; Nguyen, N.H.A.; Baheti, V.; Ashraf, M.; Militky, J.; Mansoor, T.; Noman, M.T.; Ahmad, S. Electrical conductivity and physiological comfort of silver coated cotton fabrics. *J. Text. Inst.* **2018**, *109*, 620–628. [CrossRef]
51. Noman, M.; Petrů, M.; Amor, N.; Louda, P. Thermophysiological comfort of zinc oxide nanoparticles coated woven fabrics. *Sci. Rep.* **2020**, *10*, 21080. [CrossRef]
52. Malik, S.A.; Kocaman, R.T.; Kaynak, H.; Gereke, T.; Aibibu, D.; Babaarslan, O.; Cherif, C. Analysis and prediction of air permeability of woven barrier fabrics with respect to material, fabric construction and process parameters. *Fibers Polym.* **2017**, *18*, 2005–2017. [CrossRef]
53. Malik, S.A.; Kocaman, R.T.; Gereke, T.; Aibibu, D.; Cherif, C. Prediction of the Porosity of Barrier Woven Fabrics with Respect to Material, Construction and Processing Parameters and Its Relation with Air Permeability. *Fibres Text. East. Eur.* **2018**, *26*, 71–79. [CrossRef]
54. Wong, A.; Li, Y.; Yeung, P. Predicting Clothing Sensory Comfort with Artificial Intelligence Hybrid Models. *Text. Res. J.* **2004**, *74*, 13–19. [CrossRef]
55. Xu, J.; He, Z.; Li, S.; Ke, W. Production cost optimization of enzyme washing for indigo dyed cotton denim by combining Kriging surrogate with differential evolution algorithm. *Text. Res. J.* **2020**, *90*, 1860–1871. [CrossRef]
56. Unal, P.; Üreyen, M.; Mecit, D. Predicting properties of single jersey fabrics using regression and artificial neural network models. *Fibers Polym.* **2012**, *13*, 87–95. [CrossRef]
57. Yang, C.S.; Lin, C.; Chen, W. Using deep principal components analysis-based neural networks for fabric pilling classification. *Electronics* **2019**, *8*, 474. [CrossRef]
58. Özçelik Kayseri, G.; Kirtay, E. Part II. Predicting the Pilling Tendency of the Cotton Interlock Knitted Fabrics by Artificial Neural Network. *J. Eng. Fibers Fabr.* **2015**, *10*, 62–71. [CrossRef]
59. Fayala, F.; Alibi, H.; Benltoufa, S.; Jemni, A. Neural Network for Predicting Thermal Conductivity of Knit Materials. *J. Eng. Fibers Fabr.* **2008**, *3*, 53–60. [CrossRef]
60. Majumdar, A. Modelling of thermal conductivity of knitted fabrics made of cotton–bamboo yarns using artificial neural network. *J. Text. Inst.* **2011**, *102*, 752–762. [CrossRef]
61. Kanat, Z.E.; Özdil, N. Application of artificial neural network (ANN) for the prediction of thermal resistance of knitted fabrics at different moisture content. *J. Text. Inst.* **2018**, *109*, 1247–1253. [CrossRef]
62. Mitra, A.; Majumdar, A.; Majumdar, P.K.; Bannerjee, D. Predicting thermal resistance of cotton fabrics by artificial neural network model. *Exp. Therm. Fluid Sci.* **2013**, *50*, 172–177. [CrossRef]
63. Wang, D.; Barber, J.; Lu, W.; Thouless, M. Use of wavelet analysis for an objective evaluation of the formation of pills in nonwoven fabrics. *J. Ind. Text.* **2019**, *49*, 663–675. [CrossRef]
64. Kalkanci, M.; Sinecen, M.; Kurumer, G. Prediction of dimensional change in finished fabric through artificial neural networks. *Tekst. Konfeksiyon* **2018**, *28*, 43–51. [CrossRef]
65. Majumdar, A.; Das, A.; Hatua, P.; Ghosh, A. Optimization of Woven Fabric Parameters for Ultraviolet Radiation Protection and Comfort Using Artificial Neural Network and Genetic Algorithm. *Neural Comput. Appl.* **2016**, *27*, 2567–2576. [CrossRef]
66. Farooq, A.; Irshad, F.; Azeemi, R.; Iqbal, N. Prognosticating the Shade Change after Softener Application using Artificial Neural Networks. *Autex Res. J.* **2020**, 79–84. [CrossRef]
67. Malik, S.A.; Arain, R.A.; Khatri, Z.; Saleemi, S.; Cherif, C. Neural network modeling and principal component analysis of antibacterial activity of chitosan/$AgCl$-$TiO_2$ colloid treated cotton fabric. *Fibers Polym.* **2015**, *16*, 1142–1149. [CrossRef]
68. Furferi, R.; Governi, L.; Volpe, Y. Modelling and simulation of an innovative fabric coating process using artificial neural networks. *Text. Res. J.* **2012**, *82*, 1282–1294. [CrossRef]
69. Ni, C.; Li, Z.; Zhang, X.; Sun, X.; Huang, Y.; Zhao, L.; Zhu, T.; Wang, D. Online Sorting of the Film on Cotton Based on Deep Learning and Hyperspectral Imaging. *IEEE Access* **2020**, *8*, 93028–93038. [CrossRef]
70. Hanbay, K.; Talu, M.F.; Ozguvenc, O.F. Fabric defect detection systems and methods—A systematic literature review. *Optik* **2016**, *127*, 11960–11973. [CrossRef]

71. Czimmermann, T.; Ciuti, G.; Milazzo, M.; Chiurazzi, M.; Roccella, S.; Oddo, C.M.; Dario, P. Visual-Based Defect Detection and Classification Approaches for Industrial Applications—A SURVEY. *Sensors* **2020**, *20*, 1459. [CrossRef] [PubMed]
72. Rasheed, A.; Zafar, B.; Rasheed, A.; Ali, N.; Sajid, M.; Hanif Dar, S.; Habib, U.; Shehryar, T.; Mahmood, M.T. Fabric Defect Detection Using Computer Vision Techniques: A Comprehensive Review. *Math. Probl. Eng.* **2020**, *2020*, 8189403. [CrossRef]
73. Eldessouki, M.; Hassan, M.; Qashqary, K.; Shady, E. Application of Principal Component Analysis to Boost the Performance of an Automated Fabric Fault Detector and Classifier. *Fibres Text. East. Eur.* **2014**, *22*, 51–57.
74. Liu, Z.; Wang, B.; Li, C.; Yu, M.; Ding, S. Fabric defect detection based on deep-feature and low-rank decomposition. *J. Eng. Fibers Fabr.* **2020**, *15*, 1–12. [CrossRef]
75. Sezer, O.; Ercil, A.; Ertuzun, A. Using perceptual relation of regularity and anisotropy in the texture with independent component model for defect detection. *Pattern Recognit.* **2007**, *40*, 121–133. [CrossRef]
76. Yapi, D.; Allili, M.S.; Baaziz, N. Automatic Fabric Defect Detection Using Learning-Based Local Textural Distributions in the Contourlet Domain. *IEEE Trans. Autom. Sci. Eng.* **2018**, *15*, 1014–1026. [CrossRef]
77. Li, Y.; Zhao, W.; Pan, J. Deformable Patterned Fabric Defect Detection With Fisher Criterion-Based Deep Learning. *IEEE Trans. Autom. Sci. Eng.* **2017**, *14*, 1256–1264. [CrossRef]
78. Han, Y.J.; Yu, H.J. Fabric Defect Detection System Using Stacked Convolutional Denoising Auto-Encoders Trained with Synthetic Defect Data. *Appl. Sci.* **2020**, *10*, 2511. [CrossRef]
79. Kuo, C.F.J.; Juang, Y. A study on the recognition and classification of embroidered textile defects in manufacturing. *Text. Res. J.* **2016**, *86*, 393–408. [CrossRef]
80. Huang, M.L.; Fu, C.C. Applying Image Processing to the Textile Grading of Fleece Based on Pilling Assessment. *Fibers* **2018**, *6*, 73. [CrossRef]
81. Li, Y.; Zhang, C. Automated vision system for fabric defect inspection using Gabor filters and PCNN. *SpringerPlus* **2016**, *5*, 765. [CrossRef]
82. Sacco, C.; Radwan, A.; Harik, R.; Tooren, M.V. Automated Fiber Placement Defects: Automated Inspection and Characterization. In Proceedings of the SAMPE 2018 Conference and Exhibition, Long Beach, CA, USA, 25 May 2018; pp. 1–13.
83. Liang, Z.; Xu, B.; Chi, Z.; Feng, D. Intelligent characterization and evaluation of yarn surface appearance using saliency map analysis, wavelet transform and fuzzy ARTMAP neural network. *Expert Syst. Appl.* **2012**, *39*, 4201–4212. [CrossRef]
84. Jing, J.; Wang, Z.; Rätsch, M.; Zhang, H. Mobile-Unet: An efficient convolutional neural network for fabric defect detection. *Text. Res. J.* **2020**, 0040517520928604. [CrossRef]
85. Jeyaraj, P.R.; Nadar, E.R.S. Effective textile quality processing and an accurate inspection system using the advanced deep learning technique. *Text. Res. J.* **2020**, *90*, 971–980. [CrossRef]
86. Wei, B.; Hao, K.; Tang, X.S.; Ding, Y. A new method using the convolutional neural network with compressive sensing for fabric defect classification based on small sample sizes. *Text. Res. J.* **2019**, *89*, 3539–3555. [CrossRef]
87. Hu, G.; Huang, J.; Wang, Q.; Li, J.; Xu, Z.; Huang, X. Unsupervised fabric defect detection based on a deep convolutional generative adversarial network. *Text. Res. J.* **2020**, *90*, 247–270. [CrossRef]
88. Sapuan, S.; Mujtaba, I. *Composite Materials Technology: Neural Network Applications*; CRC Press: Boca Raton, FL, USA, 200 ; pp. 1–355.
89. Müzel, S.D.; Bonhin, E.P.; Guimarães, N.; Guidi, E.S. Application of the Finite Element Method in the Analysis of Composite Materials: A Review. *Polymers* **2020**, *12*, 818. [CrossRef]
90. Dixit, A.; Mali, H.S. Modeling techniques for predicting the mechanical properties of woven-fabric textile composites: A Review. *Mech. Compos. Mater.* **2013**, *49*, 1–20. [CrossRef]
91. Schimmack, M.; Haus, B.; Leuffert, P.; Mercorelli, P. An Extended Kalman Filter for temperature monitoring of a metal-polymer hybrid fibre based heater structure. In Proceedings of the 2017 IEEE International Conference on Advanced Intelligent Mechatronics (AIM), Munich, Germany, 3–7 July 2017; pp. 376–381. [CrossRef]
92. González, C.; Fernández-León, J. A Machine Learning Model to Detect Flow Disturbances during Manufacturing of Composites by Liquid Moulding. *J. Compos. Sci.* **2020**, *4*, 71. [CrossRef]
93. Altarazi, S.; Allaf, R.; Alhindawi, F. Machine Learning Models for Predicting and Classifying the Tensile Strength of Polymeric Films Fabricated via Different Production Processes. *Materials* **2019**, *12*, 1475. [CrossRef]
94. Balcioglu, H.E.; Seckin, A.C. Comparison of machine learning methods and finite element analysis on the fracture behavior of polymer composites. *Arch. Appl. Mech.* **2020**, *91*, 223–239. [CrossRef]
95. Khan, S.; Malik, S.A.; Gull, N.; Saleemi, S.; Islam, A.; Butt, M.T.Z. Fabrication and modelling of the macro-mechanical properties of cross-ply laminated fibre-reinforced polymer composites using artificial neural network. *Adv. Compos. Mater.* **2019**, *28*, 409–423. [CrossRef]
96. Boon, Y.D.; Joshi, S.C.; Bhudolia, S.K.; Gohel, G. Recent Advances on the Design Automation for Performance-Optimized Fiber Reinforced Polymer Composite Components. *J. Compos. Sci.* **2020**, *4*, 61. [CrossRef]
97. He, M.; Wang, Y.; Ramakrishnan, K.R.; Zhang, Z. A comparison of machine learning algorithms for assessment of delamination in fiber-reinforced polymer composite beams. *Struct. Health-Monit. Int. J.* **2020**, 1997–2012. [CrossRef]
98. Matsuzaki, R.; Tachikawa, T.; Ishizuka, J. Estimation of state and material properties during heat-curing molding of composite materials using data assimilation: A numerical study. *Heliyon* **2018**, *4*, e00554. [CrossRef]

99. Ishizuka, J.; Matsuzaki, R.; Tachikawa, T. Data assimilation-based state estimation of composites during molding. *Adv. Compos. Mater.* **2019**, *28*, 225–243. [CrossRef]
100. Konstantopoulos, G.; Koumoulos, E.P.; Charitidis, C.A. Classification of mechanism of reinforcement in the fiber-matrix interface: Application of Machine Learning on nanoindentation data. *Mater. Des.* **2020**, *192*, 108705. [CrossRef]
101. Golushko, S. *Mathematical Modeling and Numerical Optimization of Composite Structures*; Intechopen: London, UK, 2019; pp. 13–34. [CrossRef]
102. Golushko, S.; Buznik, V.; Nuzhny, G. Mathematical modeling and numerical analysis of reinforced composite beams. *J. Phys. Conf. Ser.* **2019**, *1268*, 012018. [CrossRef]
103. Zhang, Y.; Xu, X. Predicting the delamination factor in carbon fibre reinforced plastic composites during drilling through the Gaussian process regression. *J. Compos. Mater.* **2020**, 2061–2067. [CrossRef]
104. Qi, Z.; Zhang, N.; Liu, Y.; Chen, W. Prediction of mechanical properties of carbon fiber based on cross-scale FEM and machine learning. *Compos. Struct.* **2019**, *212*, 199–206. [CrossRef]
105. Liang, L.; Liu, M.; Martin, C.; Sun, W. A deep learning approach to estimate stress distribution: A fast and accurate surrogate of finite-element analysis. *J. R. Soc. Interface* **2018**, *15*, 20170844. [CrossRef]
106. Shaban, Y.; Meshreki, M.; Yacout, S.; Balazinski, M.; Attia, H. Process control based on pattern recognition for routing carbon fiber reinforced polymer. *J. Intell. Manuf.* **2017**, *28*, 165–179. [CrossRef]
107. Zhao, Z.; Yua, M.; Dong, S. Damage Location Detection of the CFRP Composite Plate Based on Neural Network Regression. In Proceedings of the 7th Asia-Pacific Workshop on Structural Health Monitoring, Hong Kong, China, 12–15 November 201 ; pp. 1–10.
108. Nasser, J.; Groo, L.; Sodano, H. Artificial neural networks and phenomenological degradation models for fatigue damage tracking and life prediction in laser induced graphene interlayered fiberglass composites. *Smart Mater. Struct.* **2021**, *30*, 085010. [CrossRef]
109. Liu, K.; Ma, Z.; Liu, Y.; Yang, J.; Yao, Y. Enhanced Defect Detection in Carbon Fiber Reinforced Polymer Composites via Generative Kernel Principal Component Thermography. *Polymers* **2021**, *13*, 825. [CrossRef] [PubMed]
110. Sun, L.; Hu, S.J.; Freiheit, T. Feature-based quality classification for ultrasonic welding of carbon fiber reinforced polymer through Bayesian regularized neural network. *J. Manuf. Syst.* **2021**, *58*, 335–347. [CrossRef]
111. Schimmack, M.; McGaw, D.; Mercorelli, P. Wavelet based Fault Detection and RLS Parameter Estimation of Conductive Fibers with a Simultaneous Estimation of Time-Varying Disturbance. *IFAC-PapersOnLine* **2015**, *48*, 1773–1778. [CrossRef]
112. Liu, X.; Gasco, F.; Goodsell, J.; Yu, W. Initial failure strength prediction of woven composites using a new yarn failure criterion constructed by deep learning. *Compos. Struct.* **2019**, *230*, 111505. [CrossRef]
113. Soman, R.; Ostachowicz, W. Kalman Filter Based Load Monitoring in Beam Like Structures Using Fibre-Optic Strain Sensors. *Sensors* **2019**, *19*, 103. [CrossRef]
114. Soman, R.; Ostachowicz, W. Kalman Filter based Neutral Axis tracking for damage detection in composites structures under changing axial loading conditions. *Compos. Struct.* **2018**, *206*, 517–525. [CrossRef]
115. Hallal, A.; Younes, R.; Fardoun, F. Review and comparative study of analytical modeling for the elastic properties of textile composites. *Compos. Part B Eng.* **2013**, *50*, 22–31. [CrossRef]
116. Balokas, G.A.; Czichon, S.; Rolfes, R. Neural network assisted multiscale analysis for the elastic properties prediction of 3D braided composites under uncertainty. *Compos. Struct.* **2018**, *183*, 550–562. [CrossRef]
117. Jiang, D.; Li, Y.; Fei, Q.; Wu, S. Prediction of uncertain elastic parameters of a braided composite. *Compos. Struct.* **2015**, *126*, 123–131. [CrossRef]
118. Doucet, A.; Johansen, A.M. *Handbook of Nonlinear Filtering*; Oxford University Press: Oxford, UK, 2009; Volume 12, pp. 656–704
119. Simon, D. *Optimal State Estimation: Kalman, $H_\infty$, and Nonlinear Approaches*; Wiley: Hoboken, NJ, USA, 2006; p. 552. [CrossRef]
120. Amor, N.; Meddeb, A.; Marrouchi, S.; Chebbi, S. A comparative study of nonlinear Bayesian filtering algorithms for estimation of gene expression time series data. *Turk. J. Electr. Eng. Comput. Sci.* **2019**, *27*, 2648–2665. [CrossRef]
121. Crisan, D.; Doucet, A. A survery of convergence results on particle filtering methods for practitioners. IEEE *Trans. Signal Process.* **2002**, *50*, 736–746. [CrossRef]
122. Huang, Y.; Werner, S.; Huang, J.; Kashyap, N.; Gupta, V. State Estimation in Electric Power Grids: Meeting New Challenges Presented by the Requirements of the Future Grid. *IEEE Signal Process. Mag.* **2012**, *29*, 33–43. [CrossRef]
123. Battistello, G.; Ulmke, M. Exploitation of a-priori information for tracking maritime intermittent data sources. In Proceedings of the International Conference on Information Fusion, Chicago, IL, USA, 5–8 July 2011; pp. 1–8.
124. Yang, C.; Bakich, M.; Blasch, E. Nonlinear constrained tracking of targets on roads. In Proceedings of the International Conference on Information Fusion, Sydney, NSW, Australia, 4–7 July 2005; pp. 235–242. [CrossRef]
125. Agate, C.S.; Sullivan, K.J. Road-constrained target tracking and identification using a particle filter. In Proceedings of the Signal and Data Processing of Small Targets, San Diego, CA, USA, 5–7 August 2003; Volume 5204, pp. 532–543. [CrossRef]
126. Ko, S.; Bitmead, R.R. State estimation for linear systems with state equality constraints. *Automatica* **2007**, *43*, 1363–1368. [CrossRef]
127. Lang, L.; Chen, W.S.; Bakshi, B.R.; Goel, P.K.; Ungarala, S. Bayesian estimation via sequential Monte Carlo sampling Constrained dynamic systems. *Automatica* **2007**, *43*, 615–622. [CrossRef]
128. Ungarala, S. A direct sampling particle filter from approximate conditional density function supported on constrained state space. *Comput. Chem. Eng.* **2011**, *35*, 1110–1118. [CrossRef]

129. Shao, X.; Huang, B.; Lee, J.M. Constrained Bayesian state estimation: A comparative study and a new particle filter based approach. *J. Process Control* **2010**, *20*, 143–157. [CrossRef]
130. Papi, F.; Podt, M.; Boers, Y.; Battistello, G. On constraints exploitation for particle filtering based target tracking. In Proceedings of the International Conference on Information Fusion, Singapore, 9–12 July 2012; pp. 455–462.
131. Prakash, J.; Patwardhan, S.C.; Shah, S.L. On the choice of importance distributions for unconstrained and constrained state estimation using particle filter. *J. Process Control* **2011**, *21*, 3–16. [CrossRef]
132. Straka, O.; Dunik, J.; Simandl, M. Truncation nonlinear filters for state estimation with nonlinear inequality constraints. *Automatica* **2012**, *48*, 273–286. [CrossRef]
133. Zhao, Z.; Huang, B.; Liu, F. *Constrained Particle Filtering Methods for State Estimation of Nonlinear Process*; Wiley: Hoboken, NJ, USA, 2014; Volume 60, pp. 2072–2082. [CrossRef]
134. Amor, N.; Rasool, G.; Bouaynaya, N.C. Constrained State Estimation—A Review. *arXiv* **2018**, arXiv:eess.SP/1807.03463.
135. Amor, N.; Bouaynaya, N.C.; Shterenberg, R.; Chebbi, S. On the convergence of the constrained particle filters. IEEE *Signal Process. Lett.* **2017**, *24*, 858–862. [CrossRef]
136. Amor, N.; Rasool, G.; Bouaynaya, N.C.; Shterenberg, R. Constrained particle filtering for movement identification in forearm prosthesis. *Signal Process.* **2019**, *161*, 25–35. [CrossRef]
137. Amor, N.; Bouaynaya, N.; Georgieva, P.; Shterenberg, R.; Chebbi, S. EEG Dynamic Source Localization using Constrained Particle Filtering. In Proceedings of the International Conference on Symposium Series on Computational Intelligence (SSCI), Athens, Greece, 6–9 December 2016; pp. 1–8. [CrossRef]
138. Amor, N.; Rasool, G.; Bouaynaya, N.; Shterenberg, R. Hand Movement Discrimination Using Particle Filters. In Proceedings of the 2018 IEEE Signal Processing in Medicine and Biology Symposium (SPMB), Philadelphia, PA, USA, 1 December 201 ; pp. 1–5. [CrossRef]
139. Ali, A.; Sattar, M.; Riaz, T.; Khan, B.A.; Awais, M.; Militky, J.; Noman, M.T. Highly stretchable durable electro-thermal conductive yarns made by deposition of carbon nanotubes. *J. Text. Inst.* **2021**, 1–10. [CrossRef]
140. Jamshaid, H.; Mishra, R.; Militký, J.; Noman, M.T. Interfacial performance and durability of textile reinforced concrete. *J. Text. Inst.* **2018**, *109*, 879–890. [CrossRef]
141. Noman, M.T.; Militky, J.; Wiener, J.; Saskova, J.; Ashraf, M.A.; Jamshaid, H.; Azeem, M. Sonochemical synthesis of highly crystalline photocatalyst for industrial applications. *Ultrasonics* **2018**, *83*, 203–213. [CrossRef] [PubMed]

polymers

MDPI

*Review*

# Geopolymers and Fiber-Reinforced Concrete Composites in Civil Engineering

**Aamir Mahmood [1], Muhammad Tayyab Noman [2,*], Miroslava Pechočiaková [1], Nesrine Amor [2], Michal Petrů [2], Mohamed Abdelkader [3,4,5], Jiří Militký [1], Sebnem Sozcu [1] and Syed Zameer Ul Hassan [6]**

1 Department of Material Engineering, Faculty of Textile Engineering, Technical University of Liberec, Studentská 1402/2, 46117 Liberec, Czech Republic; aamir.mahmood@tul.cz (A.M.); miroslava.pechociakova@tul.cz (M.P.); jiri.militky@tul.cz (J.M.); sebnem.sozcu@tul.cz (S.S.)
2 Department of Machinery Construction, Institute for Nanomaterials, Advanced Technologies and Innovation (CXI), Technical University of Liberec, Studentská 1402/2, 46117 Liberec, Czech Republic; nesrine.amor@tul.cz (N.A.); michal.petru@tul.cz (M.P.)
3 Department of Advanced Materials, Institute for Nanomaterials, Advanced Technologies and Innovation (CXI), Technical University of Liberec, Studentská 1402/2, 46117 Liberec, Czech Republic; mohamed.fawzy@mena.vt.edu
4 Department of Mechanical and Materials Engineering, Vilnius Gediminas Technical University, Sauletekio al. 11, 10221 Vilnius, Lithuania
5 Department of Nanoengineering, Center for Physical Sciences and Technology (FTMC), Savanoriu Ave. 231, 02300 Vilnius, Lithuania
6 Department of Textile Engineering, Balochistan University of Information Technology, Engineering and Management Sciences, Quetta 87300, Pakistan; syed.zameer@buitms.edu.pk
* Correspondence: muhammad.tayyab.noman@tul.cz

**Citation:** Mahmood, A.; Noman, M.T.; Pechočiaková, M.; Amor, N.; Petrů, M.; Abdelkader, M.; Militký, J.; Sozcu, S.; Hassan, S.Z.U. Geopolymers and Fiber-Reinforced Concrete Composites in Civil Engineering. *Polymers* **2021**, *13*, 2099. https://doi.org/10.3390/polym13132099

Academic Editor: Enzo Martinelli

Received: 4 June 2021
Accepted: 23 June 2021
Published: 25 June 2021



**Abstract:** This paper discusses the influence of fiber reinforcement on the properties of geopolymer concrete composites, based on fly ash, ground granulated blast furnace slag and metakaolin. Traditional concrete composites are brittle in nature due to low tensile strength. The inclusion of fibrous material alters brittle behavior of concrete along with a significant improvement in mechanical properties i.e., toughness, strain and flexural strength. Ordinary Portland cement (OPC) is mainly used as a binding agent in concrete composites. However, current environmental awareness promotes the use of alternative binders i.e., geopolymers, to replace OPC because in OPC production, significant quantity of $CO_2$ is released that creates environmental pollution. Geopolymer concrete composites have been characterized using a wide range of analytical tools including scanning electron microscopy (SEM) and elemental detection X-ray spectroscopy (EDX). Insight into the physicochemical behavior of geopolymers, their constituents and reinforcement with natural polymeric fibers for the making of concrete composites has been gained. Focus has been given to the use of sisal, jute, basalt and glass fibers.

**Keywords:** jute; basalt; glass; geopolymers; composites; concrete

## 1. Introduction

The modern world is facing many challenges due to a day-by-day deterioration in the natural environment and the depletion of natural resources. New techniques and materials are being sought to combat this situation. The construction industry has been an important contribution to environmental deterioration. The last few decades have witnessed the use of geopolymeric materials in the construction industry in an attempt to address certain environmental and economic concerns. Rapid industrialization, development and innovations in almost every part of the world have brought about an alarming day-by-day deterioration in the atmospheric environment. With an exponential increase in human population, the demand for major construction has increased to fulfil the human necessity for shelter. This can be considered a threat to the depletion of natural resources. Concrete is one of the major materials utilized in the construction industry and it is estimated that the production of concrete is responsible for 5–8% of total $CO_2$ emissions in the world. Most, 95%, $CO_2$ emissions are due to the fabrication of cement, which is the main constituent of concrete [1]. This reflects the need for the development of more economical and environment friendly materials with lower $CO_2$ emissions. In this regard, efforts have been made to find alternatives that can be used with lower pollution and less energy consumption. Various methods for reducing $CO_2$ emissions during the fabrication of cement have been examined previously [2]. These methods include the use of alternative fuel [3], improving energy efficiency inside the cement kiln [4], modifying cement with optimum percentage of nanoparticles [5] and creating alternative cement like geopolymers [6]. Among these options, geopolymers with environmentally friendly characteristics are receiving more and more attention from the scientific community and are new entrants in the field of construction materials. The fabrication of geopolymers is cost-effective and requires less energy consumption in comparison with the formulation of other alternatives available. In addition, the inner portion of industrial waste such as fly ash adds many advantages and causes geopolymers to be a completely new green material for use in the construction industry [7–9]. Some studies indicate that the use of geopolymers can reduce greenhouse gas emissions from 44% to 64% in comparison to that for the production of ordinary Portland cement (OPC) [10–12]. If careful design of geopolymer cement is carried out, its production will require less energy and may reduce $CO_2$ emission by 80% compared to those from OPC production [13]. Geopolymers have performance characteristics similar to those of OPC and hence are viewed as one of the most likely alternatives to OPC. OPC can be replaced with geopolymers, which serve as a binding phase and at the same time, can enhance mechanical properties of concrete [14]. The situation described above has motivated scientists from all over the world to focus on the development of geopolymers. Figure 1 shows an increasing interest on geopolymer development due to reduced energy consumption, lower $CO_2$ emission, and a high reduction in cost compared to the generation of traditional materials.

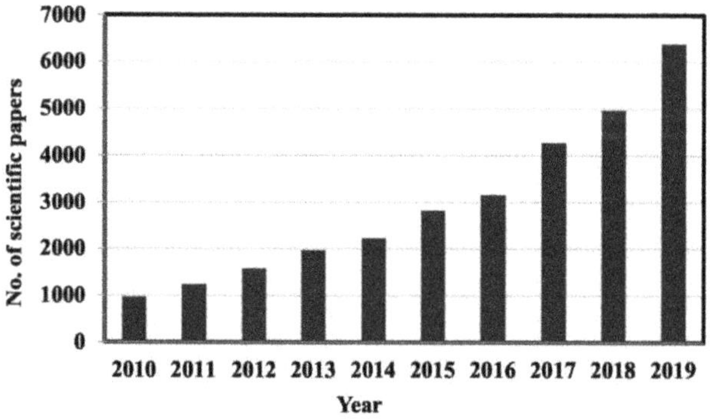


**Figure 1.** A graphical representation of data published on geopolymers from 2010 to 2020 based on its economical and energy consumption benefits. Reprinted with permission from Ref. [2]. Copyright 2021, with permission from Elsevier.

## 2. Geopolymers

Geopolymers are used in many fields ranging from aeronautics and civil engineering to the plastics industry [15–18]. The synthesis of geopolymers is based on a chemical reaction between solid aluminosilicate compounds ($Si_2O_5$, $Al_2O_2$) and a highly concentrated alkali hydroxide or polysilicate solution [19]. Kaolin and metakaolin are the examples of naturally occurring aluminosilicate compounds whereas industrial wastes i.e., fly ash and blast furnace slag, are considered to be industrial sources. The dissolution of the aluminosilicate compounds by hydrolysis forms silicate and aluminate due to the presence of alkaline activators. After dissolution, a reaction with silicates takes place promoted by the presence of activators solution. At a higher pH, the amorphous aluminosilicates are quickly dissolved and form a highly saturated aluminosilicate solution that further produces a gel like structure. Large networks are formed by oligomers in an aqueous media due to condensation. This gel-like structure is bi-phasic due to the water and aluminosilicate binder. The conversion time from a supersaturated aluminosilicate solution to gel depends on the synthesis conditions, materials composition and concentration of activators solution. After the gelation process, the system continues rearranging as the linkages of the gel network are enhanced, resulting in the three-dimensional (3D) aluminosilicate network named geopolymers [20]. A general flow diagram of the synthesis of geopolymers is illustrated in Figure 2 [21]. However, a chemical view of geopolymerization process shows that Al-O/Si-O bonds are broken in the solid alumina/silica rich binder material during the hydrolysis by hydroxyl initiator $[OH]^-$ and produces a tetrahedral aluminate and silicate intermediate species i.e., $[Al(OH)4]^-$ and $[Si(OH)4]^{4-}$. Water condenses during gelation and an overall shrinkage occurs in the structure. Therefore, the system reorganizes and rearranges itself into a 3D network as explained in Figure 3 [22].

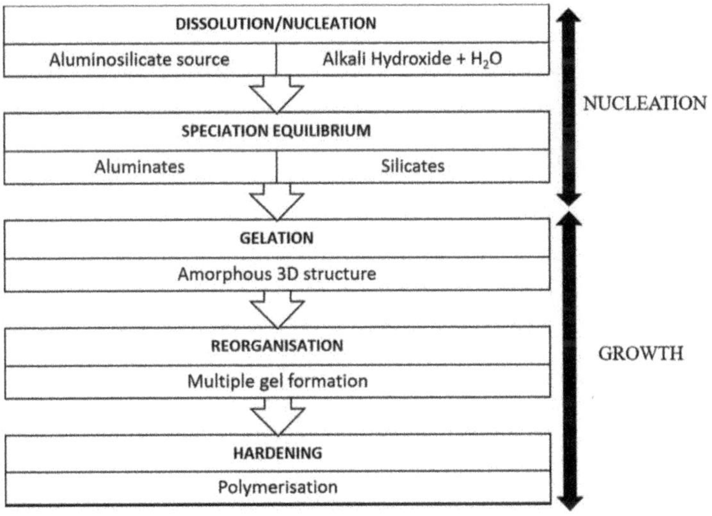


**Figure 2.** A general flow of geopolymerization process. Reprinted with permission from Ref. [21]. Copyright 2020, with permission from Elsevier.

Geopolymers are categorized on Si/Al ratios, and the three basic geopolymeric structures based on Si/Al ratios are poly (sialate), poly (sialate-siloxo) and poly (sialate-disiloxo) as illustrated in Figure 4 [23,24]. Sialate is the common name for silicon-oxo-aluminate and it ranges from amorphous to semi-crystalline in nature. Therefore, raw materials that contain rich silica and alumina content can be used for the production of geopolymers.

In general, geopolymers are defined as ceramic-like materials formed at room temperature. They are inorganic in nature and contain a aluminosilicate polymer as a building block. Geopolymers can be made from both industrial and agricultural wastes like fly ash, metakaolin, rice husk ash and wheat straw ash [25,26]. Geopolymers are prepared by mixing alkaline activators (mainly CaO) and water along with aluminosilicate-based sources. Alkali metal silicates, hydroxides or even carbonates have been used for making an alkaline solution for the activation. However, some researchers also used NaOH or KOH in this regard [27–29]. During the activation process, $SiO_4$ and $AlO_4$ species were obtained as the solid aluminosilicate oxide dissolution took place and, as a result of the polycondensation reaction, an amorphous three-dimensional geopolymeric system formed. However, the inclusion of acids involves the deterioration of Si-O-Si bonds during geopolymerization or may leave an Al depletion silica system that depends on the distribution of Si and Al in the framework. Figure 5 shows the SEM images of water-immersed geopolymer samples and acid-immersed samples after 70 days of immersion [30].

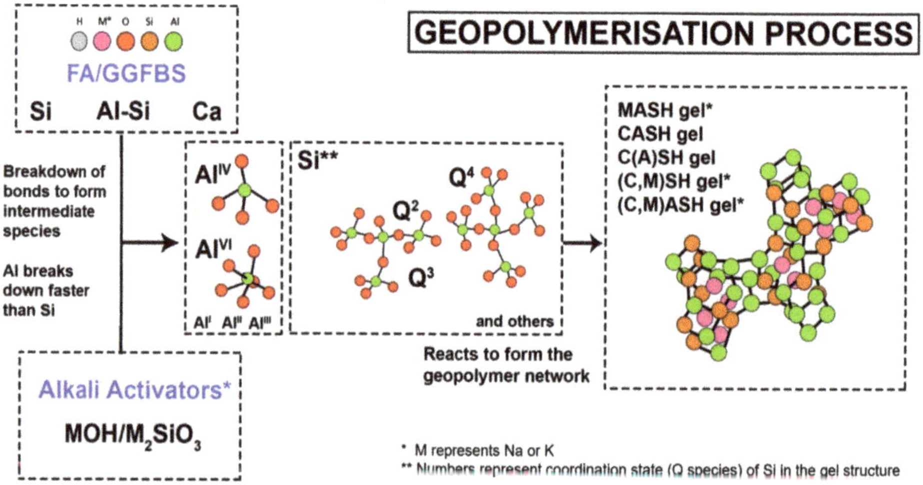


**Figure 3.** A chemical view of the geopolymerization reaction mechanism of aluminosilicate (breakdown of bonds and formation of intermediate species) dissolution in the activator solution. Reprinted with permission from Ref. [22]. Copyright 2021, with permission from MDPI.

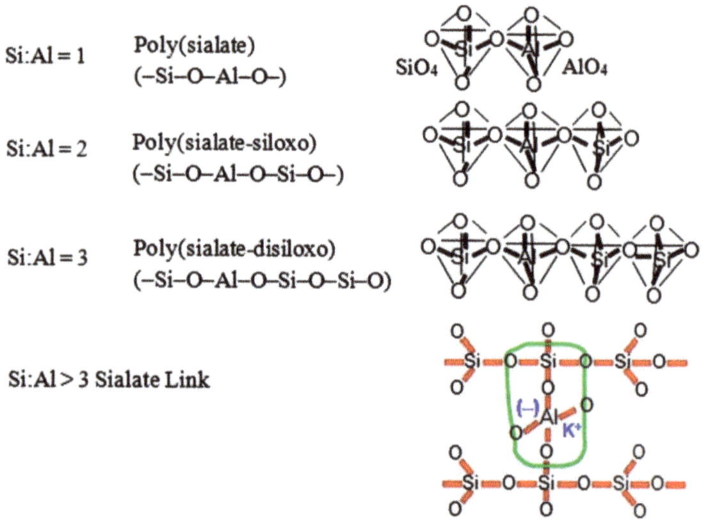


**Figure 4.** Siloxo Si-O units based on different geopolymeric systems. Reprinted with permission from Ref. [24]. Copyright 2016, with permission from Elsevier.

Geopolymer cement is reported to attain rapid hardening, and after 4 h of manufacturing at 20 °C, it achieves a compressive strength in the range of 20 MPa whereas after 28 days, it gains compressive strength from 70 to 100 MPa. Heat-treated, low-calcium content, fly-ash-based geopolymer concretes have excellent compressive strength and do not suffer from drying shrinkage and low creep [31]. Geopolymers exhibit better chemical resistance to sulphates than OPC. For low-cost ceramic materials at ambient temperatures and for pre-cost applications, drying and hardening of geopolymers after the initial setting is very important especially in applications where shrinkage is not desirable [32]. It is important to define the exact chemical nature of geopolymeric gel, which is further responsible for determining the mechanical properties of geopolymers. It is also important to note that the nanostructure and molecular nature within the gel determine the thermal and chemical stability of the binder. In this regard, correct selection of raw materials and mixing design are essential factors in achieving the enhanced properties of geopolymers for specific purposes. Geopolymers are reported to be fire resistant because of their inorganic framework and show high thermal stability in comparison with OPC [20]. It is also reported that geopolymers have excellent mechanical properties in comparison to OPC [33]. Density, compressive strength, flexural strength and modulus of elasticity are the important mechanical properties of geopolymers for further use in structural applications. Table 1 summarizes the detail of the constituents of geopolymers, their role on the properties of fiber-reinforced composites and the experimental conditions, etc.

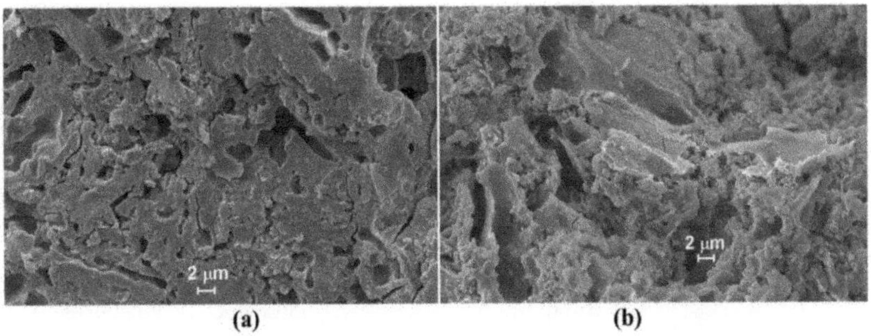


**Figure 5.** SEM micrographs of illustration of (**a**) geopolymers sample immersed in water exhibited glass-like microstructural surface and (**b**) geopolymers sample after acid exposure showed debris-like colloidal particles (could be the precipitates of silica gel) on the surface. Reprinted with permission from Ref. [30]. Copyright 2018, with permission from Elsevier.

**Table 1.** A brief summary of the constituents of geopolymers.

| Constituents | Composition | Temperature [°C] | Time [Days] | Chemicals | Compressive Strength [MPa] | References |
|---|---|---|---|---|---|---|
| Coal fly ash & metakaolin | - | 250 | 3 | NaOH | - | [34] |
| Fly ash, GGBFS & zeolite | Al/Si | 32 | 1 | NaOH | 100 | [35] |
| Coal fly ash | Al/Si | 80 | 1 | NaOH | 18 | [36] |
| Fly ash | $SiO_2/Al_2O_3$ | 85 | 3 | KOH | 19 | [37] |

### *2.1. Factors Affecting on Geopolymers*

Many factors that affect the compressive strength, flexural strength and other mechanical properties of geopolymers have been reported by different researchers [22,38–42]. These factor, such as different calcium-containing raw materials [21,43], ionic additives, curing procedures and post-curing chemical treatment, have been considered important for final properties [44–46]. An amorphous structure of geopolymers is better for realizing anticipated mechanical strength. It is important to note that the properties of geopolymers greatly depend on the $SiO_2/Al_2O_3$ ratio, $NaOH/Al_2O_3$ or $SiO_2/KOH$ ratio and the liquid–solid ratio. Research works investigating the influence of the C-S-H phase on the geopolymerization of aluminosilicates with a focus of its role on early-age strength have been made previously [47–49]. Phair and Deventer investigated the C-S-H phases and demonstrated

C-S-H phases at different pH levels. According to them, the presence of C-S-H at pH 12 did not improve the compressive strength significantly as compared to pH 14 [50].

The effect of admixtures on geopolymers is another important factor that alters the overall properties. It has been noted that sucrose and citric acid, as admixtures that perform the role of retarder in OPC, have dissimilar mechanisms in fly-ash-based geopolymers [51]. Commercial superplasticizers such as naphthalene and polycarboxylate based superplasticizers were also investigated. It is reported that a naphthalene-based superplasticizer is effective when a single activator is used, rendering a 136% increase in relative slump without disturbing the compressive strength. When a multi-compound activator is used, a modified polycarboxylate-based superplasticizer is more effective [52,53]. The retarding effect of a polycarboxylate-based superplasticizer in a fly ash/slag blended system is reported in the literature along with a significant improvement in workability compared to a naphthalene-based superplasticizer.

Many researchers have conducted studies on the properties of geopolymer pastes based on various curing conditions. For achieving complete geopolymerization, researchers reported curing of samples at temperatures between 40 °C and 85 °C. Palomo et al. reported that the alkali-activated fly ash, when cured at a temperature 85 °C for 24 h, gave significant compressive strength of the geopolymer as compared to the same composition at 65 °C [54]. However, no significant increase in compressive strength was noted with curing time extended beyond 24 h. Heah et al. showed that metakaolin-based geopolymers cured at higher temperatures result in an increase of strength after 1 to 3 days. They also discovered that samples cured at a higher temperature for a longer time period result in sample failure. This failure was described by the thermolysis of the -Si-O-Al-O- bond [55]. According to Rovnanik, a metakaolin-based geopolymer cured at a higher temperature (40–80 °C) showed deterioration of mechanical properties when compared with the results obtained for slightly decreased temperature. In order to achieve better mechanical and durability properties of geopolymers, suitable curing is a must [56]. Previous research has shown that the mechanical properties of geopolymer are strongly influenced by its Si/Al ratio. Additionally, the liquid/solid ratio, which is linked to the water content in the reactive mixture, has an effect on the geopolymers formulation. Zuhua et al. demonstrated that an increased liquid/solid ratio supported the transfer of ions and thus boosted the dissolution of the aluminosilicate source. On the other hand, this increase slowed the polycondensation reactions and resulted in an increased porosity rate, thus leading to poor mechanical performance [57].

### 2.2. *Cementitious Materials for Geopolymers*

Commonly cementitious materials are created by alkaline activation either based on Si and Ca or based on Si and Al. In recent times, efforts have been made to replace OPC with other cementitious materials containing Si and Al. Various activated natural materials and industrial by-products are being used to produce alkaline-activated binders for further use in developing cementless mortar and concrete. Supplementary cementitious materials for geopolymers have less environmental impacts compared to OPC. Research work on geopolymers has indicated that the inclusion of raw materials such as fly ash and slag in geopolymers fabrication is gaining interest worldwide, and these industrial waste materials are responsible for deciding the final properties of geopolymers [58,59]. In this regard, most of the researchers have focused on employing fly ash/blast furnace slag for geopolymers systems. Microstructure, physical, mechanical, chemical and thermal properties of geopolymers, in contrast to their macroscopic characteristics, mainly depend on the raw materials. Materials that are rich in aluminum and silicone can be used in the fabrication of geopolymers. These materials include fly ash, slag, waste glass and some pure Al-Si minerals and clays (kaolinite and metakaolinite). Among them, fly ash and slag are the most widely used ones.

### 2.2.1. Fly Ash

Fly ash is a residue generated in coal-fired power plants for electricity generation and is viewed as solid waste material [60]. Fly ash is produced at an amount of about 800 million tons annually, worldwide. China is the largest country in fly ash production followed by India, USA and EU [61]. It has been reported that the fly ash utilization rate is estimated as 50% for the USA, more than 90% for EU, 60% for India and 67% for China. Coal fly ash usually consists of coarse bottom ash and fine fly ash, of which coarse bottom ash is about 5–15% by weight and fine fly ash is about 85–95% by weight of total coal ash generated. Bottom ash accumulates on the bottom of the boiler by air flow whereas the fly ash is caught from the flue gas and collected by electrostatic or mechanical precipitation. Fly ash particles are mainly spherical and comprise solid spheres, cenospheres, irregular shaped waste and porous unburnt carbon [62]. Fly ash is considered to be a pozzolan-like material at room temperature when mixed with water and calcium hydroxide form cementitious products. The color of fly ash is generally grey, and the amount of unburned coal in the ash is responsible for defining its color from dark to dull to black, with fine, powdery particles mostly spherical in shape. These particles are either solid or hollow and predominantly amorphous [63]. The type of coal used at power plants significantly influences the physical and chemical characteristics of fly ash [64]. Generally, fly ash contains quartz, hematite, mullite and amorphous particles. Chemical composition of fly ash is greatly influenced by factors such as the type of coal used for burning and conditions under which this process takes place, as well as the removal effectiveness of the air-pollution-control device [65]. A complete production process of fly ash (from coal and after pulverization, boiling and precipitation) is thoroughly explained in Figure 6 [41].

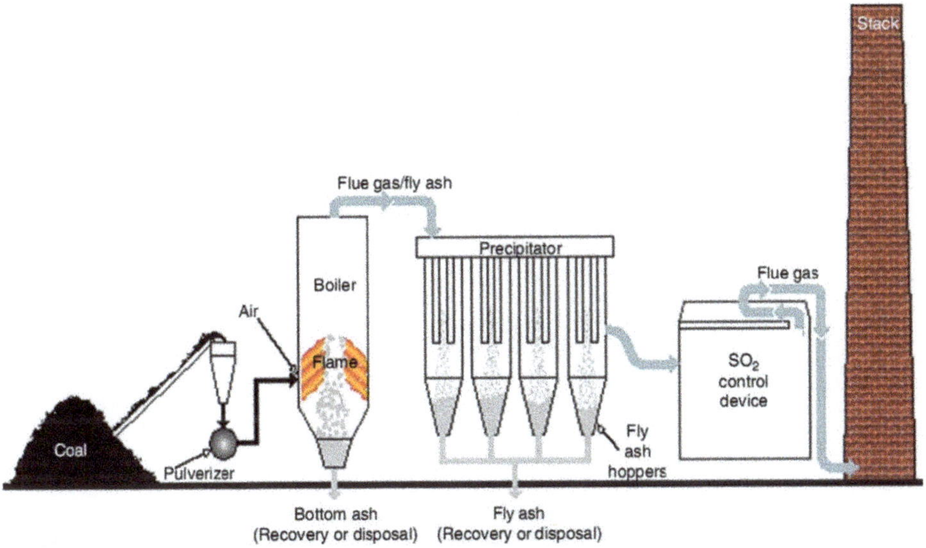


**Figure 6.** A schematic depiction of a clean production process of fly ash from coal. Reprinted with permission from Ref. [41]. Copyright 2021, with permission from Elsevier.

Fly ash can be classified based on chemical and mineralogical composition into two types, that is, class C and class F, depending on the coal used for combustion. Class C fly ash is obtained during the combustion of lignite and sub-bituminous coals and contains $SiO_2$, $Al_2O_3$ and $Fe_2O_3$, together less than 50%, whereas CaO ranges between 20% and 30%. On the contrary, class F fly ash is obtained when anthracite and bituminous coals are burned, and contains more than 70% of $SiO_2$, $Al_2O_3$ and $Fe_2O_3$ with CaO content less than 5% [66]. Generally bituminous coal fly ash particle size is less than 0.075 mm and is normally similar to silt particle size, whereas sub-bituminous coal fly ash is also similar to that of silt particles, but is a little coarser compared to bituminous coal fly ash. The specific gravity of fly ash is found to be between 2.1 and 3 and specific surface area between 170 and 1000 $m^2/kg$. The bulk density of fly ash is in the range of 0.54–0.86 $g/cm^3$. Properties of fly ash of classes C and F differ significantly and hence both classes have different uses. Class C fly ash is characterized as high-calcium fly ash and can be considered as a cementitious material if CaO content is higher than 20%. High-calcium fly ash with CaO between 10% and 20% is said to be a cementitious and pozzolanic material [67]. Past years have witnessed increasing research on fly ash and its industrial utilization. Fly ash has been investigated mainly in areas such as cement fabrication, ceramics, paints, plastics, agriculture and construction industry [68]. Low-calcium fly ash represents the properties of normal pozzolan (a material with silicate glass and modified with aluminum and iron). Pozzolanic activity to form strength-developing products with low-calcium fly ash takes place when it interacts with $Ca(OH)_2$. For this purpose, low-calcium fly ash is used in combination with OPC to produce $Ca(OH)_2$ during the hydration process. On the other hand, the suitability of high-calcium fly ash in concrete is viewed with doubts. This can be explained by high free CaO and sulphur content in such a fly ash chemical composition that can disturb concrete volume stability and durability. Yet, high-calcium fly ash can cause early strength development in concrete and, if proportioned accurately, can increase the quality of concrete.

Fly ash as a replacement for cement in concrete is limited to 15–20% by mass of the total cementitious material [69]. However, Malhotra reported that more than 50% of fly ash replacement in concrete can be used, subject to the fact that acceptable material responses, such as strength, durability, permeability and shrinkage, are ensured [70]. Fly ash increases the workability and lessens the bleeding of fresh concrete. It also increases the durability properties of concrete and, if designed properly, exhibits enhanced strength and low permeability. Concretes with partial replacement of OPC with fly ash show considerable increase in workability compared to OPC concrete. Fly ash containing concretes show increased workability with increasing levels of fly ash replacement thus reducing the water demand for the system [71]. Fly ash can be used for controlling sulphate attack and both classes of fly ash behave differently in such a corrosive environment. High-calcium fly ash contains considerable amounts of soluble calcium, aluminum and sulphur-bearing minerals as well as a substantial amount of calcium aluminate glass, which can release calcium and aluminum into the solution slowly. It increases the pH of the solution when it comes in contact with water and the expansion of cement occurs, thus leading to cracking. Class C fly ash increases the exposure to sulphate attack as it is rich in lime and hydrates independently. On the other hand, class F fly ash hinders the sulphate attack by hampering the formation of alumino silicate hydrate compounds. It has been reported that low-calcium fly ash concrete is more resistant compared to concrete with high-calcium fly ash [72]. A flow of fly ash and its components in the utilization of real-life applications in various industries is presented in Figure 7 [41].

### 2.2.2. Ground-Granulated Blast Furnace Slag (GGBFS)

Ground-granulated blast furnace slag (GGBFS) is among the other by-products during the production of iron and steel. During the process of iron and steel production in blast furnaces, quick removal of slag on the top of molten iron is carried out and then grounded to obtain GGBFS. It is a powder-like material and white in color. Physical properties of GGBFS depend on the cooling process and chemical properties depend on the selection of raw materials for iron production. Its fineness specific surface area is between 300 $m^2/kg$ and 500 $m^2/kg$ with specific gravity ranging from 2.4 to 3.0 [73]. GGBFS is pozzolanic in nature, and for decades, has been used as cementitious component for making cement/concrete composites. The effectiveness of GGBFS in cementitious composites depends on many factors, such as chemical composition, fineness and hydraulic reactivity. SEM and EDX analysis for GGBFS explains that GGBFS contains denser particles and those particles have different size grades that work as pore filtration as well as higher silicon contents than OPC, as illustrated in Figure 8 [74].

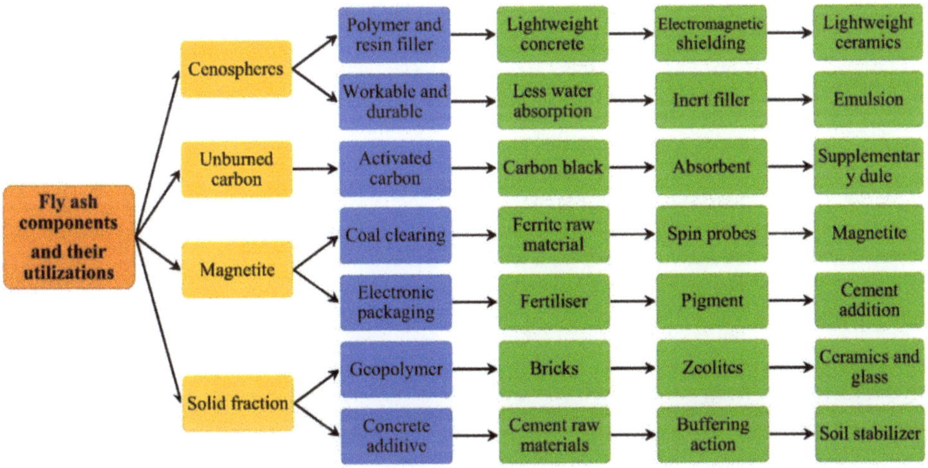


**Figure 7.** The utilization of fly ash and its components into different real-life applications. Reprinted with permission from Ref. [41]. Copyright 2021, with permission from Elsevier.

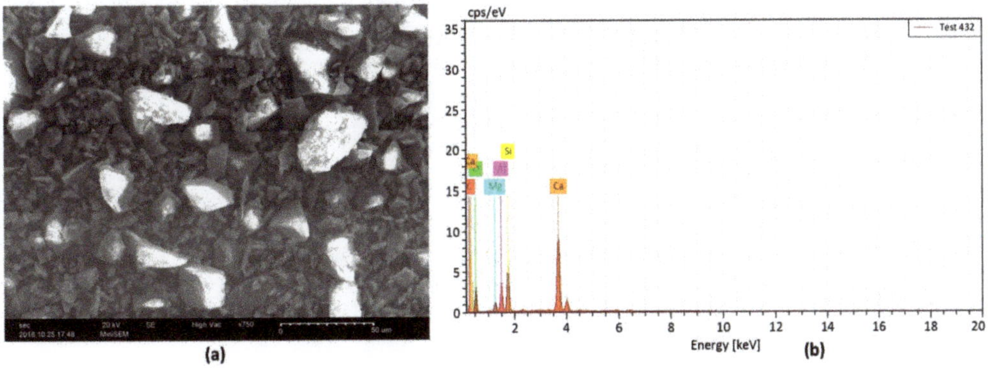


**Figure 8.** Results of ground granulated basalt furnace slag from (**a**) SEM analysis shows different grades of denser particles and (**b**) EDX analysis shows higher silicon content than OPC. Reprinted with permission from Ref. [74]. Copyright 2021, with permission from Elsevier.

GGBFS is non-metallic in nature and basically contains oxides of calcium, silica, alumina and magnesia along with some other oxides in small quantities. Both glassy and crystalline phases are present in GGBFS with glass content in the range of 85–90% [75]. The amount of glassy phase in GGBFS is very much affected by the cooling process. Slow cooling produces about 50% to 70% of the crystalline phase in GGBFS, which in turn reduces the hydraulic activity. On the other hand, quick cooling normally develops smaller and uniform particle sizes in GGBFS and contributes to advanced hydraulic activity. A number of formulas have been developed to predict the hydraulic activity of GGBFS. These formulas do not represent the exact strength performance of GGBFS because the hydration reactions that take place are very complex in nature and these formulas do not indicate them exactly [76].

Chemical composition is vital for achieving good hydraulic activity of GGBFS. GGBFS are classified based on their basicity index. A $CaO/SiO_2$ ratio more than 1.0 is one such basicity index described in the literature. Basicity defines the hydraulic activity of the slag, and if the slag is more basic, its hydraulic activity in the presence of some alkaline activator is better. With constant basicity, an increase in $Al_2O_3$ content increases the strength, whereas variations in MgO content reaching 8–10% can insignificantly affect the strength development while exceeding 10% can affect it negatively. Furthermore, increasing the amount of CaO increases the hydraulic activity of GGBFS whereas it decreases with an increasing content of $SiO_2$. In order to assure high alkalinity, without which GGBFS is hydraulically inactive, European Standard 197-1 suggests a ratio of the mass of CaO and MgO together to the mass of $SiO_2$ ratio greater than 1.0. It has been observed that initial hydration is considerably slower when GGBFS is hydrated with water only. To overcome this, OPC, alkalis or lime are used as activators to accelerate the hydration process [77]. Like other cementitious materials, surface area is used to determine the reactivity of GGBFS. Increased fineness gives better strength development. Setting time, shrinkage and economic considerations are the factors that limit fineness practically. Fineness of GGBFS affects the reactivity of slag in concrete, early strength development and water requirement. On the other hand, it is dependent on energy saving and economic considerations. It has been reported that the fineness of GGBFS should be two to three times greater than that of OPC in order to have advantages over slag in certain engineering properties like bleeding, setting time, high strength and excellent durability.

A number of research works have been carried out to study the workability of concrete and mortar with the inclusion and replacement of GGBFS. It has been reported that 60% replacement of OPC with GGBFS in concrete improved its workability. Generally, replacing OPC with an increasing percent of GGBFS results in prolonged setting times of concrete whereas replacement equal to or more than 40% has an extremely retarding effect. The use of GGBFS for applications where early-age strength is required is not recommended. Concrete with the inclusion of GGBFS under a normal temperature curing condition attains considerably slow strength development as compared to OPC concrete. However, the early strength development of GGBFS included considerable concrete increases at higher early-age temperatures [78]. In a study carried out by Oner and Akyuz, it was observed that the compressive strength of GGBFS concrete increased with an increase in GGBFS content and an optimum level for the efficient use of GGBFS content was around 55–59%. Flexural strength of concrete is prone to micro-cracking, and the inclusion of GGBFS can improve its flexural strength at later stages [79]. The influence of GGBFS on the flexural strength of concrete showed that concrete containing 60% of GGBFS was higher than the flexural strength of the same concrete without GGBFS whereas a noticeable decrease was observed with 80% replacement.

#### 2.2.3. Metakaolin

Less energy-intensive processed materials and wastes with pozzolanic behavior have been using by mankind for thousands of years. One such example in recent times is metakaolin [80]. Metakaolin belongs to the calcined clay family and is obtained from the calcination of kaolin clay. The use of kaolin as an industrial mineral in various industries largely depends on its mineralogical composition, geological conditions under which it formed and its physical and chemical properties. Kaolin is found as sedimentary, residual or hydrothermal and has unique properties in every of these cases, and therefore is subject to proper testing and evaluation for a specific use. Kaolin deposits are found in southwestern England, Georgia and South Carolina in the USA as well as in the lower Amazon of Brazil. which are known as the most utilized ones. Kaolinite crystals that cover most of the kaolin deposits are pseudo-hexagonal along with plates, some larger books and vermicular stacks [81]. Metakaolin is a highly reactive pozzolana type and needs to be processed like cement in burning kilns at temperatures around 700–900 °C and mainly contains silica and alumina. Metakaolin is produced under these temperatures, when kaolin-clay rich in kaolinite undergoes thermal activation and by dehydroxylation, which leads to breaking down or partial breakdown of the structure resulting in a transition phase with high reactivity [82]. The morphologies of fly ash and metakaolin particles are observed by SEM analysis and illustrated in Figure 9 [83]. It is observed that fly ash particles are quasi-spherical and smooth with a variation in sizes. However, metakaolin particles are mostly irregular in terms of shape. The quasi-spherical-shaped particles of fly ash are more suitable for lower water demand than metakaolin particles.

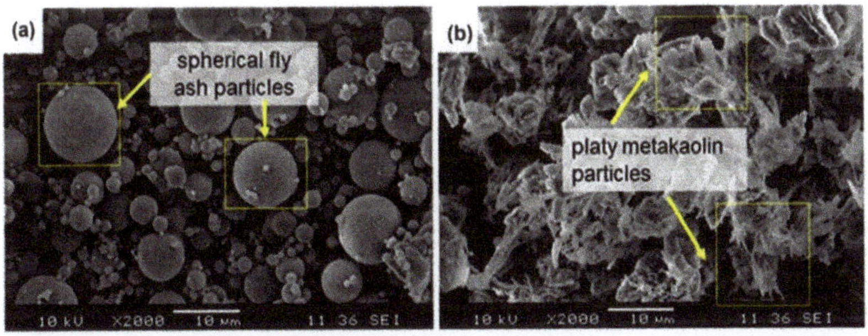


**Figure 9.** SEM images of (**a**) fly ash and (**b**) metakaolin. Reprinted with permission from Ref. [83]. Copyright 2021, with permission from Elsevier.

Metakaolin in the presence of water reacts chemically with $Ca(OH)_2$ produced by cement hydration and this generates additional gel containing secondary calcium-silicate-hydrate (C-S-H) along with crystalline products including aluminate hydrates and aluminosilicate hydrates. The hydration reaction is associated with the metakaolin reactivity level and is influenced by the processing conditions and purity of the clay type used [84]. Metakaolin is an off-white powder type of aluminosilicate material and has porous, angular-shaped, platy, smaller particle size of about 1–2 μm with a surface area in the range of 10,000–29,000 $m^2/kg$. Its specific gravity is in the range of 2.20–2.60 whereas the bulk density of metakaolin has been reported in the range of 300–400 $kg/m^3$ [85–87]. The chemical composition of metakaolin varies according to the type of kaolin used. The main components of metakaolin are silicon dioxide and alumina oxide with other components as well including ferric oxide, calcium oxide and potassium oxide. As per ASTM standard, raw or calcined natural pozzolan is characterized as class N with $SiO_2$, $Al_2O_3$ and $Fe_2O_3$, together equal or greater than 70%, $SO_3$ less than or equal to 4%, limiting oxygen index (LOI) content less than or equal to 10% and moisture content less than or equal to 3%.

Metakaolin as a pozzolanic addition has attracted much interest in recent years. The main interest regarding using metakaolin in mortar or concrete is the removal of C-H produced by the hydration of cement responsible for poor durability. Furthermore, C-H removal guarantees improved strength and has a key influence on resistance to sulphate attack and alkali silica reaction. Purity of metakaolin, OPC composition, water/binder ratio and curing conditions need to be addressed for the metakaolin content in a system for complete elimination of C-H [88]. Concretes with partial replacement of OPC with metakaolin shows a considerable reduction in workability and that the degree of workability ranges from low to very low with the increase in replacement, thus increasing the water demand of concrete. Despite a lower degree of workability, metakaolin concretes can be compacted well without any difficulty. An increase in compressive strength is caused by the filling effect, speeding up of cement hydration and pozzolanic reactivity of metakaolin in concrete containing metakaolin. Previous research showed that metakaolin-based concrete with the partial replacement of OPC with metakaolin ranging 5–15% achieved a cube compressive strength of 92–104 MPa at 28 days, whereas for the same samples, static and dynamic moduli of elasticity were observed as 45.73–46.26 and 51.78–52.86 GPa, respectively. Generally, metakaolin geopolymers set and harden within 24 h, with the short setting time of 4 h.

With high $Al_2O_3$ content, metakaolin geopolymers' setting time is short but it weakens the strength with a low content of $SiO_2$.

Metakaolin as a partial replacement in concrete acts as a filler and significantly improves early strength and increases long-term strength of concrete. Porosity and pore size distribution measurements observed previously show that the early strength development is linked with considerable pore refinement [89]. Th partial replacement of OPC with up to 20% metakaolin causes improvement in the pore structure of mortar paste, thus increasing the compressive strength. It has been observed that the filler effect is immediate and the speeding up of OPC hydration occurs significantly within the first 24 h whereas the maximum pozzolanic effect takes place between 7 and 14 days. Sulphate resistance of mortar increases by replacing cement with at least 30% metakaolin due to the reduction in C-H content in the paste, which reduces the gypsum and ettringite formation. On the other hand, it is also believed that refinement in pore structure hinders the entrance of suphate ions into the system. The inclusion of metakaolin also improves the resistance of concrete against chloride ions. Concrete with 10% metakaolin has been reported to be effective under freezing and thawing [90]. It has been observed that pozzolanic constituent materials like metakaoline affect the fire resistance of concrete. Metakaolin-based concretes show poor fire resistance at elevated temperatures as compared to OPC concretes, and this is because of the compacted micro-structure and poor porosity of the metakaolin-based concretes. Therefore, careful investigation of metakaolin-based concretes should be carried out for elements that are prone to high temperatures. Figure 10 shows the metakaolin and fly ash blended geopolymer composites with different formulations of monoaluminium phosphate and aluminum dihydrogen triphosphate that provided excellent mechanical and physical properties after a 28-day curing process [83]. It is observed that both fly ash and metakaolin particles provide more compact and homogenous microstructure of the composite that leads to higher compressive strength. In simple words, the addition of fly ash and metakaolin particles into the composite matrix increases the bonding between the particles and the composite matrix. Table 2 elucidates the detail of chemical composition and physical properties i.e., specific surface area, specific gravity, bulk density of fly ash produced with different coal types, GGBFS and metakaolin.

### 2.3. *Activating Chemical Solutions for Geopolymers*

Chemical activation is used to enhance the reactivity process and improve the fresh and or hardened state. Geopolymers are synthesized when aluminosilicate powders are mixed and activated in highly concentrated alkaline medium [91]. A strong alkaline medium is necessary to dissolve the quantity of aluminosilicates. Usually, alkali earth metal hydroxides and carbonates are used for this purpose. Sodium hydroxide is the most widely used one in this regard. In order to activate the aluminosilicates for geopolymerization, alkali hydroxides or carbonates with pH values more than 13 are required. Alkali metal hydroxides can significantly increase the solution viscosity at concentrations of more than 10 M. When sodium hydroxide is used in excess for activation, the formation of white crystalline sodium carbonates is a well-known issue. There are two types of activating mechanism in geopolymers, the liquid-activated geopolymer binder and the powder-activated geopolymer binder. The former relates to the alkali metal hydroxides, carbonates or silica fume and water whereas the latter relates to the mixing of alkali metal hydroxide and silicates in dry form. Liquid-activated geopolymer binders have been used in most of the studies and have some problems, such as high pH and variations in molarity leading to inconsistent performance [92]. As stated above, sodium hydroxide is the most commonly used alkaline medium for geopolymerization. Sodium hydroxide is commonly known as caustic soda and is a white solid inorganic material and is produced by 50% mass saturated in solution with water. Sodium hydroxide produces a huge amount of heat when it reacts with water. Sodium carbonate is an inorganic material white in color comprising sodium, carbon and oxygen elements, often called washing soda. It is a basic salt with a strong alkaline taste. Potassium hydroxide is an inorganic compound comprising potassium, oxygen and hydrogen elements also known as caustic potash. It is soluble in water and exothermic in nature.

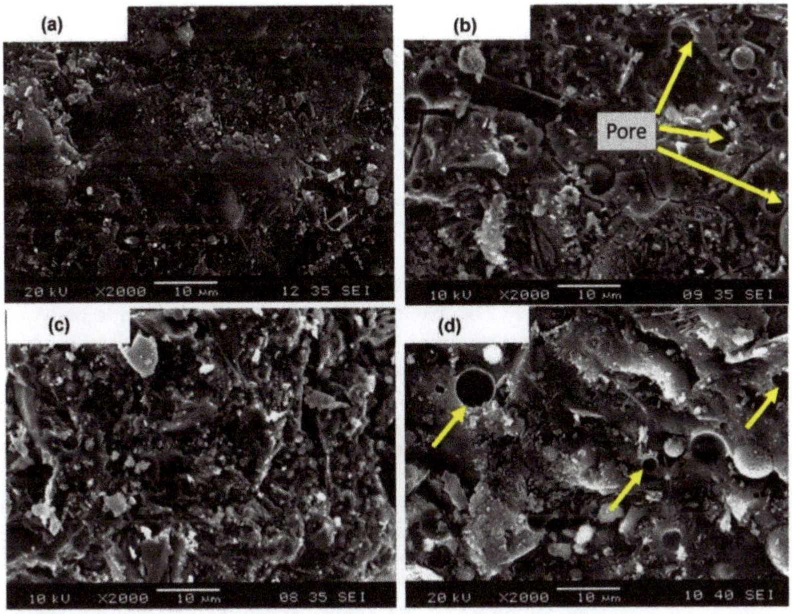


**Figure 10.** The morphologies of fly ash and metakaolin blended geopolymers composites with (**a**) 1.0 wt% monoaluminium phosphate, (**b**) 3.0 wt.% monoaluminium phosphate, (**c**) 1.0 wt.% aluminum dihydrogen triphosphate and (**d**) 3.0 wt.% aluminum dihydrogen triphosphate. Reprinted with permission from Ref. [83]. Copyright 2021, with permission from Elsevier.

**Table 2.** Chemical composition and physical properties of cementitious materials used in geopolymers concrete composites.

| | | Chemical Composition (wt.%) | | | | | | | | | Physical Properties | | | Ref. |
|---|---|---|---|---|---|---|---|---|---|---|---|---|---|---|
| | Coal Type | $SiO_2$ | CaO | $Al_2O_3$ | MgO | $SO_3$ | $Fe_2O_3$ | $K_2O$ | $Na_2O$ | $TiO_2$ | Specific Surface Area ($m^2$/kg) | Specific Gravity | Bulk Density (kg/$m^3$) | |
| Fly Ash | Sub-bituminous | 40–60 | 5–30 | 20–30 | 1–6 | 0–2 | 4–10 | 0–4 | 0–2 | - | 170–1000 | 2.1–3.0 | 540–860 | [61,62] |
| | Lignite | 15–45 | 15–40 | 10–25 | 3–10 | 0–10 | 4–15 | 0–4 | 0–6 | - | | | | |
| | Bituminous | 20–60 | 1–12 | 5–35 | 0–5 | 0–4 | 10–40 | 0–3 | 0–4 | - | | | | |
| | Anthracite | 28–57 | 1–27 | 18–36 | 1–4 | 0–9 | 3–16 | 0–4 | 0–1 | - | | | | |
| GGBFS | | 28–40 | 30–50 | 8–24 | 1–18 | 0.23–1.3 | - | - | - | - | 300–500 | 2.4–3.0 | 1200 | [73,75] |
| Metakaolin | | 51.9 | 0.11 | 45.39 | - | - | 0.92 | 0.45 | - | 0.76 | 10,000–29,000 | 2.2–2.6 | 300–400 | [82,85,86] |

Görhan and Kürklü found that fly-ash-based geopolymers prepared with low and high concentrations of sodium hydroxide exhibited poor compressive strength. They used 3, 6 and 9 M sodium hydroxide concentrations for the samples and noticed that samples with the 6 M concentration exhibited higher compressive strength of 22 MPa [93]. High-calcium, fly-ash-based geopolymers at 70%, 80%, 90% and 100% by mass of binders in combination with OPC and activated by sodium silicate and sodium hydroxide (alkali liquid/binder = 0.65 and $Na_2SiO_3$/NaOH = 0.67) cured at 60 °C showed that samples had increased compressive strength [94]. Fly-ash-based geopolymer mortar with sodium hydroxide as an activator showed increased compressive strength as compared to the same fly-ash-based geopolymer with a potassium hydroxide activator. The compressive strength recorded was 65.28 MPa for the sodium-hydroxide-activated, fly-ash-based geopolymer and 28.73 MPa for the potassium-hydroxide-activated, fly-ash-based geopolymer [95]. Zhang et al. carried out a study on a metakaolin-based geopolymer using sodium hydroxide, potassium hydroxide and sodium silicate as alkali activators with 8 M and 12 M, respectively. The highest compressive strength was observed as 40 MPa with 12 M concentration and with Si/Al = 1.9:1 [96]. In recent research, geopolymers with a combination of metakaolin and low-calcium fly ash (five different combinations of metakaolin and fly ash with metakaolin to fly ash mass ratios of 100:0, 80:20, 50:50, 20:80, 0:100) were activated by sodium hydroxide and potassium hydroxide solutions of 6 M separately. Researchers found that samples activated by sodium hydroxide showed improved compressive strength as compared to the samples activated by potassium hydroxide. GGBFS-based geopolymer concrete with sodium hydroxide and sodium silicate as alkali activators showed that the compressive strength of geopolymer concrete cured at room temperature for 28 days had a maximum value of ~44 MPa when 19 M NaOH solution with 50% $Na_2SiO_3$ concentration was used [97].

## 3. Geopolymers and Natural Fiber-Reinforced Composites

Geopolymers exhibit good thermal and durability properties, and at the same time, they are brittle in nature, show poor resistance to tensile and flexural loadings and undergo sudden failure, hence are not suitable for several structural applications [98,99]. To address this issue, research works have been focused on reinforcing geopolymers with synthetic and natural fibers in order to increase their ductility and resistance to tensile stresses. The incorporation of natural fibers in geopolymers gives a feasible solution to counter its initial brittle behavior [100]. Fiber can be defined as a hair-like material that is either a continuous filament or discrete elongated piece similar to thread. Fibers can be broadly divided into natural and man-made ones [101,102]. Figure 11 describes an overview of fiber classification.

In order to increase the flexural strength and energy absorption, fibers can be used as reinforcement in geopolymer composites in the form of threads, filaments, whiskers and nanoparticles. The inclusion of random short fibers in a cementitious medium enhances toughness, ductility and strength by bridging and reducing the cracks [103,104]. Moreover, the addition of fibers in the geopolymer matrix also increases its energy absorption and resistance to deformation. Geopolymers reinforced with any type of fiber show better toughness results in comparison with OPC-based composites. Several factors influence fiber performance in geopolymer composites, including inherent properties of the fiber used, its content, precursors, curing and age of the composites. Yet, the main role for overall mechanical properties is the interface between the fiber and matrix, and with a strong contact interface, high loads can easily be transferred from the matrix to fibers. Most research works concerning fiber-reinforced geopolymers have been done using steel fibers, carbon fibers, glass fibers, polypropylene fibers, polyvinyl alcohol fibers and basalt fibers [105–107]. Using cellulosic fibers as reinforcement in recent times has been witnessed as well. In this paper, our focus is on sisal, jute, basalt and glass fiber-reinforced geopolymers composites.

### 3.1. Cellulosic Fiber Reinforced Geopolymer Composites

During the past years, natural cellulosic fibers composites have gained considerable attention due to their low cost and low density and their use as a renewable source [108,109]. Cellulosic fibers have been used as an alternative instead of steel or synthetic fibers within cementitious composites as reinforcement [110]. Cement-based geopolymers reinforced with cellulosic fibers show enhanced toughness, ductility, flexural capacity and crack resistance in comparison to cement-based composites without fiber reinforcement. Fiber reinforcement presents its main benefit during cracking where fibers bridge the matrix cracks and transfer the loads, thus presenting more demanding use of such materials in the construction industry [111–113]. However, they have some disadvantages including low durability, efficiency at high fiber content that reduces the workability of a fresh composite, inconsistent material properties and poor interaction with the matrix [114,115]. As mentioned earlier, using cellulosic fibers has the problem of low durability. The use of cellulosic fibers as reinforcement in cement-based materials is limited due to relatively low degradation resistance in alkaline environments. Lignin and hemicellulose phases of natural fibers placed in OPC dissolve, therefore weakening the fiber structure and leading to their degradation in a highly alkaline environment. It has been reported that with an increase in fiber diameter, the corresponding mechanical strength and modulus of fibers decrease. Al-Oraimi and Seibi reported that a small percentage of natural fibers caused an improvement in mechanical properties and impact resistance of concrete, and has the same performance as that of synthetic fiber concrete [116]. Ramakrishna and Sundarajan reported that the addition of fibers increased impact resistance 3–18 times greater than those of without any fibers [117].

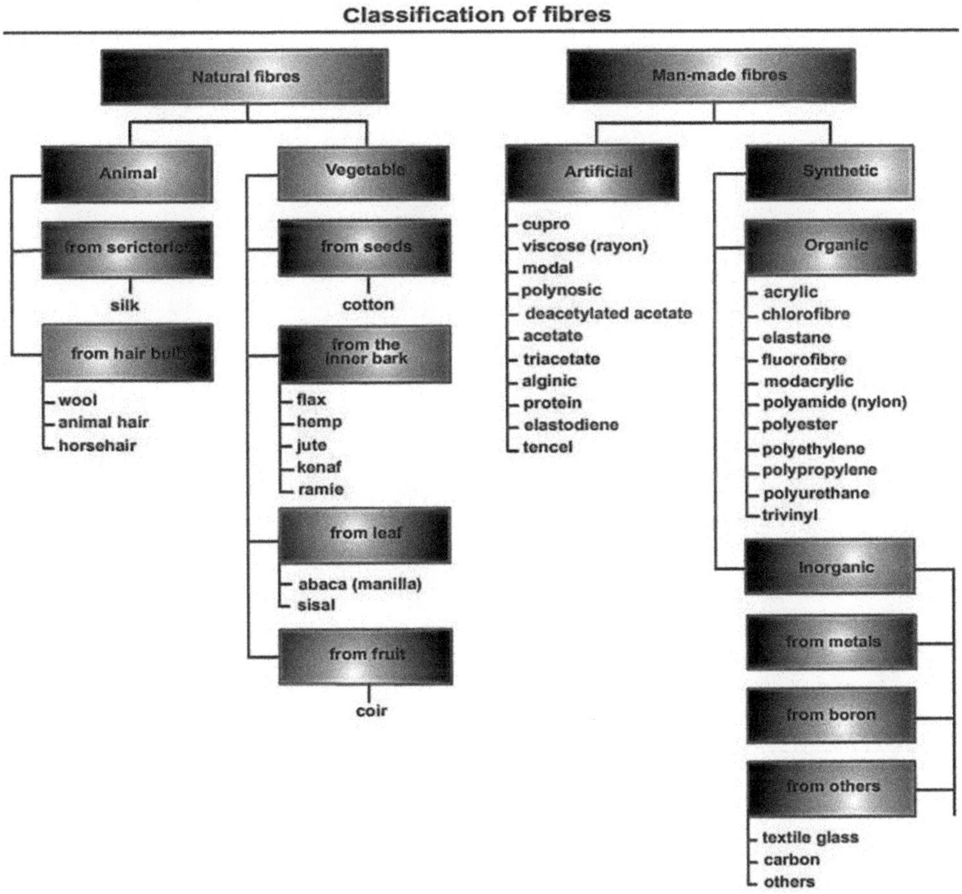


**Figure 11.** Classification of textile fibers.

3.1.1. Sisal and Its Composite

Sisal fiber is among the most commonly used cellulosic natural fibers. Plant Agave sisalana is the source of sisal fiber. The name sisal fiber comes from Mexico where Maya Indians used it for making ropes, clothes and carpets. Sisal fiber is mainly grown in Brazil, Haiti, East Africa, India, Indonesia and China. The current estimated production of sisal fiber is around 0.30 million tons per year [118–121]. In every sisal plant, there are 200–250 leaves with 1000–1200 fiber bundles in each leaf comprising 4% fiber, 0.75% cutile,

8% dry matter and 87.25% water. Each leaf of sisal contains three forms of fibers, namely mechanical, ribbon and xylem, of which the mechanical one is the most commercially useful and is extracted from the periphery of the leaf. Sisal fibers can be extracted by either retting or by mechanical means through decorticators. The mechanical method is more suitable for obtaining good-quality fibers extracted with average yield 2–4% (15 kg/8 h). The chemical composition of sisal fiber constitutes 67–78% cellulose, 10–14.2% hemicellulose, 8–11% lignin and 2% waxes [122]. Sisal fibers have low density and relatively high tensile strength. The density of sisal fiber is in the range of 1.03–1.45 $g/cm^3$, the tensile strength of sisal fiber is in the range of 347–700 MPa, its Young's modulus is reported in the range of 15.4–18.79 GPa whereas moisture content is reported as 11%. Elongation at break for sisal fiber was reported to be 2.0–2.5% [123].

Sisal fiber possesses a higher percentage of cellulose among the different plant-leaves cellulosic fibers and hence is thought to increase tensile properties. Moreover, sisal fibers have higher resistance to water permeability, thus making it attractive for use in construction applications. Sisal-reinforced composites represent higher impact strength with moderate tensile and flexural strengths [124]. In recent years, there has been a continuous effort for the development of sisal-fiber-reinforced composites with high strength in tension and compression. Liang et al. worked with long sisal fibe-reinforced and short sisal fiber-reinforced biocomposites and observed significantly high mechanical properties for long sisal fiber-reinforced composites. The results reveal holes and a fractured and uneven surface of composites for short fibers due to fibers debonding from the matrix during composite destruction. However, the fractured surface was clean, and the polymer matrix had a good wrapping effect on longer sisal fibers. The results of short sisal fiber-reinforced composites and long sisal fiber-reinforced composites after alkali treatment are illustrated in Figures 12 and 13, respectively [125]. It has been reported that reinforcing sisal with thermosets and thermoplastics results in increased mechanical properties. Prasad and Rao prepared composites using sisal fiber as a reinforcement in polyester resin matrix. The results reveal that the tensile strength of sisal/polyester increased by around 52%, tensile modulus increased by 67%, flexural strength increased by around 45% and flexural modulus increased by around 38% [126]. Hashmi et al. prepared the sisal/polypropylene composite and observed that the tensile strength increased by around 36%, flexural strength increased by 25.54% and impact strength increased by 56.52% [127]. In a recent study, Castro et al. investigated mechanical properties of untreated woven sisal fiber-reinforced green high-density polyethylene composites. They produced the composites using woven sisal fibers with a mass percentage proportion of 30:70 (fiber/matrix) and arranging 0°/90° and ±45° stacking sequences and under low-cost manufacturing process based on hot compression molding. Their results showed a 39% increase in tensile strength, a 13% increase in flexural strength, a 35% increase in flexural modulus and a 68% increase in ultimate strain in comparison with traditional polyethylene without sisal fibers [128]. In another study, sisal fiber was reinforced with poly-lactic acid (a biodegradable polymer). Results showed that with sisal fiber inclusion, tensile strength, flexural modulus and impact strength increased by 13.33%, 58.57% and 56.66%, respectively, whereas tensile modulus and flexural strength decreased by 17.24% and 2%, respectively [129]. Savastano et al. used 8% sisal pulp and blast furnace slag as a binder along with OPC in order to prepare composites and observed a maximum flexural strength in the range of 18–20 MPa with an improvement of at least 58% as opposed to that of the composites with fiber inclusion. Using kraft pulps from sisal and banana waste and from Eucalyptus grandis pulp mill residue as fiber reinforcement in cement-based composites showed an optimum performance with a 12% mass fiber content, a flexural strength of around 20 MPa and fracture toughness in the range of 1–1.5 $kJ/m^2$ [130].

Baloyi et al. prepared sisal–glass composites by layering methods with varying sisal fiber content treated with NaOH. They found that sisal fiber treated with 20% NaOH placed in nine layers with four layers of glass showed a tensile strength of 57.6 MPa and a flexural strength of 36 MPa [131]. Bahja et al. studied the effect of different treatments on the morphological properties of the sisal fiber and their effect in cement mortar. Their results showed that sisal fiber treatment results in an increased thermal resistance. The addition of sisal fibers by 4% mass of cement decreased the density of mortar and increased its porosity. A reduction in compressive and flexural strengths of the mortar with treated sisal fibers was noticed with satisfactory results for the elastic modulus of mortar [132]. Ren et al. conducted a study on the mechanical properties of sisal fiber-reinforced ultra-high-performance concrete. Samples were prepared using 1, 2 and 3% sisal fibers with lengths of 6, 12 and 18 mm. They noticed that sisal fiber had little effect on compressive strength of ultra-high-performance concrete and reduced the flowability of the concrete as well. Their results revealed that with 2% content of sisal fibers of 18 mm length, the flexural strength and toughness of the ultra-high performance concrete increased by 16.7% and 540%, respectively, compared to the control sample [133].

**Figure 12.** SEM images of (**a**) short sisal fiber-reinforced composites and (**b**) long sisal fiber-reinforced composites. Reprinted with permission from Ref. [125]. Copyright 2021, with permission from MDPI.

It is a known fact that, after cracking, the inclusion of short fibers as a mass reinforcement in fiber-reinforced concrete (FRC) mainly provides crack control due to the tensile stress transfer capability across the crack surfaces of the fibers known as crack bridging. It also improves the energy absorption capacity of the composites structures [134]. In this manner, fibers provide significant resistance to shear across developing cracks and, therefore, FRC demonstrates a pseudo-ductile response, increased residual strength (especially in tension) and enhanced energy dissipations capacities, relative to the brittle behavior of plain concrete mixtures. Furthermore, the advantageous characteristics of FRC under tension are also very important for the shear response of concrete structural members that are governed by the tensile response of the fibrous material. Thus, fibers have proved to be a promising non-conventional reinforcement in concrete elements under shear stresses due to the beneficial cracking performance of FRC, and under specific circumstances, could alter the brittle shear failures to ductile flexural ones. Different researchers worked on the shear critical analysis and reported findings that are useful on the state of the practice and for real-scale constructions [135]. Kytinou et al. reported the flextural properties and enhanced short-term behavior of steel-fiber-reinforced composites through the finite element method. The results explain that type of fibers, their volume fraction and aspect ratio play a significant role in achieving better compressive and tensile properties. Samples with higher amounts of reinforced fibers showed lower deformation at the same applied load than with the lower amount and vice versa. Moreover, a higher amount of steel fibers in reinforced composites show higher post cracking stress and vice versa [136]. Choi et al. also reported the crack mechanism of high-performance FRC that ultimately explain the strain hardening response based on the length of fibers. The results show that tensile properties were significantly overestimated for FRC when individual segments were estimated during a comparison of probabilistic and experimental results. This indicates that fiber tension should not be applied directly in order to abstain from multiple cracks and strain hardening behavior [137].

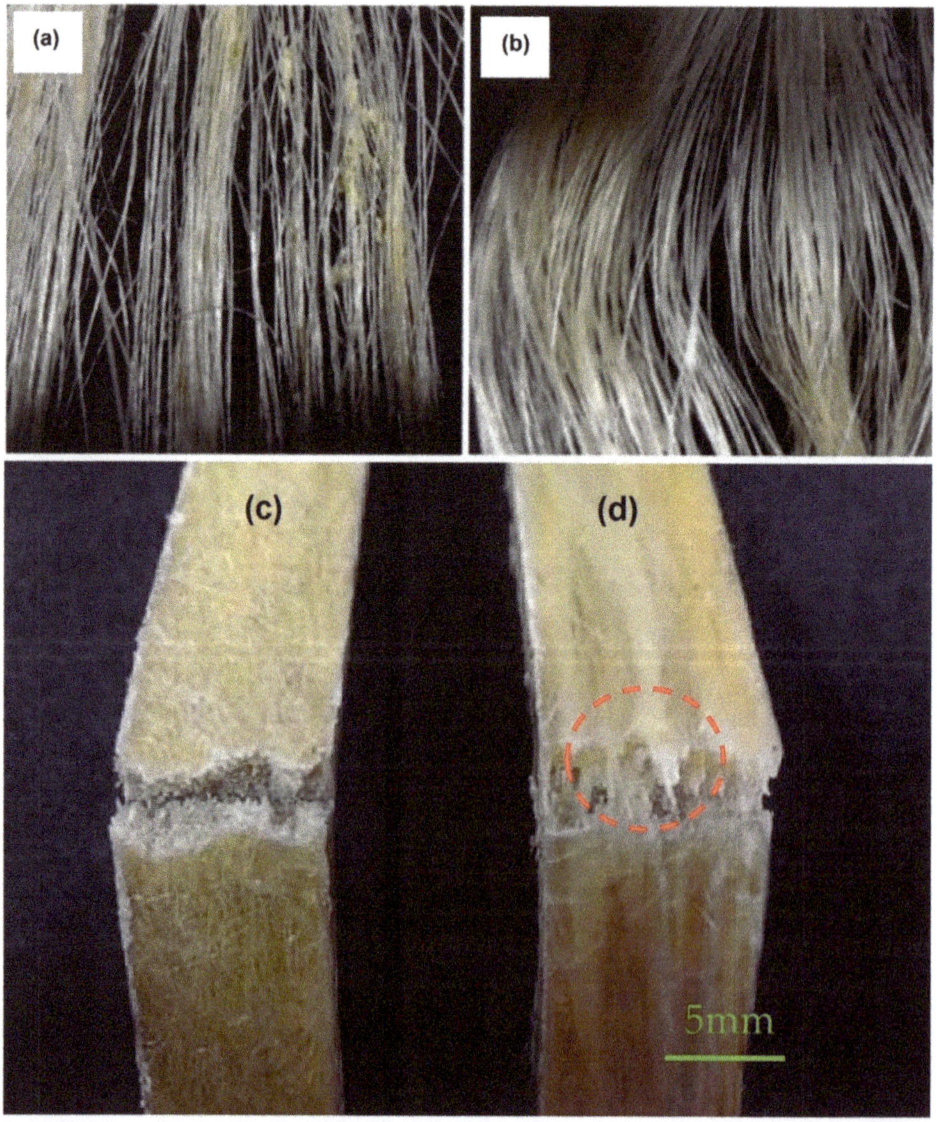


**Figure 13.** Sisal fibers (**a**) before alkali treatment, (**b**) after alkali treatment, (**c**) flexural strength test for short sisal fibers reinforced composites and (**d**) flexural strength test for long sisal fibers reinforced composites. Reprinted with permission from Ref. [125]. Copyright 2021, with permission from MDPI.

The study of flowability is another important topic to discuss here. In an experimental study, flowability of the cement paste with three different fibers, polypropylene, polyvinyl alcohol and sisal, was examined. Results revealed that the increase in length and dosage of each fiber decreased the flowability of cement paste. It was further revealed that the flowability of cement paste with sisal fiber was the lowest [138]. Recently, the behavior of sisal-reinforced concrete in exterior beam–column joints under monotonic loads was studied. The authors replaced cement in the concrete mix by 0.5, 1, 1.5 and 2% sisal fibers. Under gradually increasing loads, exterior beam–column joints with sisal-reinforced concrete showed lower deflection and enhanced shear strength [139]. Silva et al. studied the potential use of long aligned sisal fibers in thin cement-based laminates for semi-structural and structural applications. Samples were prepared replacing OPC with 30% metakaolin and 20% calcined waste crushed clay bricks. They prepared calcium-hydroxide-free samples by replacing 50% cement by the calcined clays at 28 days of age. They observed the ultimate tensile strength of calcium hydroxide free composites to be 13.95 MPa with an increase of around 34% and toughness under tensile loads twice that of OPC composites. They further noticed that quicker aging of calcium-hydroxide-free composites through hot water immersion revealed an ultimate strength 3.8 times higher and toughness 42.4 times greater than OPC composites under the same conditions [140].

### 3.1.2. Jute and Its Composites

Jute is one of the natural organic cellulosic fibers finding application in green composites and is the most used cellulosic fiber after cotton [141–144]. Jute is a type of bast fiber obtained from two species, namely Corchorus capsularis and Corchorus olitorius, and is mainly grown in tropical regions of the world including India, Bangladesh and China. Currently, jute production around the world is estimated as $2300 \times 10^3$ tons per year. Jute fiber represents a complex mixture of chemical compounds that are produced during the growth of fiber in the plant stem by photosynthesis process. Soil condition, climate, development of the plant and retting process greatly influence the constituents of fiber. In general, jute fiber is made up of 60% cellulose, 22% hemicellulose, 12% lignin, 1% fatty and waxy matter, 1% nitrogenous matter, 1% mineral matter and 3% miscellaneous. The main constituents i.e., cellulose, hemicellulose and lignin, essentially have influence on the fiber's structure, as others are very minor in proportion [145]. Jute fiber has a relatively low density and high strength and stiffness. In general, jute fiber has a density of around 1.3 $g/cm^3$, tensile strength of 393–773 MPa, Young's modulus of 26.5 GPa and 1.5–1.8% elongation. Jute fiber at 65% relative humidity and at 21 °C has a value of 12 for equilibrium moisture content [146]. Jute is among the most-studied fibers for reinforcement in thermoset and thermoplastic polymers. Alshaaer worked with the synthesis and recyclability of jute-reinforced geopolymers composites and reported four times higher flexural strength of jute-reinforced geopolymers composites as compared to non-reinforced geopolymers composites. Their results are illustrated in Figure 14 [147].

**Figure 14.** SEM micrographs of (**a**) jute fibers, (**b**) jute-fiber-reinforced geopolymers composites (lower magnification) where 1 shows jute fibril and 2 shows geopolymer aggregates and (**c**) jute fibers reinforced geopolymers composites (higher magnification) where 1 shows jute fibril and 2 shows geopolymer aggregates. Reprinted with permission from Ref. [147]. Copyright 2021, with permission from Frontiersin.

Khondker et al. studied the effects of molding temperature and pressure on mechanical and interfacial properties of polylactic acid and homo-polypropylene-based thermoplastic composites using untreated and treated unidirectional jute yarns. Their results indicated that molding condition at 175 °C and 2.7 MPa pressure was more appropriate for optimized properties of the unidirectional jute fiber/polylactic acid composites. They noticed that in the case of jute fiber/polylactic acid microbraid composites, maximum tensile stress and modulus increased with an increasing fiber volume fraction. In case of jute/homo-polypropylene composites, jute reinforcement caused increase tensile and bending properties of the composites, and with a 20% jute-fiber inclusion, jute/homo-polypropylene composites showed remarkable improvement in tensile and bending properties [148]. Ramakrishnan et al. prepared jute/nano-clay/epoxy hybrid composites employing various ratios of untreated and treated jute fiber and nano-clay and used the compression molding technique to evaluate the dynamic mechanical and free vibration behaviors. They used various NaOH concentrations of 2.5%, 5% and 7.5% to modify the jute fiber surface. In order to improve the dynamic properties of jute-fiber-reinforced epoxy composites, nano-clay 1, 3, 5 and 7 wt.% was added, along with the primary 5% NaOH treated jute fibers. Their research revealed that the dynamic mechanical behavior is greatly influenced by NaOH solution concentration and nano-clay content. Their results indicated that the jute treated with 5% NaOH epoxy composites showed the peak storage modulus value of 3884.56 MPa with a peak loss modulus value of 484.07 MPa. On the other hand, epoxy composites with jute treated with 5% NaOH and 5 wt.% nano-clay showed a storage modulus peak value of 4446.38 MPa and a loss modulus peak value of 544.04 MPa [149]. Yao et al. studied the impact of jute fiber and polyvinyl alcohol (PVA) on flexural and fracture performance of soil-cement column. Their results showed that presence of fibers considerably enhanced the flexural performance and fracture energy of soil-cement. Furthermore, the inclusion of fibers also successfully controlled the formation and propagation of plastic shrinkage cracks at an early age and reduced the flexural strength loss under wet–dry cycles [150].

In a recent study, four fibers, namely piassava, tucum palm, razor grass and jute with 1.5, 3 and 4.5% mass addition of the composite binder, were used as reinforcement in cement mortars to study the mechanical properties with 50% OPC with 40% metakaolin and 10% fly ash. Fibers were treated using four techniques, namely washing in hot water, hornification, 8% sodium hydroxide treatment and hybridization in order to obtain improved physical and mechanical properties. In this respect, sodium hydroxide treatment was selected for jute fiber. Their results showed that treated fibers increased the performance of the composites during mechanical testing. It was further noticed that the inclusion of treated fibers above 3% increased the flexural strength than those without fibers [151]. A study on the mechanical properties of the jute-reinforced geopolymer composite was carried out by Sankar and Kriven. They revealed that the jute-reinforced geopolymer composite showed a flexural strength of 20.5 MPa with an improvement in tensile strength from 8.8 MPa to 14.5 MPa with alkali-treated jute weaves. It was also revealed that the jute-reinforced geopolymer composite could absorb an impact energy of 9.64 J on average [152]. Trindade et al. conducted a study on the mechanical properties of natural fibers using metakaolin as an aluminosilicate source for a geopolymer with sodium hydroxide and sodium silicate as an alkaline activator. A total volumetric fraction of 10% for the Jute and sisal fibers were used for reinforcing the composite, and five layers each of jute and sisal fibers were embedded. Jute- and sisal-fiber-reinforced geopolymer composites showed a significant compressive strength at an average of 72.70 MPa at 7 days. They observed an ultimate tensile strength of 6.31 MPa and an ultimate flexural strength of 15.21 MPa whereas the modulus of elasticity was recorded at 14.92 GPa. They revealed that the jute-fiber-reinforced geopolymer composite exhibited a strain capacity of 28.34 mm [153].

Bheel et al. studied the effect of jute fiber and wheat straw ash on the mechanical properties of concrete. They used 0.25, 0.5, 0.75 and 1% jute fiber as reinforcement and 20, 30 and 40% wheat straw ash as a replacement for fine aggregates. Their results revealed that composites reinforced with 0.5% jute fiber along with 30% wheat straw ash at 28 days were enhanced with values of 32.88 MPa, 3.8 MPa and 5.3 MPa for compressive, splitting tensile and flexural strengths, respectively. Similarly, the modulus of elasticity for all compositions increased as well, reaching 30.45 GPa for composites with 1% jute fiber reinforcement at 28 days. Authors also observed a decrease in permeability and workability of concrete with increasing values of jute fiber and wheat straw ash in concrete [154]. Fonseca et al. prepared fiber-cement-reinforced composites with NaOH-treated jute fibers, cellulose nanofibrils and a hybrid of both Jute fibers and cellulose nanofibrils at 0.5 and 2% mass of cement to study the physical and mechanical properties using an extrusion process. The samples were subjected to natural weathering for 5 months before they were analyzed. Their results revealed that composites reinforced with cellulose nanofabrils, the hybrid (0.5% jute fiber and 1.5% cellulose nanofibrils) and 0.5% jute fiber had the highest apparent density values. With an increase in the proportion of jute fibers and cellulose nanofabrils from 0.5 to 2%, composites showed a reduction in apparent porosity values. The sample reinforced with 2% cellulose nanofabrils showed a 75% reduction in apparent porosity. The composite reinforced with 1.5% cellulose nanofabrils and 0.5% jute fibers showed the strongest mechanical properties. Composites with all compositions showed a reduction in the modulus of elasticity after natural weathering. On the other hand, hybrid reinforced composites showed an increase, on average, of 1 MPa for the modulus of rupture and limit of proportionality [155]. Table 3 summarizes the chemical composition and mechanical properties of sisal and jute fibers.

**Table 3.** Chemical composition and properties of cellulosic fibers.

| Fiber Name | Chemical Composition (wt.%) | | | Physico-Mechanical Properties | | | | | Ref. |
|---|---|---|---|---|---|---|---|---|---|
| | Cellulose | Hemicellulose | Lignin | Density (g/cm$^3$) | Tensile Strength (MPa) | Young's Modulus (GPa) | Elongation at Break (%) | Equilibrium Moisture Content (%) | |
| Sisal | 67–78 | 10–14.2 | 8–11 | 1.03–1.45 | 347–700 | 15.4–18.79 | 2–2.5 | 12 | [119,122,123] |
| Jute | 60 | 22 | 12 | 1.3 | 393–773 | 26.5 | 1.5–1.8 | 12 | [145,146] |

### 3.2. *Inorganic Fiber-Reinforced Geopolymer Composites*

The use of inorganic fibers like asbestos dates back to prehistoric time. During the past decades, efforts have been focused on developing high-performance materials that could meet the requirements of improved tensile strength and modulus values. Among them, inorganic fibers with improved mechanical properties have been employed in polymers to produce composites of enhanced properties. Glass, basalt, boron, boron carbide, boron nitride, zirconia, silicon carbide and silicon nitride are among the many inorganic fibers being used for producing composites. Due to their low cost, inorganic fibers are being replaced with carbon fibers in high temperature-resistance applications. The addition of inorganic fibers in polymers significantly improves physical, structural, thermal and rheological behaviors of the composites. Inorganic fibers contain mainly alumina and silica and have high melting temperatures. Low cost, chemical stability and good insulating properties along with enhanced mechanical properties make them attractive for use in different industries including civil engineering. The percentage increase of silica to alumina enhances the tensile strength but it negatively affects the modulus with 100% alumina content. Inorganic fibers with 52% $Al_2O_3$ can withstand 1250 °C, and with a further increase in alumina content, can resist even higher temperatures. Metakaolin-based geopolymers reinforced with high alumina fibers at 600–1000 °C show improved mechanical strength, high energy absorption and reduced shrinkage [156]. However, Welter et al. indicated the weaknesses of some inorganic fiber composites in high-temperature applications [157].

#### 3.2.1. Basalt and Its Composites

Basalt, being an eco-friendly, nontoxic and affordable material, is gaining popularity for reinforcing composites. Basalt fiber, among the new materials of the 21st century, is broadly employed in many fields including aerospace, construction, the chemical industry, agriculture, medicine and electronics. Basalt fiber was extensively used in military and aeronautical applications during the second world war. Basalt is obtained from volcanic rocks that are dark in color and form as a result of the solidification of volcanic lava. Basalt flows mainly contain $SiO_2$ along with $Al_2O_3$, $Fe_2O_3$, FeO, CaO and MgO. Basalt rocks are classified into alkaline, mildy acidic and acidic according to the $SiO_2$ % content present; of which acidic basalts satisfy the conditions for fiber preparation [158]. Basalt fibers are obtained in a fine-fiber shape of 9–13 μm diameter by melting and beating through the centrifugal blowing process of basalt rocks above 1500 °C [159]. Basalt fibers are receiving more attention for use in civil engineering because of their low density, high fatigue strength, low water absorption, good heat and insulation properties, good processability along with cheap fabrication process compared to glass and carbon fibers and high chemical resistance [160–162]. Basalt fibers have higher strength values than other inorganic fibers; for example, basalt fibers have a higher tensile strength than E-glass and their strain to failure is greater than that of carbon fiber. Another advantage of basalt fiber is its reasonable resistance to acid attack; however, basalt corrodes in an alkaline environment [163]. Basalt fibers have a high melting point ~1000 °C and are more suitable for high-temperature-resistant applications than cellulosic fibers. Basalt fibers have moisture absorption less than 0.02% for 24 h with a moisture regain value of 1. Basalt fibers show a significant resistance at high temperatures for a short time period, up to 750 °C, and for longer exposure in 260–700 °C and even in some cases up to 1000 °C. Basalt fibers lose 20–25% of their initial strength without losing insulation properties [164]. Chopped basalt fibers with different length and the effect of alkali or acid treatment are very important factors during composite preparation. Acid or base treatments are obliged to add abundant functional groups on the surface of basalt fibers. It is reported that during alkali treatment, basalt fibers provides more rough surface than untreated samples, whereas a very smooth and clean surface is observed for acid treatment as shown in Figure 15 [165].

**Figure 15.** (**A**) Different types of chopped basalt fibers from left to right: With fiber length of 24 mm, 12 mm and 6 mm, respectively, (**B**) morphologies of basalt fibers (**a**) untreated, (**b**) HCl treated and (**c**) NaOH treated. Reprinted with permission from Refs. [165,166]. Copyright 2021, with permission from MDPI and Elsevier.

The inclusion of basalt fibers into carbon-epoxy composites increases the absorbed impact energy of the composites. In a previous study, vinylester and epoxy reinforced with basalt fibers were tested for structural applications. The results showed that the ultimate tensile strength of basalt/epoxy composites increased by 29% and compressive strength increased by 85% as compared to basalt/vinylester composites. It was observed that the failure mode in compression was the same for both types of composites. Kim et al. investigated the effects of modified carbon nanotube/epoxy/basalt on the flexural and fracture properties. Carbon nanotubes were modified with silane treatment and acid treatment. Their results showed that silane-treated carbon nanotube/epoxy/basalt composites had an increase in flexural strength and flexural modulus by 14% and 10%, respectively, compared to acid-treated composites. The fracture toughness of silane-treated composites showed an increase of 40% compared to acid-treated carbon nanotube/epoxy/basalt composites [167]. Szabo and Czigany studied the static properties of short fiber-reinforced polypropylene composites using basalt and ceramic as a reinforcement with 5, 15 and 25 wt.%. Their results showed that fracture toughness greatly depends on the type and direction of the load and thickness. It had been noticed that the characteristic damage form was pull out in transverse and debonding in longitudinal directions [168]. Zhang et al. studied the tensile, flexural and impact properties of basalt-fiber-reinforced poly(butylene succinate) composites. Their results revealed that tensile strength and modulus gradually increased with increasing basalt fiber content. It was observed that tensile strength of the composite increased from 31 MPa to 46 MPa with basalt content increase from 3 vol.% to 15 vol.%. Moreover, the increase in tensile strength at higher loadings of basalt fiber was comparatively smaller than those at lower loadings. Flexural strength of the composite was observed to increase from 18 MPa to 71 MPa with an increase of basalt fiber from 0 to 15 vol.%. Similarly, the flexural modulus also increased from 551 MPa to 3.8 GPa. It was also observed that the impact strength of the basalt-reinforced poly(butylene succinate) did not change at 3 vol.% fiber loading, and with further loading, impact strength increased linearly between 5 and 15 vol.% with its highest value of 7.5 kJ/m$^2$ at 15 vol.% [169].

Punurai et al. studied the mechanical properties, microstructure and drying shrinkage of hybrid fly-ash–basalt fiber-reinforced geopolymer paste by replacing fly ash with basalt fiber at 0, 10, 20, 30, 40 and 100%. Their results showed that the inclusion of basalt fiber resulted in increased setting times and the initial, and final setting times with 100% basalt fiber inclusion were 280% and 110% higher than that of 100% fly ash geopolymer paste. The 7-day compressive strength of the geopolymer paste with 40% basalt fiber inclusion was recorded as 112% higher than that of the 100% fly ash geopolymer paste, whereas the 28-day compressive strength of the same sample was 118% higher than the 100% fly ash geopolymer paste. The flexural strength also increased with an increasing content of basalt fiber. The 28-day flexural strength of the sample with 40% basalt fiber content was noted as 64% higher than the 100% fly ash geopolymer paste. Their results indicated that the drying shrinkage of geopolymer paste decreased with increasing basalt fiber content [170]. Figure 16 shows the four-point flexural test results for basalt-fiber-reinforced geopolymers composites [166].

Li and Xu studied the dynamic compressive strength, deformation and energy absorption of basalt-fiber-reinforced geopolymeric concrete. They observed that the inclusion of 0.1 and 0.2% of basalt fiber resulted in 10.1 and 30.9% reduction in dynamic compressive strength whereas with 0.3% basalt fiber, there was no substantial change in dynamic compressive strength. With 0.3% basalt fiber addition to geopolymer concrete, critical strain at the strain rate of 100 s$^{-1}$ increased by 7.7% resulting in improved deformation capacity of the geopolymer concrete. Furthermore, the addition of 0.3% basalt fiber results in an increase of 8.9–13.2% in specific energy absorption at the strain rate from 40 to 100 s$^{-1}$ showing a noticeable improvement in energy absorption capacity [171]. Yang et al. studied the effects of basalt fiber content on the uniaxial compressive mechanical properties of concrete. Their research showed that 6 kg/m$^3$ of basalt fiber in concrete could improve the compressive strength and could reduce the density and intensity of acoustic emission. They concluded that with the increase in basalt fiber content, local damage could be effectively weakened. Their results showed that the proper amount of basalt fiber in concrete delayed the early cracking and reduced the transverse strain of concrete [172]. Shen et al. conducted a study on reinforced concrete beam-column joints by strengthening them with basalt-fiber-reinforced polymer sheets in different ways under cyclic loads. Their results showed that overall seismic performance of the joints strengthened with basalt-fiber-reinforced polymer sheets was enhanced noticeably, and there was good interface behavior between the concrete and basalt-reinforced polymer sheets. They also observed that load-bearing capacity, ductility and stiffness of joints increased by strengthening with basalt-fiber-reinforced polymer sheets. Likewise, the energy dissipation capacity of the joints with basalt-fiber-reinforced polymer sheets also increased [173].

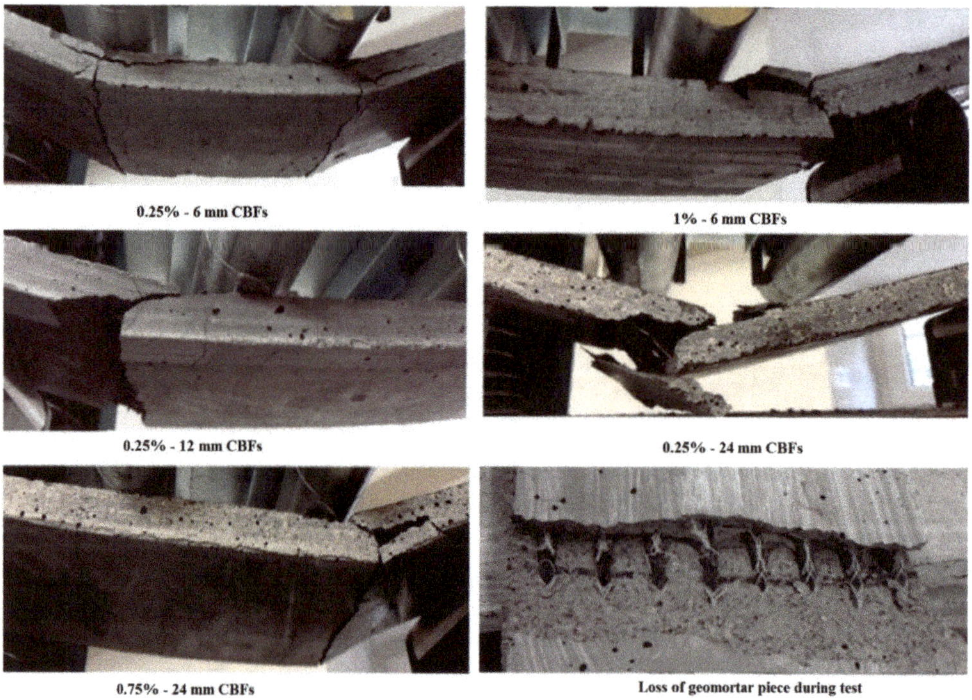


**Figure 16.** Four-point flexural strength test with varying fiber length of basalt chopped fibers. Reprinted with permission from Ref. [166]. Copyright 2021, with permission from MDPI.

### 3.2.2. Glass and Its Composites

Glass fiber is one of the most common reinforcement in polymer composites. It is a strong, less brittle, lightweight and cost-effective material. The application areas of glass fibers include automobile, marine, sports, leisure goods, aerospace and the construction industry [174–176]. In the construction industry, glass fibers are mainly used for the production of fibrocement-based objects and for external strengthening of existing buildings. The global production of glass fiber accounts for about 5 million metric tons annually and it is estimated that the global market value will reach more than $21 billion in 2025. The use of glass dates back to ancient times e.g., many ancient Egyptian ships were made by winding glass fibers on a rim of clay of appropriate form. Commercial production of glass fiber started in 1930s by the Owens-Illinois Glass Company. There exist many groups of glasses including silica, oxynitride, phosphate and halide of which silica glasses are used for composites reinforcement. Glass fibers are manufactured by using different types of broken glass that contain silica along with other components like alumina, oxides of calcium, magnesium and boron. The manufacturing process of glass fiber involves high-temperature conversion of raw materials into a homogeneous melt that is later converted into glass fibers. The production process has three phases, namely raw material handling, glass melting and refining and fiber formation. Glass fibers are produced in many forms, including continuous fiber, rovings, staple fiber and chopped strand. The mixing of continuous and chopped-strand glass fibers with resin is more common. Depending on the chemical composition and end use, glass fibers are characterized into many classes, namely C-glass, D-glass, R-glass, E-glass and S-glass [177]. E-glass, S-glass and C-glass are the leading glass fibers, with E-glass being most widely used in composites because of its low cost and relatively low moduli. S-glass fibers are stiffer and stronger than E-glass and have better resistance to fatigue and creep. E-glass are alumino-borosilicate and are mainly used for glass-reinforced plastics, while S-glass are also alumino-silicate with no CaO content and the highest value of tensile strength among all glass fibers and are mainly used in aircraft components and missile casings [178]. AR-glass fibers represent good resistance in alkaline media and are being used in cement substrates and concrete. C-glass fibers have good resistance to chemical resistance [179]. Figure 17 gives a random orientation of AR-glass fibers [180].

**Figure 17.** AR glass fibers. Reprinted with permission from Ref. [180]. Copyright 2021, with permission from Elsevier.

The main advantages of glass fibers are excellent high tensile strength and low cost of production. The structure of glass fibers is amorphous and the Young's modulus of glass fiber is same as in the bulk form of glass. While the strength-to-weight ratio of glass fibers is high and elastic modulus is low, they increase stiffness and reduce elongation of plastic composites. Glass fibers have a high ratio of surface area to weight, and this makes them vulnerable to chemical attacking. On the other hand, glass fibers have a good thermal insulation property. The main drawbacks of glass fibers are their relatively low elasticity modulus, reduced long-term strength and weak resistance to moisture and alkaline mediums. Other drawbacks of glass fibers are their high sensitivity to abrasion during handling, poor fatigue resistance and high hardness.

Faizal et al. studied the tensile behavior of a plane-woven E-glass fiber-reinforced polyester composite. Composites were prepared using a symmetrical and non-symmetrical lay-up of glass fibers and were cured at different curing pressures. Their results revealed that for both symmetrical and non-symmetrical lay-ups, the tensile modulus decreased with increasing curing pressure. They observed a common stiffness characteristic for both symmetrical and non-symmetrical arrangements at a curing pressure of about 87.1 $kg/m^2$.

For the symmetrical lay-up arrangement, ductility decreased with an increasing curing pressure whereas it increased with increasing curing pressure in the case of the non-symmetrical lay-up arrangement [181]. Kushwaha and Kumar investigated the mechanical properties of the bamboo glass mat (strand and woven)-reinforced epoxy and polyester laminate composites. They found that in the glass strand mat and bamboo epoxy composites, improved tensile strength and tensile modulus could be achieved with a comparatively lower weight percentage of glass fibers along with bamboo fiber reinforcement. The same behavior was noticed for flexural strength as well; however, the flexural modulus increased with higher percentages of glass fiber, whereas in woven glass mat and bamboo-reinforced epoxy composites, tensile strength and modulus and flexural strength and modulus increased with the increase in glass fiber content. The same results were observed for the polyester-based composites [182]. Devendra and Rangaswamy investigated the mechanical properties of E-glass fiber-reinforced epoxy composites using varying concentrations of fly ash, aluminum oxide, magnesium hydroxide and hematite powder as fillers. Their results revealed that the composite filled by 10 vol.% of magnesium hydroxides showed the maximum ultimate strength when compared with other filled composites. High impact strength was observed with 10 vol.% of fly ash, whereas aluminum oxide and magnesium hydroxide showed good impact strength at 10 vol.% and their further increase led to a reduction in impact strength. Results showed that impact strength increased with the increasing amount of hematite powder. Furthermore, the hardness of composites increased with an increasing amount of magnesium hydroxide, aluminum oxide and hematite and decreased with the increasing amount of fly ash. Magnesium hydroxide exhibited the highest hardness number when compared to other fillers [183].

Etcheverry and Barbosa studied the glass fiber/polypropylene adhesion improvements. They revealed that in-situ metallocenic polymerization of propylene on the glass fiber surface increased the adhesion between polypropylene matrix and glass fiber reinforcement [184]. In a previous study, the mechanical behavior of E-chopped strand glass fiber-reinforced wood sawdust/polyvinyl chloride composites with 50% wood were examined. The varying percentages of glass fiber were 10, 20 and 30% with initial fiber lengths of 3, 6 and 12 mm. The results showed that the stiffness and strength of wood sawdust/polyvinyl chloride composites improved with increasing glass fiber content. Composites reinforced with 30% glass fiber with a final length showed greater tensile and flexural strength and moduli. Increasing the glass fiber content led to an increased impact strength of wood/polyvinyl chloride composites, whereas elongation at break slightly decreased with an increasing content of glass fiber. A reduction in percentage shrinkage of the composites occurred with increasing glass fiber content [185]. Chen et al. studied the mechanical properties of a polyamide66/polyphenylene sulphide blend reinforced with 5, 10, 20 and 30% volume content of glass fiber. Their results showed that the inclusion of glass fiber significantly improved the tensile strength, flexural strength and hardness of the composites but decreased the impact strength of the blend. The maximum tensile strength and flexural strength were achieved with 30 vol.% and with 25 vol.% of glass fiber, respectively [186]. Cheng et al. investigated the mechanical properties of glass-fiber-reinforced cement with fly ash or slag after natural curing for 28, 180 and 360 days and accelerated aging at 80 °C for 8 days. They used 3% glass fiber content by weight of the total solid mix by replacing an equal quantity of sand, whereas fly ash or slag were added at a ratio of 0%, 20% and 40%, respectively. Their results showed that regardless of admixture, the modulus of rupture of glass-reinforced cement increased with an increase in curing time and its value was noted as being higher than that of the mortar without glass fiber. They further observed that after 8 days of accelerated aging at 80 °C, the modulus of rupture of glass-reinforced cement without any admixture decreased radically and had a lower value than that of glass-reinforced cement with natural curing for 360 days and mortar without glass-fiber accelerated curing for 880 °C. They concluded that fly ash and slag could improve the long-term strength of glass-reinforced cement but could not constrain the toughness degradation of glass reinforced cement mortars [187]. Fang et al. investigated the compressive, flexural strength and water resistance of fiber-glass-reinforced magnesium phosphate cement mortar with fiber volume fractions of 1.5%, 2.5%, 3% and 3.5%, respectively. Their results revealed the optimal volume fraction of glass fiber at 2.5% and that the glass fibers had more noticeable effects on the flexural strength than on compressive strength. They further observed that the water-resistance performance in the compressive and flexural strength might not be improved with glass fiber magnesium phosphate cement mortars [188]. Gese et al. investigated the performance of AR-glass-reinforced mortar composite in flexural strengthening of RC beams considering three factors, namely age (3, 7 and 28 days) of AR-glass reinforced mortar, number (2, 3 and 4) of AR-glass-reinforced mortar layers and the pre-cracking level (no pre-cracking, 50% and 100% of the yielding load). It was observed that the AR-glass-reinforced mortar external reinforcement decreased the ductility of the beams. Their results showed a 49% increase in yield load for 28-day beams and 33–30% for 3–7-day beams, respectively. The ultimate load was considerably improved by AR-glass reinforcement and increased by 31%, 54% and 72% for the beams strengthened with two, three and four layers, respectively, whereas AR-glass-reinforced mortar ages and pre-cracking had no significant differences in the ultimate load for beams. Although pre-cracking affected the crack load, it modified the behavior of stage II of the flexural test. Higher stiffness noted in stage II implied better performance in pre-cracked beams for yield loads than that of crack load increase [189]. Table 4 summarizes the chemical composition and the mechanical properties of basalt and different types of glass fibers.

**Table 4.** Chemical composition and properties of inorganic fibers.

| Fiber Name | Chemical Composition (wt.%) | | | | | | | | Physico-Mechanical Properties | | | | Ref. |
|---|---|---|---|---|---|---|---|---|---|---|---|---|---|
| | $SiO_2$ | $Al_2O_3$ | CaO | MgO | $Fe_2O_3$ | $Na_2O$ | $B_2O_3$ | Others | Density ($g/cm^3$) | Tensile Strength (MPa) | Young's Modulus (GPa) | Elongation at Break (%) | |
| Basalt | 52.8 | 17.5 | 8.59 | 4.63 | 10.3 | 3.34 | - | ~3.34 | 2.65–2.83 | 3000–4840 | 89–110 | 3–3.15 | [99,158] |
| E-glass | 52–56 | 12–16 | 16–25 | 0–5 | - | - | 5–10 | - | 2.58 | 1.7–3.5 | 69–72 | 4.8 | [177] |
| S-glass | 65 | 25 | - | 10 | - | - | - | - | 2.48 | 2–4.5 | 85 | 5.7 | |
| AR-glass | 55–75 | 0–5 | 1–10 | - | - | - | 0–8 | - | 2.7 | 3.24 | 73.1 | 4.4 | |
| C-glass | 65 | 4 | 14 | 3 | - | - | 5.5 | - | 2.52 | 1.7–2.8 | 68.9 | 4.8 | |

## 4. Summary and Future Direction

- This paper represents the inclusion of selective cellulosic and non-cellulosic fibers in geopolymers-based, fiber-reinforced concrete composites from a construction and civil engineering perspective. Geopolymers are the relatively new materials being employed in the construction industry to replace the use of traditional concrete materials. Due to a number of advantages, interest in developing, characterizing and implementing the use of geopolymers in the construction industry is growing.
- Geopolymer cement-based materials are developed using alumina silicate sources, with fly ash, metakaolin and GGBFS being the most-used ones. First, the properties and uses of these alumino-silicate materials were briefly discussed and represented in this paper. Moreover, the second part discussed the inclusion of fibers as a reinforcement in concrete composites. It is well-known that geopolymers alone cannot respond adequately to certain mechanical properties and hence need to be employed in combination with other suitable materials.
- As discussed in this paper, geopolymers are weak in tension and possess brittle behavior that represents poor tensile/flexural properties. To overcome this problem, one such solution is the inclusion of fibers in geopolymers-based composites. Natural fibers are gaining attention regarding their use in composites due to a number of reasons, including their relatively low density, excellent strength and environmental friendliness.

- This paper examines the use of cellulosic and non-cellulosic fibers in composites for the construction industry. A brief description of sisal, jute, basalt and glass fibers are discussed in this context and represent some recent works conducted in the area.

**Author Contributions:** M.T.N. and A.M. conceived, designed, performed experiments, analyzed the results and wrote manuscript. M.P. (Miroslava Pechociakova), N.A., M.A., S.S. and S.Z.U.H. performed data curation and experiments. M.P. (Michal Petru) and J.M. analyzed the results, supervised and acquired funding. All of the authors participated in critical analysis and preparation of the manuscript. All authors have read and agreed to the published version of the manuscript.

**Funding:** This work was supported by the Ministry of Education, Youth and Sports of the Czech Republic and the European Union (European Structural and Investment Funds-Operational Programme Research, Development and Education) in the frames of the project "Modular platform for autonomous chassis of specialized electric vehicles for freight and equipment transportation", Reg. No. CZ.02.1.01/0.0/0.0/16_025/0007293. This work was also supported by the Ministry of Education, Youth and Sports of the Czech Republic and the European Union-European Structural and Investment Funds in the Frames of Operational Programme Research, Development and Education-project Hybrid Materials for Hierarchical Structures (HyHi, Reg. No. CZ.02.1.01/0.0/0.0/16_019/0000843) and students grant competition SGS-2021-6025.

**Institutional Review Board Statement:** Not applicable.

**Informed Consent Statement:** Not applicable.

**Data Availability Statement:** Not applicable.

**Conflicts of Interest:** The authors declare no conflict of interest.

## References

1. Huntzinger, D.N.; Eatmon, T.D. A life-cycle assessment of Portland cement manufacturing: Comparing the traditional process with alternative technologies. *J. Clean. Prod.* **2009**, *17*, 668–675. [CrossRef]
2. Shehata, N.; Sayed, E.T.; Abdelkareem, M.A. Recent progress in environmentally friendly geopolymers: A review. *Sci. Total Environ.* **2020**, *762*, 143166. [CrossRef]
3. El-Salamony, A.-H.R.; Mahmoud, H.M.; Shehata, N. Enhancing the efficiency of a cement plant kiln using modified alternative fuel. *Environ. Nanotechnol. Monit. Manag.* **2020**, *14*, 100310.
4. Hasanbeigi, A.; Morrow, W.; Masanet, E.; Sathaye, J.; Xu, T. Energy efficiency improvement and $CO_2$ emission reduction opportunities in the cement industry in China. *Energy Policy* **2013**, *57*, 287–297. [CrossRef]
5. Shafeek, A.M.; Khedr, M.H.; El-Dek, S.I.; Shehata, N. Influence of ZnO nanoparticle ratio and size on mechanical properties and whiteness of White Portland Cement. *Appl. Nanosci.* **2020**, *10*, 3603–3615. [CrossRef]
6. Bajpai, R.; Choudhary, K.; Srivastava, A.; Sangwan, K.S.; Singh, M. Environmental impact assessment of fly ash and silica fume based geopolymer concrete. *J. Clean. Prod.* **2020**, *254*, 120147. [CrossRef]
7. Amran, Y.M.; Alyousef, R.; Alabduljabbar, H.; El-Zeadani, M. Clean production and properties of geopolymer concrete; A review. *J. Clean. Prod.* **2020**, *251*, 119679. [CrossRef]
8. Verma, M.; Dev, N. Effect of ground granulated blast furnace slag and fly ash ratio and the curing conditions on the mechanical properties of geopolymer concrete. *Struct. Concr.* **2021**. [CrossRef]
9. Shekhawat, P.; Sharma, G.; Singh, R.M. Microstructural and morphological development of eggshell powder and flyash-based geopolymers. *Constr. Build. Mater.* **2020**, *260*, 119886. [CrossRef]
10. McLellan, B.C.; Williams, R.P.; Lay, J.; Van Riessen, A.; Corder, G.D. Costs and carbon emissions for geopolymer pastes in comparison to ordinary portland cement. *J. Clean. Prod.* **2011**, *19*, 1080–1090. [CrossRef]
11. Kathirvel, P.; Sreekumaran, S. Sustainable development of ultra high performance concrete using geopolymer technology. *J. Build. Eng.* **2021**, *39*, 102267. [CrossRef]
12. Matalkah, F.; Ababneh, A.; Aqel, R. Efflorescence Control in Calcined Kaolin-Based Geopolymer Using Silica Fume and OPC. *J. Mater. Civ. Eng.* **2021**, *33*, 04021119. [CrossRef]
13. Tchadjie, L.N.; Ekolu, S.O. Enhancing the reactivity of aluminosilicate materials toward geopolymer synthesis. *J. Mater. Sci.* **2018**, *53*, 4709–4733. [CrossRef]
14. Ribeiro, R.A.S.; Ribeiro, M.S.; Kriven, W.M. A review of particle-and fiber-reinforced metakaolin-based geopolymer composites. *J. Ceram. Sci. Technol.* **2017**, *8*, 307.
15. Ali, A.; Sattar, M.; Riaz, T.; Alam Khan, B.; Awais, M.; Militky, J.; Noman, M.T. Highly stretchable durable electro-thermal conductive yarns made by deposition of carbon nanotubes. *J. Text. Inst.* **2021**, 1–10. [CrossRef]
16. Ashraf, M.A.; Wiener, J.; Farooq, A.; Saskova, J.; Noman, M.T. Development of Maghemite Glass Fibre Nanocomposite for Adsorptive Removal of Methylene Blue. *Fibers Polym.* **2018**, *19*, 1735–1746. [CrossRef]

17. Behera, P.; Noman, M.T.; Petrů, M. Enhanced Mechanical Properties of Eucalyptus-Basalt-Based Hybrid-Reinforced Cement Composites. *Polymers* **2020**, *12*, 2837. [CrossRef]
18. Noman, M.T.; Amor, N.; Petru, M. Synthesis and applications of ZnO nanostructures (ZONSs): A review. *Crit. Rev. Solid State Mater. Sci.* **2021**, 1–43. [CrossRef]
19. Davidovits, J. Geopolymers: Inorganic polymeric new materials. *J. Therm. Anal. Calorim.* **1991**, *37*, 1633–1656. [CrossRef]
20. Duxson, P.; Fernández-Jiménez, A.; Provis, J.L.; Lukey, G.C.; Palomo, A.; Van Deventer, J.S.J. Geopolymer technology: The current state of the art. *J. Mater. Sci.* **2007**, *42*, 2917–2933. [CrossRef]
21. Nawaz, M.; Heitor, A.; Sivakumar, M. Geopolymers in construction—Recent developments. *Constr. Build. Mater.* **2020**, *260*, 120472. [CrossRef]
22. Wong, V.; Jervis, W.; Fishburn, B.; Numata, T.; Joe, W.; Rawal, A.; Sorrell, C.C.; Koshy, P. Long-Term Strength Evolution in Ambient-Cured Solid-Activator Geopolymer Compositions. *Minerals* **2021**, *11*, 143. [CrossRef]
23. Ng, C.; Alengaram, U.J.; Wong, L.S.; Mo, K.H.; Jumaat, M.Z.; Ramesh, S. A review on microstructural study and compressive strength of geopolymer mortar, paste and concrete. *Constr. Build. Mater.* **2018**, *186*, 550–576. [CrossRef]
24. Liew, Y.-M.; Heah, C.-Y.; Mustafa, A.B.M.; Kamarudin, H. Structure and properties of clay-based geopolymer cements: A review. *Prog. Mater. Sci.* **2016**, *83*, 595–629. [CrossRef]
25. Toniolo, N.; Boccaccini, A.R. Fly ash-based geopolymers containing added silicate waste. A review. *Ceram. Int.* **2017**, *43*, 14545–14551. [CrossRef]
26. Mousavinejad, S.H.G.; Sammak, M. Strength and chloride ion penetration resistance of ultra-high-performance fiber reinforced geopolymer concrete. *Structures* **2021**, *32*, 1420–1427. [CrossRef]
27. Silva, G.; Kim, S.; Aguilar, R.; Nakamatsu, J. Natural fibers as reinforcement additives for geopolymers—A review of potential eco-friendly applications to the construction industry. *Sustain. Mater. Technol.* **2020**, *23*, e00132. [CrossRef]
28. Noman, M.; Petrů, M. Functional Properties of Sonochemically Synthesized Zinc Oxide Nanoparticles and Cotton Composites. *Nanomaterials* **2020**, *10*, 1661. [CrossRef] [PubMed]
29. Noman, M.T.; Petru, M.; Amor, N.; Louda, P. Thermophysiological comfort of zinc oxide nanoparticles coated woven fabrics. *Sci. Rep.* **2020**, *10*, 21080. [CrossRef] [PubMed]
30. Sturm, P.; Gluth, G.; Jäger, C.; Brouwers, H.; Kühne, H.-C. Sulfuric acid resistance of one-part alkali-activated mortars. *Cem. Concr. Res.* **2018**, *109*, 54–63. [CrossRef]
31. Liu, H.; He, H.; Li, Y.; Hu, T.; Ni, H.; Zhang, H. Coupling effect of steel slag in preparation of calcium-containing geopolymers with spent fluid catalytic cracking (FCC) catalyst. *Constr. Build. Mater.* **2021**, *290*, 123194. [CrossRef]
32. Singh, B.; Ishwarya, G.; Gupta, M.; Bhattacharyya, S. Geopolymer concrete: A review of some recent developments. *Constr. Build. Mater.* **2015**, *85*, 78–90. [CrossRef]
33. Kriven, W.M.; Bell, J.L.; Gordon, M. Microstructure and Microchemistry of Fully-Reacted Geopolymers and Geopolymer Matrix Composites. *Ceram. Transact.* **2003**, *153*, 1994.
34. Peng, X.; Shuai, Q.; Li, H.; Ding, Q.; Gu, Y.; Cheng, C.; Xu, Z. Fabrication and Fireproofing Performance of the Coal Fly Ash-Metakaolin-Based Geopolymer Foams. *Materials* **2020**, *13*, 1750. [CrossRef]
35. Değirmenci, F.N. Utilization of Natural and Waste Pozzolans as an Alternative Resource of Geopolymer Mortar. *Int. J. Civ. Eng.* **2018**, *16*, 179–188. [CrossRef]
36. Król, M.; Rożek, P.; Mozgawa, W. Synthesis of the Sodalite by Geopolymerization Process Using Coal Fly Ash. *Pol. J. Environ. Stud.* **2017**, *26*, 2611–2617. [CrossRef]
37. Andini, S.; Cioffi, R.; Colangelo, F.; Grieco, T.; Montagnaro, F.; Santoro, L. Coal fly ash as raw material for the manufacture of geopolymer-based products. *Waste Manag.* **2008**, *28*, 416–423. [CrossRef]
38. Zafar, I.; Rashid, K.; Ju, M. Synthesis and characterization of lightweight aggregates through geopolymerization and microwave irradiation curing. *J. Build. Eng.* **2021**, *42*, 102454. [CrossRef]
39. Nuaklong, P.; Wongsa, A.; Boonserm, K.; Ngohpok, C.; Jongvivatsakul, P.; Sata, V.; Sukontasukkul, P.; Chindaprasirt, P. Enhancement of mechanical properties of fly ash geopolymer containing fine recycled concrete aggregate with micro carbon fiber. *J. Build. Eng.* **2021**, *41*, 102403. [CrossRef]
40. Muraleedharan, M.; Nadir, Y. Factors affecting the mechanical properties and microstructure of geopolymers from red mud and granite waste powder: A review. *Ceram. Int.* **2021**, *47*, 13257–13279. [CrossRef]
41. Amran, M.; Debbarma, S.; Ozbakkaloglu, T. Fly ash-based eco-friendly geopolymer concrete: A critical review of the long-term durability properties. *Constr. Build. Mater.* **2021**, *270*, 121857. [CrossRef]
42. Simão, L.; De Rossi, A.; Hotza, D.; Ribeiro, M.J.; Novais, R.M.; Montedo, O.R.K.; Raupp-Pereira, F. Zeolites-containing geopolymers obtained from biomass fly ash: Influence of temperature, composition, and porosity. *J. Am. Ceram. Soc.* **2020**, *104*, 803–815. [CrossRef]
43. Yip, C.K.; Van Deventer, J.S.J. Microanalysis of calcium silicate hydrate gel formed within a geopolymeric binder. *J. Mater. Sci.* **2003**, *38*, 3851–3860. [CrossRef]
44. van Jaarsveld, J.; van Deventer, J.; Lukey, G. The effect of composition and temperature on the properties of fly ash- and kaolinite-based geopolymers. *Chem. Eng. J.* **2002**, *89*, 63–73. [CrossRef]
45. Nguyen, T.T.; Goodier, C.I.; Austin, S.A. Factors affecting the slump and strength development of geopolymer concrete. *Constr. Build. Mater.* **2020**, *261*, 119945. [CrossRef]

46. Zhang, H.; Li, L.; Sarker, P.K.; Long, T.; Shi, X.; Wang, Q.; Cai, G. Investigating Various Factors Affecting the Long-Term Compressive Strength of Heat-Cured Fly Ash Geopolymer Concrete and the Use of Orthogonal Experimental Design Method. *Int. J. Concr. Struct. Mater.* **2019**, *13*, 63. [CrossRef]
47. Zhang, B.; MacKenzie, K.J.D.; Brown, I.W.M. Crystalline phase formation in metakaolinite geopolymers activated with NaOH and sodium silicate. *J. Mater. Sci.* **2009**, *44*, 4668–4676. [CrossRef]
48. De Vargas, A.S.; Molin, D.C.D.; Vilela, A.; da Silva, F.J.; Pavão, B.; Veit, H.M. The effects of Na2O/SiO2molar ratio, curing temperature and age on compressive strength, morphology and microstructure of alkali-activated fly ash-based geopolymers. *Cem. Concr. Compos.* **2011**, *33*, 653–660. [CrossRef]
49. Noman, M.T.; Petru, M.; Militký, J.; Azeem, M.; Ashraf, M.A. One-Pot Sonochemical Synthesis of ZnO Nanoparticles for Photocatalytic Applications, Modelling and Optimization. *Materials* **2019**, *13*, 14. [CrossRef]
50. Phair, J.; Van Deventer, J. Effect of the silicate activator pH on the microstructural characteristics of waste-based geopolymers. *Int. J. Miner. Process.* **2002**, *66*, 121–143. [CrossRef]
51. Kusbiantoro, A.; Ibrahim, M.S.; Muthusamy, K.; Alias, A. Development of Sucrose and Citric Acid as the Natural based Admixture for Fly Ash based Geopolymer. *Proc. Environ. Sci.* **2013**, *17*, 596–602. [CrossRef]
52. Nematollahi, B.; Sanjayan, J. Effect of different superplasticizers and activator combinations on workability and strength of fly ash based geopolymer. *Mater. Des.* **2014**, *57*, 667–672. [CrossRef]
53. Jang, J.; Lee, N.; Lee, H. Fresh and hardened properties of alkali-activated fly ash/slag pastes with superplasticizers. *Constr. Build. Mater.* **2014**, *50*, 169–176. [CrossRef]
54. Palomo, A.; Grutzeck, M.; Blanco-Varela, M.T. Alkali-activated fly ashes: A cement for the future. *Cem. Concr. Res.* **1999**, *29*, 1323–1329. [CrossRef]
55. Heah, C.; Kamarudin, H.; Al Bakri, A.M.; Binhussain, M.; Luqman, M.; Nizar, I.K.; Ruzaidi, C.; Liew, Y. Effect of Curing Profile on Kaolin-based Geopolymers. *Phys. Proc.* **2011**, *22*, 305–311. [CrossRef]
56. Rovnaník, P. Effect of curing temperature on the development of hard structure of metakaolin-based geopolymer. *Constr. Build. Mater.* **2010**, *24*, 1176–1183. [CrossRef]
57. Zuhua, Z.; Xiao, Y.; Huajun, Z.; Yue, C. Role of water in the synthesis of calcined kaolin-based geopolymer. *Appl. Clay Sci.* **2009**, *43*, 218–223. [CrossRef]
58. Van Jaarsveld, J.; van Deventer, J.; Lukey, G. The characterisation of source materials in fly ash-based geopolymers. *Mater. Lett.* **2003**, *57*, 1272–1280. [CrossRef]
59. Isaia, G.C.; Gastaldini, A.L.G. Concrete sustainability with very high amount of fly ash and slag. *Rev. IBRACON Estrut. Mater.* **2009**, *2*, 244–253. [CrossRef]
60. Panias, D.; Giannopoulou, I.P.; Perraki, T. Effect of synthesis parameters on the mechanical properties of fly ash-based geopolymers. *Colloids Surf. A Physicochem. Eng. Asp.* **2007**, *301*, 246–254. [CrossRef]
61. Belviso, C. State-of-the-art applications of fly ash from coal and biomass: A focus on zeolite synthesis processes and issues. *Prog. Energy Combust. Sci.* **2018**, *65*, 109–135. [CrossRef]
62. Yao, Z.; Ji, X.; Sarker, P.; Tang, J.; Ge, L.; Xia, M.; Xi, Y. A comprehensive review on the applications of coal fly ash. *Earth Sci. Rev.* **2015**, *141*, 105–121. [CrossRef]
63. Ahmaruzzaman, M. A review on the utilization of fly ash. *Prog. Energy Combust. Sci.* **2010**, *36*, 327–363. [CrossRef]
64. Cho, Y.K.; Jung, S.H.; Choi, Y.C. Effects of chemical composition of fly ash on compressive strength of fly ash cement mortar. *Constr. Build. Mater.* **2019**, *204*, 255–264. [CrossRef]
65. Temuujin, J.; Van Riessen, A.; Williams, R. Influence of calcium compounds on the mechanical properties of fly ash geopolymer pastes. *J. Hazard. Mater.* **2009**, *167*, 82–88. [CrossRef]
66. Sarkar, A.; Rano, R.; Udaybhanu, G.; Basu, A. A comprehensive characterisation of fly ash from a thermal power plant in Eastern India. *Fuel Process. Technol.* **2006**, *87*, 259–277. [CrossRef]
67. Papadakis, V.G. Effect of fly ash on Portland cement systems: Part II. High-calcium fly ash. *Cem. Concr. Res.* **2000**, *30*, 1647–1654. [CrossRef]
68. Çiçek, T.; Çinçin, Y. Use of fly ash in production of light-weight building bricks. *Constr. Build. Mater.* **2015**, *94*, 521–527. [CrossRef]
69. Bendapudi, S.C.K.; Saha, P. Contribution of fly ash to the properties of mortar and concrete. *Int. J. Earth Sci. Eng.* **2011**, *4*, 1017–1023.
70. Malhotra, V. Durability of concrete incorporating high-volume of low-calcium (ASTM Class F) fly ash. *Cem. Concr. Compos.* **1990**, *12*, 271–277. [CrossRef]
71. Johari, M.A.M.; Brooks, J.; Kabir, S.; Rivard, P. Influence of supplementary cementitious materials on engineering properties of high strength concrete. *Constr. Build. Mater.* **2011**, *25*, 2639–2648. [CrossRef]
72. Hemalatha, T.; Ramaswamy, A. A review on fly ash characteristics—Towards promoting high volume utilization in developing sustainable concrete. *J. Clean. Prod.* **2017**, *147*, 546–559. [CrossRef]
73. Cheah, C.B.; Tan, L.E.; Ramli, M. Recent advances in slag-based binder and chemical activators derived from industrial by-products—A review. *Constr. Build. Mater.* **2020**, *272*, 121657. [CrossRef]
74. Kumar, V.P.; Gunasekaran, K.; Shyamala, T. Characterization study on coconut shell concrete with partial replacement of cement by GGBS. *J. Build. Eng.* **2019**, *26*, 100830. [CrossRef]

75. Siddique, R.; Bennacer, R. Use of iron and steel industry by-product (GGBS) in cement paste and mortar. *Resour. Conserv. Recycl.* **2012**, *69*, 29–34. [CrossRef]
76. Pal, S.; Mukherjee, A.; Pathak, S. Investigation of hydraulic activity of ground granulated blast furnace slag in concrete. *Cem. Concr. Res.* **2003**, *33*, 1481–1486. [CrossRef]
77. Özbay, E.; Erdemir, M.; Durmuş, H.I. Utilization and efficiency of ground granulated blast furnace slag on concrete properties—A review. *Constr. Build. Mater.* **2016**, *105*, 423–434. [CrossRef]
78. Barnett, S.; Soutsos, M.; Millard, S.; Bungey, J. Strength development of mortars containing ground granulated blast-furnace slag: Effect of curing temperature and determination of apparent activation energies. *Cem. Concr. Res.* **2006**, *36*, 434–440. [CrossRef]
79. Oner, A.; Akyuz, S. An experimental study on optimum usage of GGBS for the compressive strength of concrete. *Cem. Concr. Compos.* **2007**, *29*, 505–514. [CrossRef]
80. Sabir, B.; Wild, S.; Bai, J. Metakaolin and calcined clays as pozzolans for concrete: A review. *Cem. Concr. Compos.* **2001**, *23*, 441–454. [CrossRef]
81. Murray, H.H. Traditional and new applications for kaolin, smectite, and palygorskite: A general overview. *Appl. Clay Sci.* **2000**, *17*, 207–221. [CrossRef]
82. Siddique, R.; Klaus, J. Influence of metakaolin on the properties of mortar and concrete: A review. *Appl. Clay Sci.* **2009**, *43*, 392–400. [CrossRef]
83. Zulkifly, K.; Cheng-Yong, H.; Yun-Ming, L.; Abdullah, M.M.A.B.; Shee-Ween, O.; Bin Khalid, M.S. Effect of phosphate addition on room-temperature-cured fly ash-metakaolin blend geopolymers. *Constr. Build. Mater.* **2021**, *270*, 121486. [CrossRef]
84. Cai, R.; He, Z.; Tang, S.; Wu, T.; Chen, E. The early hydration of metakaolin blended cements by non-contact impedance measurement. *Cem. Concr. Compos.* **2018**, *92*, 70–81. [CrossRef]
85. Rashad, A.M. Metakaolin as cementitious material: History, scours, production and composition—A comprehensive overview. *Constr. Build. Mater.* **2013**, *41*, 303–318. [CrossRef]
86. Paiva, H.; Yliniemi, J.; Illikainen, M.; Rocha, F.; Ferreira, V.M. Mine Tailings Geopolymers as a Waste Management Solution for A More Sustainable Habitat. *Sustainability* **2019**, *11*, 995. [CrossRef]
87. Wang, M.-R.; Jia, D.-C.; He, P.-G.; Zhou, Y. Microstructural and mechanical characterization of fly ash cenosphere/metakaolin-based geopolymeric composites. *Ceram. Int.* **2011**, *37*, 1661–1666. [CrossRef]
88. Poon, C.-S.; Azhar, S.; Anson, M.; Wong, Y.-L. Performance of metakaolin concrete at elevated temperatures. *Cem. Concr. Compos.* **2003**, *25*, 83–89. [CrossRef]
89. Khatib, J.; Wild, S. Sulphate Resistance of Metakaolin Mortar. *Cem. Concr. Res.* **1998**, *28*, 83–92. [CrossRef]
90. Khatib, J.; Hibbert, J. Selected engineering properties of concrete incorporating slag and metakaolin. *Constr. Build. Mater.* **2005**, *19*, 460–472. [CrossRef]
91. Neupane, K. Fly ash and GGBFS based powder-activated geopolymer binders: A viable sustainable alternative of portland cement in concrete industry. *Mech. Mater.* **2016**, *103*, 110–122. [CrossRef]
92. Assi, L.N.; Carter, K.; Deaver, E.; Ziehl, P. Review of availability of source materials for geopolymer/sustainable concrete. *J. Clean. Prod.* **2020**, *263*, 121477. [CrossRef]
93. Görhan, G.; Kürklü, G. The influence of the NaOH solution on the properties of the fly ash-based geopolymer mortar cured at different temperatures. *Compos. Part. B Eng.* **2014**, *58*, 371–377. [CrossRef]
94. Sanchindapong, S.; Narattha, C.; Piyaworapaiboon, M.; Sinthupinyo, S.; Chindaprasirt, P.; Chaipanich, A. Microstructure and phase characterizations of fly ash cements by alkali activation. *J. Therm. Anal. Calorim.* **2020**, *142*, 1–8. [CrossRef]
95. Rahim, R.A.; Rahmiati, T.; Azizli, K.A.; Man, Z.; Nuruddin, M.F.; Ismail, L. Comparison of Using NaOH and KOH Activated Fly Ash-Based Geopolymer on the Mechanical Properties. *Mater. Sci. Forum* **2014**, *803*, 179–184. [CrossRef]
96. Zhang, F.; Zhang, L.; Liu, M.; Mu, C.; Liang, Y.N.; Hu, X. Role of alkali cation in compressive strength of metakaolin based geopolymers. *Ceram. Int.* **2017**, *43*, 3811–3817. [CrossRef]
97. Fu, C.; Ye, H.; Zhu, K.; Fang, D.; Zhou, J. Alkali cation effects on chloride binding of alkali-activated fly ash and metakaolin geopolymers. *Cem. Concr. Compos.* **2020**, *114*, 103721. [CrossRef]
98. Ranjbar, N.; Talebian, S.; Mehrali, M.; Kuenzel, C.; Metselaar, H.S.C.; Jumaat, M.Z. Mechanisms of interfacial bond in steel and polypropylene fiber reinforced geopolymer composites. *Compos. Sci. Technol.* **2016**, *122*, 73–81. [CrossRef]
99. Ranjbar, N.; Zhang, M. Fiber-reinforced geopolymer composites: A review. *Cem. Concr. Compos.* **2020**, *107*, 103498. [CrossRef]
100. Silva, G.; Kim, S.; Bertolotti, B.; Nakamatsu, J.; Aguilar, R. Optimization of a reinforced geopolymer composite using natural fibers and construction wastes. *Constr. Build. Mater.* **2020**, *258*, 119697. [CrossRef]
101. Amor, N.; Noman, M.T.; Petru, M. Prediction of functional properties of nano $TiO_2$ coated cotton composites by artificial neural network. *Sci. Rep.* **2021**, *11*, 1–11. [CrossRef]
102. Noman, M.T.; Petru, M.; Louda, P.; Kejzlar, P. Woven textiles coated with zinc oxide nanoparticles and their thermophysiological comfort properties. *J. Nat. Fiber* **2021**, *18*, 1–14. [CrossRef]
103. Arisoy, B.; Wu, H.-C. Material characteristics of high performance lightweight concrete reinforced with PVA. *Constr. Build. Mater.* **2008**, *22*, 635–645. [CrossRef]
104. Shaikh, F.U.A. Review of mechanical properties of short fibre reinforced geopolymer composites. *Constr. Build. Mater.* **2013**, *43*, 37–49. [CrossRef]

105. Jamshaid, H.; Mishra, R.; Militký, J.; Noman, M.T. Interfacial performance and durability of textile reinforced concrete. *J. Text. Inst.* **2017**, *109*, 879–890. [CrossRef]
106. Jamshaid, H.; Mishra, R.; Militky, J.; Pechociakova, M.; Noman, M.T. Mechanical, thermal and interfacial properties of green composites from basalt and hybrid woven fabrics. *Fibers Polym.* **2016**, *17*, 1675–1686. [CrossRef]
107. Yang, T.; Hu, L.; Xiong, X.; Petrů, M.; Noman, M.T.; Mishra, R.; Militký, J. Sound Absorption Properties of Natural Fibers: A Review. *Sustainability* **2020**, *12*, 8477. [CrossRef]
108. Noman, M.T.; Ashraf, M.A.; Ali, A. Synthesis and applications of nano-TiO2: A review. *Environ. Sci. Pollut. Res.* **2019**, *26*, 3262–3291. [CrossRef]
109. Noman, M.T.; Ashraf, M.A.; Jamshaid, H.; Ali, A. A Novel Green Stabilization of TiO2 Nanoparticles onto Cotton. *Fibers Polym.* **2018**, *19*, 2268–2277. [CrossRef]
110. Yan, L.; Kasal, B.; Huang, L. A review of recent research on the use of cellulosic fibres, their fibre fabric reinforced cementitious, geo-polymer and polymer composites in civil engineering. *Compos. Part. B Eng.* **2016**, *92*, 94–132. [CrossRef]
111. Savastano, H., Jr.; Warden, P.G.; Coutts, R. Mechanically pulped sisal as reinforcement in cementitious matrices. *Cem. Concr. Compos.* **2003**, *25*, 311–319. [CrossRef]
112. Morton, J.; Cooke, T.; Akers, S. Performance of slash pine fibers in fiber cement products. *Constr. Build. Mater.* **2010**, *24*, 165–170. [CrossRef]
113. Noman, M.; Amor, N.; Petru, M.; Mahmood, A.; Kejzlar, P. Photocatalytic Behaviour of Zinc Oxide Nanostructures on Surface Activation of Polymeric Fibres. *Polymers* **2021**, *13*, 1227. [CrossRef]
114. Azwa, Z.; Yousif, B.; Manalo, A.; Karunasena, W. A review on the degradability of polymeric composites based on natural fibres. *Mater. Des.* **2013**, *47*, 424–442. [CrossRef]
115. Ardanuy, M.; Claramunt, J.; Filho, R.T. Cellulosic fiber reinforced cement-based composites: A review of recent research. *Constr. Build. Mater.* **2015**, *79*, 115–128. [CrossRef]
116. Al-Oraimi, S.; Seibi, A. Mechanical characterisation and impact behaviour of concrete reinforced with natural fibres. *Compos. Struct.* **1995**, *32*, 165–171. [CrossRef]
117. Ramakrishna, G.; Sundararajan, T. Impact strength of a few natural fibre reinforced cement mortar slabs: A comparative study. *Cem. Concr. Compos.* **2005**, *27*, 547–553. [CrossRef]
118. Naveen, J.; Jawaid, M.; Amuthakkannan, P.; Chandrasekar, M. Mechanical and physical properties of sisal and hybrid sisal fiber-reinforced polymer composites. In *Mechanical and Physical Testing of Biocomposites, Fibre-Reinforced Composites and Hybrid Composites*; Elsevier BV: Amsterdam, The Netherlands, 2019; pp. 427–440.
119. Kumre, A.; Rana, R.; Purohit, R. A Review on mechanical property of sisal glass fiber reinforced polymer composites. *Mater. Today Proc.* **2017**, *4*, 3466–3476. [CrossRef]
120. Senthilkumar, K.; Saba, N.; Rajini, N.; Chandrasekar, M.; Jawaid, M.; Siengchin, S.; Alotman, O.Y. Mechanical properties evaluation of sisal fibre reinforced polymer composites: A review. *Constr. Build. Mater.* **2018**, *174*, 713–729. [CrossRef]
121. Yan, L.; Chouw, N.; Jayaraman, K. Flax fibre and its composites–A review. *Compos. Part B Eng.* **2014**, *56*, 296–317. [CrossRef]
122. Kumar, P.S.S.; Allamraju, K.V. A Review of Natural Fiber Composites [Jute, Sisal, Kenaf]. *Mater. Today Proc.* **2019**, *18*, 2556–2562. [CrossRef]
123. Wei, J.; Meyer, C. Degradation mechanisms of natural fiber in the matrix of cement composites. *Cem. Concr. Res.* **2015**, *73*, 1–16. [CrossRef]
124. Idicula, M.; Neelakantan, N.R.; Oommen, Z.; Joseph, K.; Thomas, S. A study of the mechanical properties of randomly oriented short banana and sisal hybrid fiber reinforced polyester composites. *J. Appl. Polym. Sci.* **2005**, *96*, 1699–1709. [CrossRef]
125. Liang, Z.; Wu, H.; Liu, R.; Wu, C. Preparation of Long Sisal Fiber-Reinforced Polylactic Acid Biocomposites with Highly Improved Mechanical Performance. *Polymers* **2021**, *13*, 1124. [CrossRef]
126. Prasad, A.R.; Rao, K.M. Mechanical properties of natural fibre reinforced polyester composites: Jowar, sisal and bamboo. *Mater. Des.* **2011**, *32*, 4658–4663. [CrossRef]
127. Hashmi, S.; Rajput, R.S.; Naik, A.; Chand, N.; Singh, R. Investigations on weld joining of sisal CSM-thermoplastic composites. *Polym. Compos.* **2014**, *36*, 214–220. [CrossRef]
128. De Castro, B.D.; Fotouhi, M.; Vieira, L.M.G.; De Faria, P.E.; Rubio, J.C.C. Mechanical Behaviour of a Green Composite from Biopolymers Reinforced with Sisal Fibres. *J. Polym. Environ.* **2021**, *29*, 429–440. [CrossRef]
129. Rajesh, G.; Prasad, A.R.; Gupta, A. Mechanical and degradation properties of successive alkali treated completely biodegradable sisal fiber reinforced poly lactic acid composites. *J. Reinf. Plast. Compos.* **2015**, *34*, 951–961. [CrossRef]
130. Savastano, H., Jr.; Warden, P.; Coutts, R. Brazilian waste fibres as reinforcement for cement-based composites. *Cem. Concr. Compos.* **2000**, *22*, 379–384. [CrossRef]
131. Baloyi, R.B.; Ncube, S.; Moyo, M.; Nkiwane, L.; Dzingai, P. Analysis of the properties of a glass/sisal/polyester composite. *Sci. Rep.* **2021**, *11*, 1–10. [CrossRef]
132. Bahja, B.; Elouafi, A.; Tizliouine, A.; Omari, L. Morphological and structural analysis of treated sisal fibers and their impact on mechanical properties in cementitious composites. *J. Build. Eng.* **2021**, *34*, 102025. [CrossRef]
133. Ren, G.; Yao, B.; Huang, H.; Gao, X. Influence of sisal fibers on the mechanical performance of ultra-high performance concretes. *Constr. Build. Mater.* **2021**, *286*, 122958. [CrossRef]

134. Guerini, V.; Conforti, A.; Plizzari, G.; Kawashima, S. Influence of Steel and Macro-Synthetic Fibers on Concrete Properties. *Fibers* **2018**, *6*, 47. [CrossRef]
135. Chalioris, C.E.; Panagiotopoulos, T.A. Flexural analysis of steel fibre-reinforced concrete members. *Comput. Concr.* **2018**, *22*, 11–25.
136. Kytinou, V.K.; Chalioris, C.E.; Karayannis, C.G. Analysis of Residual Flexural Stiffness of Steel Fiber-Reinforced Concrete Beams with Steel Reinforcement. *Materials* **2020**, *13*, 2698. [CrossRef]
137. Choi, S.-W.; Choi, J.; Lee, S.-C. Probabilistic Analysis for Strain-Hardening Behavior of High-Performance Fiber-Reinforced Concrete. *Materials* **2019**, *12*, 2399. [CrossRef] [PubMed]
138. Zhang, K.; Pan, L.; Li, J.; Lin, C. What is the mechanism of the fiber effect on the rheological behavior of cement paste with polycarboxylate superplasticizer? *Constr. Build. Mater.* **2021**, *281*, 122542. [CrossRef]
139. Okeola, A.A.; Mwero, J.; Bello, A. Behavior of sisal fiber-reinforced concrete in exterior beam-column joint under monotonic loading. *Asian J. Civ. Eng.* **2021**, *22*, 627–636. [CrossRef]
140. De Andrare Silva, F.; Toledo Filho, R.D.; de Almeida Melo Filho, J.; Fairbairn, E.D.M.R. Physical and mechanical properties of durable sisal fiber–cement composites. *Constr. Build. Mater.* **2010**, *24*, 777–785. [CrossRef]
141. La Mantia, F.P.; Morreale, M. Green composites: A brief review. *Compos. Part A Appl. Sci. Manuf.* **2011**, *42*, 579–588. [CrossRef]
142. Sever, K.; Sarikanat, M.; Seki, Y.; Erkan, G.; Erdogan, U.H.; Erden, S. Surface treatments of jute fabric: The influence of surface characteristics on jute fabrics and mechanical properties of jute/polyester composites. *Ind. Crops Prod.* **2012**, *35*, 22–30. [CrossRef]
143. Ramamoorthy, S.K.; Skrifvars, M.; Persson, A. A Review of Natural Fibers Used in Biocomposites: Plant, Animal and Regenerated Cellulose Fibers. *Polym. Rev.* **2015**, *55*, 107–162. [CrossRef]
144. Noman, M.T.; Wiener, J.; Saskova, J.; Ashraf, M.A.; Vikova, M.; Jamshaid, H.; Kejzlar, P. In-situ development of highly photocatalytic multifunctional nanocomposites by ultrasonic acoustic method. *Ultrason. Sonochem.* **2018**, *40*, 41–56. [CrossRef]
145. Kerni, L.; Singh, S.; Patnaik, A.; Kumar, N. A review on natural fiber reinforced composites. *Mater. Today Proc.* **2020**, *28*, 1616–1621. [CrossRef]
146. Faruk, O.; Bledzki, A.K.; Fink, H.-P.; Sain, M. Biocomposites reinforced with natural fibers: 2000–2010. *Prog. Polym. Sci.* **2012**, *37*, 1552–1596. [CrossRef]
147. Alshaaer, M. Synthesis, Characterization, and Recyclability of a Functional Jute-Based Geopolymer Composite. *Front. Built Environ.* **2021**, *7*, 38. [CrossRef]
148. Khondker, O.; Ishiaku, U.; Nakai, A.; Hamada, H. A novel processing technique for thermoplastic manufacturing of unidirectional composites reinforced with jute yarns. *Compos. Part A Appl. Sci. Manuf.* **2006**, *37*, 2274–2284. [CrossRef]
149. Ramakrishnan, S.; Krishnamurthy, K.; Rajeshkumar, G.; Asim, M. Dynamic Mechanical Properties and Free Vibration Characteristics of Surface Modified Jute Fiber/Nano-Clay Reinforced Epoxy Composites. *J. Polym. Environ.* **2021**, *29*, 1076–1088. [CrossRef]
150. Yao, X.; Liu, K.; Huang, G.; Wang, M.; Dong, X. Mechanical Properties and Durability of Deep Soil–Cement Column Reinforced by Jute and PVA Fiber. *J. Mater. Civ. Eng.* **2021**, *33*, 04021021. [CrossRef]
151. Da Fonseca, R.P.; Rocha, J.C.; Cheriaf, M. Mechanical Properties of Mortars Reinforced with Amazon Rainforest Natural Fibers. *Materials* **2020**, *14*, 155. [CrossRef]
152. Sankar, K.; Kriven, W.M. Sodium geopolymer reinforced with jute weave. In *Developments in Strategic Materials and Computational Design V*; Kriven, W.M., Zhou, D., Moon, K., Hwang, T., Wang, J., Lewinssohn, C., Zhou, Y., Eds.; John Wiley & Sons Inc.: Hoboken, NJ, USA, 2015; pp. 39–60.
153. Trindade, A.C.; Arêas, I.O.; Almeida, D.C.; Alcamand, H.A.; Borges, P.H.; Silva, F.A. Mechanical behavior of geopolymeric composites reinforced with natural fibers. In Proceedings of the International Conference on Strain-Hardening Cement-Based Composites, Dresden, Germany, 18–20 September 2017; pp. 383–391.
154. Bheel, N.; Sohu, S.; Awoyera, P.; Kumar, A.; Abbasi, S.A.; Olalusi, O.B. Effect of Wheat Straw Ash on Fresh and Hardened Concrete Reinforced with Jute Fiber. *Adv. Civ. Eng.* **2021**, *2021*, 1–11. [CrossRef]
155. Fonseca, C.S.; Scatolino, M.V.; Silva, L.E.; Martins, M.A.; Júnior, M.G.; Tonoli, G.H.D. Valorization of Jute Biomass: Performance of Fiber–Cement Composites Extruded with Hybrid Reinforcement (Fibers and Nanofibrils). *Waste Biomass Valorizat* **2021**, 19. [CrossRef]
156. Bernal, S.A.; Bejarano, J.; Garzón, C.; de Gutiérrez, R.M.; Delvasto, S.; Rodríguez, E.D. Performance of refractory aluminosilicate particle/fiber-reinforced geopolymer composites. *Compos. Part B Eng.* **2012**, *43*, 1919–1928. [CrossRef]
157. Welter, M.; Schmücker, M.; MacKenzie, K. Evolution of the fibre-matrix interactions in basalt-fibre-reinforced geopolymer-matrix composites after heating. *J. Ceram. Sci. Technol.* **2015**, *6*, 17–24.
158. Dhand, V.; Mittal, G.; Rhee, K.Y.; Park, S.-J.; Hui, D. A short review on basalt fiber reinforced polymer composites. *Compos. Part B Eng.* **2015**, *73*, 166–180. [CrossRef]
159. Colombo, C.; Vergani, L.; Burman, M. Static and fatigue characterisation of new basalt fibre reinforced composites. *Compos. Struct.* **2012**, *94*, 1165–1174. [CrossRef]
160. Monaldo, E.; Nerilli, F.; Vairo, G. Basalt-based fiber-reinforced materials and structural applications in civil engineering. *Compos. Struct.* **2019**, *214*, 246–263. [CrossRef]
161. Dehkordi, M.T.; Nosraty, H.; Shokrieh, M.M.; Minak, G.; Ghelli, D. The influence of hybridization on impact damage behavior and residual compression strength of intraply basalt/nylon hybrid composites. *Mater. Des.* **2013**, *43*, 283–290. [CrossRef]

162. Noman, M.T.; Petru, M.; Amor, N.; Yang, T.; Mansoor, T. Thermophysiological comfort of sonochemically synthesized nano $TiO_2$ coated woven fabrics. *Sci. Rep.* **2020**, *10*, 1–12. [CrossRef]
163. Wei, B.; Cao, H.; Song, S. Tensile behavior contrast of basalt and glass fibers after chemical treatment. *Mater. Des.* **2010**, *31*, 4244–4250. [CrossRef]
164. Masi, G.; Rickard, W.; Bignozzi, M.C.; van Riessen, A. The effect of organic and inorganic fibres on the mechanical and thermal properties of aluminate activated geopolymers. *Compos. Part B Eng.* **2015**, *76*, 218–228. [CrossRef]
165. Hou, X.; Yao, S.; Wang, Z.; Fang, C.; Li, T. Enhancement of the mechanical properties of polylactic acid/basalt fiber composites via in-situ assembling silica nanospheres on the interface. *J. Mater. Sci. Technol.* **2021**, *84*, 182–190. [CrossRef]
166. Le, C.; Louda, P.; Buczkowska, K.E.; Dufkova, I. Investigation on Flexural Behavior of Geopolymer-Based Carbon Textile/Basalt Fiber Hybrid Composite. *Polymers* **2021**, *13*, 751. [CrossRef]
167. Kim, M.; Rhee, K.; Park, S.; Hui, D. Effects of silane-modified carbon nanotubes on flexural and fracture behaviors of carbon nanotube-modified epoxy/basalt composites. *Compos. Part B Eng.* **2012**, *43*, 2298–2302. [CrossRef]
168. Szabó, J.; Czigány, T. Static fracture and failure behavior of aligned discontinuous mineral fiber reinforced polypropylene composites. *Polym. Test.* **2003**, *22*, 711–719. [CrossRef]
169. Zhang, Y.; Yu, C.; Chu, P.K.; Lv, F.; Zhang, C.; Ji, J.; Zhang, R.; Wang, H. Mechanical and thermal properties of basalt fiber reinforced poly(butylene succinate) composites. *Mater. Chem. Phys.* **2012**, *133*, 845–849. [CrossRef]
170. Punurai, W.; Kroehong, W.; Saptamongkol, A.; Chindaprasirt, P. Mechanical properties, microstructure and drying shrinkage of hybrid fly ash-basalt fiber geopolymer paste. *Constr. Build. Mater.* **2018**, *186*, 62–70. [CrossRef]
171. Li, W.; Xu, J. Mechanical properties of basalt fiber reinforced geopolymeric concrete under impact loading. *Mater. Sci. Eng. A* **2009**, *505*, 178–186. [CrossRef]
172. Yang, L.; Xie, H.; Fang, S.; Huang, C.; Chao, Y.J. Experimental study on mechanical properties and damage mechanism of basalt fiber reinforced concrete under uniaxial compression. *Structures* **2021**, *31*, 330–340. [CrossRef]
173. Shen, D.; Li, M.; Kang, J.; Liu, C.; Li, C. Experimental studies on the seismic behavior of reinforced concrete beam-column joints strengthened with basalt fiber-reinforced polymer sheets. *Constr. Build. Mater.* **2021**, *287*, 122901. [CrossRef]
174. Ashik, K.P.; Sharma, R.S. A Review on Mechanical Properties of Natural Fiber Reinforced Hybrid Polymer Composites. *J. Miner. Mater. Charact. Eng.* **2015**, *3*, 420–426. [CrossRef]
175. Noman, M.T.; Militky, J.; Wiener, J.; Saskova, J.; Ashraf, M.A.; Jamshaid, H.; Azeem, M. Sonochemical synthesis of highly crystalline photocatalyst for industrial applications. *Ultrasonics* **2018**, *83*, 203–213. [CrossRef]
176. Noman, M.T.; Petru, M. Effect of Sonication and Nano TiO2 on Thermophysiological Comfort Properties of Woven Fabrics. *ACS Omega* **2020**, *5*, 11481–11490. [CrossRef]
177. Sathishkumar, T.; Satheeshkumar, S.; Naveen, J. Glass fiber-reinforced polymer composites—A review. *J. Reinforc. Plast. Compos.* **2014**, *33*, 1258–1275. [CrossRef]
178. Zhang, M.; Matinlinna, J.P. E-Glass Fiber Reinforced Composites in Dental Applications. *Silicon* **2012**, *4*, 73–78. [CrossRef]
179. XiaoChun, Q.; Xiaoming, L.; Xiaopei, C. The applicability of alkaline-resistant glass fiber in cement mortar of road pavement: Corrosion mechanism and performance analysis. *Int. J. Pavement Res. Technol.* **2017**, *10*, 536–544. [CrossRef]
180. Arslan, M.E.; Aykanat, B.; Subaşı, S.; Maraşlı, M. Cyclic behavior of autoclaved aerated concrete block infill walls strengthened by basalt and glass fiber composites. *Eng. Struct.* **2021**, *240*, 112431. [CrossRef]
181. Faizal, M.A.; Beng, Y.K.; Dalimin, M.N. Tensile property of hand lay-up plain-weave woven e-glass/polyester composite: Curing pressure and Ply arrangement effect. *Borneo Sci.* **2006**, *19*, 27–34.
182. Kushwaha, P.K.; Kumar, R. The studies on performance of epoxy and polyester-based composites reinforced with bamboo and glass fibers. *J. Reinforc. Plast. Compos.* **2010**, *29*, 1952–1962. [CrossRef]
183. Devendra, K.; Rangaswamy, T. Strength Characterization of E-glass Fiber Reinforced Epoxy Composites with Filler Materials. *J. Miner. Mater. Charact. Eng.* **2013**, *01*, 353–357. [CrossRef]
184. Etcheverry, M.; Barbosa, S.E. Glass fiber reinforced polypropylene mechanical properties enhancement by adhesion improvement. *Materials* **2012**, *5*, 1084–1113. [CrossRef] [PubMed]
185. Tungjitpornkull, S.; Chaochanchaikul, K.; Sombatsompop, N.; Tungjitpornkull, S.; Chaochanchaikul, K.; Sombatsompop, N. Mechanical Characterization of E-Chopped Strand Glass Fiber Reinforced Wood/PVC Composites. *J. Thermoplast. Compos. Mater.* **2007**, *20*, 535–550. [CrossRef]
186. Chen, Z.; Liu, X.; Lü, R.; Li, T. Mechanical and tribological properties of PA66/PPS blend. III. Reinforced with GF. *J. Appl. Polym. Sci.* **2006**, *102*, 523–529. [CrossRef]
187. Cheng, C.; He, J.; Zhang, J.; Yang, Y. Study on the time-dependent mechanical properties of glass fiber reinforced cement (GRC) with fly ash or slag. *Constr. Build. Mater.* **2019**, *217*, 128–136. [CrossRef]
188. Fang, Y.; Chen, B.; Oderji, S.Y. Experimental research on magnesium phosphate cement mortar reinforced by glass fiber. *Constr. Build. Mater.* **2018**, *188*, 729–736. [CrossRef]
189. Giese, A.C.H.; Giese, D.N.; Dutra, V.F.P.; Da Silva Filho, L.C.P. Flexural behavior of reinforced concrete beams strengthened with textile reinforced mortar. *J. Build. Eng.* **2021**, *33*, 101873. [CrossRef]

MDPI
St. Alban-Anlage 66
4052 Basel
Switzerland
Tel. +41 61 683 77 34
Fax +41 61 302 89 18
www.mdpi.com

*Polymers* Editorial Office
E-mail: polymers@mdpi.com
www.mdpi.com/journal/polymers

www.ingramcontent.com/pod-product-compliance
Lightning Source LLC
LaVergne TN
LVHW070405100526
838202LV00014B/1397